Discrete Mathematics and Theoretical Computer Science

Springer
*London
Berlin
Heidelberg
New York
Barcelona
Hong Kong
Milan
Paris
Singapore
Tokyo*

Series Editors
Douglas S. Bridges, *Canterbury University, NZ*
Cristian S. Calude, *University of Auckland, NZ*

Advisory Editorial Board
J. Casti, *Sante Fe Institute, USA*
G. J. Chaitin, *IBM Research Center, USA*
E. W. Dijkstra, *University of Texas at Austin, USA*
J. Goguen, *University of California at San Diego, USA*
R. L. Graham, *University of California at San Diego, USA*
J. Hartmanis, *Cornell University, USA*
H. Jürgensen, *University of Western Ontario, Canada*
A. Nerode, *Cornell University, USA*
G. Rozenberg, *Leiden University, The Netherlands*
A. Salomaa, *Turku University, Finland*

I. Antoniou, C.S. Calude and M.J. Dinneen (Eds)

Unconventional Models of Computation, UMC'2K

Proceedings of the Second International Conference on Unconventional Models of Computation, (UMC'2K)

Springer

I. Antoniou
Solvay Institutes, Free University of Brussels, Belgium

C.S. Calude
M.J. Dinneen
Department of Computer Science, University of Auckland, Auckland, New Zealand

ISBN 1-85233-415-0 Springer-Verlag London Berlin Heidelberg

British Library Cataloguing in Publication Data
Unconventional models of computation, UMC'2K : proceedings of the second International Conference on Unconventional Modelsof Computation, (UMC'2K). - (Discrete mathematics and theoretical computer science)
1. Soft computing – Congresses   2. Computer science – Mathematics - Congresses   3. Algorithms – Congresses   I. Antoniou, I. (Ioannis), 1955-   II. Calude, Cristian, 1952-   III. Dinneen, M.J.   IV. International Conference on Unconventional Models of Computation, (UMC'2K) (2nd : 2000 : Brussels)
511.3
ISBN 1852334150

Library of Congress Cataloging-in-Publication Data
A catalog record for this book is available from the Library of Congress

Apart from any fair dealing for the purposes of research or private study, or criticism or review, as permitted under the Copyright, Designs and Patents Act 1988, this publication may only be reproduced, stored or transmitted, in any form or by any means, with the prior permission in writing of the publishers, or in the case of reprographic reproduction in accordance with the terms of licences issued by the Copyright Licensing Agency. Enquiries concerning reproduction outside those terms should be sent to the publishers.

© Springer-Verlag London Limited 2001
Printed in Great Britain

The use of registered names, trademarks etc. in this publication does not imply, even in the absence of a specific statement, that such names are exempt from the relevant laws and regulations and therefore free for general use.

The publisher makes no representation, express or implied, with regard to the accuracy of the information contained in this book and cannot accept any legal responsibility or liability for any errors or omissions that may be made.

Typesetting: Camera ready by editors
Printed and bound at the Athenæum Press Ltd., Gateshead, Tyne & Wear
34/3830-543210 Printed on acid-free paper   SPIN 10790047

Preface

The Second International Conference on **Unconventional Models of Computation, UMC'2K**, organized by the Centre for Discrete Mathematics and Theoretical Computer Science, the International Solvay Institutes for Physics and Chemistry and the Vrije Universiteit Brussel Theoretical Physics Division was held at Solvay Institutes from 13 to 16 December, 2000.

The computers as we know them today, based on silicon chips, are getting better and better, cheaper and cheaper, and are doing more and more for us. Nonetheless, they still give rise to frustrations because they are unable to cope with many tasks of practical interest: Too many problems are effectively intractable. A simple example: cyber movie networks face the near impossible task of building a brand in a computing and communication almost vacuum.

Fortunately, for billions of years nature itself has been "computing" with molecules and cells. These natural processes form the main motivation for the construction of radically new models of computation, the core interest of our conference.

The ten invited speakers at the conference were: L. Accardi (Rome, Italy), S. Bozapalidis (Thessaloniki, Greece), K. Gustafson (Boulder, USA), T. Head (Binghamton, USA), T. Hida (Nagoya, Japan), V. Ivanov (Dubna, Russia), G. Păun (Bucharest, Romania), G. Rozenberg (Leiden, the Netherlands). H. Siegelmann (Haifa, Israel), and E. Winfree (Caltech, USA).

The Programme Committee consisting of M. Amos (Liverpool, UK), I. Antoniou (Co-chair, Brussels, Belgium), S. Bozapalidis (Thessaloniki, Greece), G. Brassard (Montreal, Canada), C.S. Calude (Co-chair, Auckland, New Zealand), M.J. Dinneen (Secretary, Auckland, New Zealand), A. Ekert (Oxford, UK), T. Head (Binghamton, NY, USA), S. Lloyd (Boston, USA), G. Păun (Bucharest, Romania), B. Pavlov (Auckland, New Zealand), I. Prigogine (Brussels, Belgium), G. Rozenberg (Leiden, Holland), A. Salomaa (Turku, Finland), K. Svozil (Vienna, Austria), H.T. Wareham (St. John's, Canada), E. Winfree (Caltech, USA) had a difficult task selecting 11 papers out of 26 submissions covering all major areas of unconventional computation, especially quantum computing, computing using organic molecules (DNA), membrane computing, cellular computing, dynamical system and nonlinear computing, and evolutionary algorithms.

The conference would not have been possible without the assistance of the following referees:

M. Amos	M.J. Dinneen	A. Salomaa
I. Antoniou	A. Ekert	K. Svozil
H. Bordihn	K.E. Gustafson	P. Evans
G. Brassard	M. Hallett	G. Păun
D.S. Bridges	T. Head	R. Sheehan
C.S. Calude	C. Moore	E. Yarevsky
E. Calude	R. Nicolescu	H.T. Wareham
K. Cattell	R. Polozov	E. Winfree
G.A. Creek	G. Rozenberg	M. Zimand
J. Dassow		

We extend our thanks to all members of the Conference Committee, particularly to P. Akritas, I. Antoniou (Chair), P. Barry, F. Bosco, C.S. Calude, U. Guenther (Registration), A. De Nayer-Westcott, E. Karpov, F. Lambert, I. Melnikov, S. Reddy (Registration), Z. Suchanecki, E. Yarevsky for their invaluable organizational work.

We extend our gratitude to Professors Ilya Prigogine and Franklin Lambert for their encouragement and support.

We thank the International Solvay Institutes for Physics and Chemistry, the Vrije Universiteit Brussel Theoretical Physics Division and the Centre for Discrete Mathematics of the University of Auckland for their technical support. The hospitality of our hosts, the International Solvay Institutes for Physics and Chemistry, the Vrije Universiteit Brussel Theoretical Physics Division, was marvelous.

We thank the Interuniversity Poles of Attraction Program Contract P4/08 Belgian State, the Belgium Prime Minister's Office, the Federal Office for Scientific, Technical and Cultural Affairs and Pukekohe Travel for their financial support.

It is a great pleasure to acknowledge the ideal cooperation with Springer-Verlag, London team, led by Beverley Ford and Rebecca Mowat, for producing this beautiful volume in time for the conference.

I. Antoniou
C.S. Calude
M.J. Dinneen
October 2000

Foreward

It is a little surprising to ask a non computer scientist to write the forward for a monograph which deals with computational models. However, I am glad to do so. My interest started with irreversibility, the breaking of time symmetry, in the formulation of the Laws of Dynamics, be they Classical or Quantum.

It is a well known fact that irreversibility cannot be understood and explained within the conventional Hilbert Space framework. Moreover, the trajectory complexity gives rise to probabilistic simplicity and predictability. We have shown that unconventional probabilistic formulations of the evolution of complex systems in terms of Rigged Hilbert Spaces break the time symmetry and give rise to new possibilities for prediction and control of complex systems.

These ideas have been successfully applied to 3 types of systems:

1. Large extended systems described in terms of the thermodynamic limit leading to singular distribution functions.
2. Chaos being intractable in the conventional mathematical framework but more tractable in an extended probabilistic framework beyond the conventional trajectories or wavefunctions descriptions.
3. Dressing of unstable Quantum states and decoherence. The Zeno era for the formation of the unstable state marks the onset of decoherence which is a strategic issue in Quantum Computation and can now be understood and estimated in terms of the extended formulation without artificial assumptions like Heat Baths.

The exploration of unconventional computation methods for intractable problems which is discussed in this conference is very timely and is expected to lead to new computational possibilities and software.

Ilya Prigogine
Director, International Solvay Institutes for Physics and Chemistry, Brussels

Table of Contents

Invited papers

L. Accardi and R. Sabbadini
On the Ohya–Masuda Quantum SAT Algorithm 1

I. Antoniou and V.V. Ivanov
Computational Methods and Tools for Modeling and Analysis of Complex Processes 10

S. Bozapalidis
Quantum Recognizable Tree Functions 25

K. Gustafson
An Unconventional Computational Linear Algebra: Operator Trigonometry 48

T. Head
Splicing Systems, Aqueous Computing, and Beyond 68

T. Hida
Some Methods of Computation in White Noise Calculus 85

G. Păun
Computing with Membranes: Attacking NP-Complete Problems 94

G. Rozenberg
DNA Processing in Ciliates – the Wonders of DNA Computing *in vivo* 116

H.T. Siegelmann and A. Ben-Hur
Macroscopical Molecular Computation with Gene Networks 119

E. Winfree
In-vitro Transcriptional Circuits 121

Contributed papers

H. Abe and S.C. Sung
Parallelizing with Limited Number of Ancillae 123

M.L. Campagnolo and C. Moore
Upper and Lower Bounds on Continuous-Time Computation 135

C. Martin-Vide and V. Mitrana
P Systems with Valuations 154

A.B. Mikhaylova and B.S. Pavlov
The Quantum Domain As a Triadic Relay 167

A. Păun
On P Systems with Active Membranes 187

F. Peper
Spatial Computing on Self-Timed Cellular Automata 202

A. Saito and K. Kaneko
Inaccessibility in Decision Procedures 215

Y. Sato, M. Taiji and T. Ikagami
On the Power of Nonlinear Mappings in Switching Map Systems 234

K. Svozil
Quantum Information: The New Frontier 248

C. Tamon and T. Yamakami
Quantum Computation Relative to Oracles 273

C. Zandron, C. Ferretti, and G. Mauri
Solving NP-Complete Problems Using P Systems with Active Membranes 289

On the Ohya–Masuda Quantum SAT Algorithm

Luigi Accardi and Ruben Sabbadini

Centro Vito Volterra, Università degli Studi di Roma "Tor Vergata", Rome, Italy

Abstract. Using a new explicit form of some of the gates involved, we propose a variant of the Ohya–Masuda quantum SAT algorithm which allows to reduce the dimensions of the dust-q-bit spaces and to perform a constructive analysis of its complexity. We use this algorithm to introduce a kind of *absolute measure of truth* for different sets of clauses, measured in terms of the cardinality of the set of truth functions which gives a value 1 to the given set of clauses.

1 Clauses Space

Definition 1. *A clause on the set*
$$X = \{x_1, \ldots, x_n, x'_1, \ldots, x'_n\}$$
$n \in \mathbf{N}$, *is a subset of* X.

Now consider a subset $I \subseteq \{x_1, \ldots, x_n\}$ and a subset $I' \subseteq \{x'_1, \ldots, x'_n\}$. Therefore a clause is uniquely fixed by an ordered pair (I, I') of subsets of $\{1, \ldots, n\}$. A clause (I, I') is said to be *minimal* if $I \cap I' = \emptyset$.

Remark 1. Introducing the set
$$N = \{1, \ldots, n\} \backslash (I \cup I')$$
it is easy to see that minimal clauses are in an one-to-one correspondence with the partitions
$$I \cup I' \cup N = \{1, \ldots, n\}$$
where the possibility that any of the sets of the partition is an empty set is not forbidden.

Therefore the number of minimal clauses is:
$$M = \binom{n}{3} \sim \frac{n^3}{6} \ .$$

Nevertheless deciding if a set of clauses is minimal could be an exponential problem.

1.1 The space of the clauses truth functions

The set $\{0, 1\}$ is a Boolean algebra with the operations

$$\varepsilon \vee \varepsilon' = \max\{\varepsilon, \varepsilon'\}$$

$$\varepsilon \wedge \varepsilon' = \min\{\varepsilon, \varepsilon'\}$$

($\varepsilon, \varepsilon' \in \{0, 1\}$). A clause truth function on the clauses on the set $\{x_1, \ldots, x_n, x'_1, \ldots, x'_n\}$ is a Boolean algebra homomorphism

$$t : \text{Parts of } \{x_1, \ldots, x_n, x'_1, \ldots, x'_n\} \to \{0, 1\}$$

with the property (*principle of the excluded third*):

$$t(x_j) \vee t(x'_j) = 1 ; \qquad \forall j = 1, \ldots, n$$

Because of this principle, such a function is uniquely determined by the values

$$\{t(x_1), \ldots, t(x_n)\},$$

hence there are 2^n such functions. For this reason, in the following, we will simply say *truth function on* $\{x_1, \ldots, x_n\}$ meaning a truth function on the clauses of the set $\{x_1, \ldots, x_n, x'_1, \ldots, x'_n\}$. On the contrary, given any n-ple

$$\varepsilon = (\varepsilon_1, \ldots, \varepsilon_n) \in \{0, 1\}^n$$

there exists only one truth function on $\{x_1, \ldots, x_n\}$, with the property that

$$t(x_j) = \varepsilon_j ; \qquad \forall j = 1, \ldots, n$$

In the following, given a truth function t, we will denote ε_t the string in $\{t(x_1), \ldots, t(x_n)\}$ uniquely associated to that function.

Lemma 1. *Let $\mathcal{T}$ be the set of truth functions on $\{x_1, \ldots, x_n\}$. The function*

$$t \in \mathcal{T} \mapsto |t(x_1), \ldots, t(x_n)\rangle \in \otimes^n \mathbf{C}^2$$

defines a one-to-one correspondence between $\mathcal{T}$ and the computational basis of $\otimes^n \mathbf{C}^2$.

Proof. From what was said above, the function

$$t \in \mathcal{T} \mapsto (t(x_1), \ldots, t(x_n)) \in \{0, 1\}^n$$

is a one-to-one correspondence. $\square$

Lemma 2. *Let $C \subseteq X$ be a clause and I, I' be the sets associated to it with the procedure explained in Section 1. Let t be a truth function on $\{x_1, \ldots, x_n\}$. Then*

$$t(C) = \left[\bigvee_{i \in I} t(x_i)\right] \vee \left[\bigvee_{j \in I'} (1 - t(x_j))\right]$$

Proof. By definition

$$C = \left(\bigcup_{i \in I} \{x_i\}\right) \cup \left(\bigcup_{j \in I'} \{x'_j\}\right).$$

The thesis follows because t is a Boolean algebra homomorphism. □

Definition 2. *A set of clauses $\mathcal{C}_0$ is said to be SAT if there exists a truth function t, on $\{x_1, \ldots, x_n\}$ such that*

$$\Lambda_{C \in \mathcal{C}_0} t(C) = \prod_{C \in \mathcal{C}_0} t(C) = 1$$

2 The Ohya and Masuda Technique

Let

$$\mathcal{C}_0 = \{C_1, \ldots, C_m\}$$

be a set of clauses on X. Let t be a truth function and $\varepsilon_t \in \{0,1\}^n$ be the corresponding binary string. Ohya and Masuda proposed a technique for constructing the following objects:
(i) A Hilbert space

$$\mathcal{H} = \otimes^{n+\mu} \mathbf{C}^2$$

where μ is a number to be determined as small as possible. Ohya and Masuda prove that μ can be chosen linearly in mn.
(ii) A unitary operator:

$$U_{\mathcal{C}_0} : \mathcal{H} \to \mathcal{H}$$

with the following property: for any truth function t

$$U_{\mathcal{C}_0} |\varepsilon_t, 0_\mu\rangle = |\varepsilon_t, x^{\varepsilon_t}_{\mu-1}, t(\mathcal{C}_0)\rangle$$

where 0_μ (resp. $x^\varepsilon_{\mu-1}$) is a μ (resp. $\mu - 1$ binary symbols depending on ε) string of zero

$$t(\mathcal{C}_0) = \prod_{C_j \in \mathcal{C}_0} t(C_j)$$

Furthermore $U_{\mathcal{C}_0}$ is a product of *gates*, namely of unitary operators that act on at most two q-bits each time.

Definition 3. *Any operator U that obeys the above conditions will be called an Ohya–Masuda unitary operator (OM unitary operator) for the SAT problem $(\mathcal{C}_0)$.*

The practical implementation of the Ohya–Masuda technique causes some problems in case the set of clauses considered becomes true with a very small probability. This problem has been recently studied by Ohya and Volovich [5] who made use of chaotic dynamics to amplify this probability.

3 Application of the Ohya–Masuda Technique to SAT

Let C_0 be and U_{C_0} as in Section 2 and, for every $\varepsilon \in \{0,1\}^n$, let t_ε be the corresponding truth function. Define the vector:

$$\psi_0 = \frac{1}{2^{n/2}} \sum_{\varepsilon \in \{0,1\}^n} |\varepsilon, 0_\mu\rangle$$

and apply the unitary operator U_{C_0} to it:

$$U_{C_0}\psi_0 = \frac{1}{2^{n/2}} \sum_{\varepsilon \in \{0,1\}^n} |\varepsilon, x^\varepsilon_{\mu-1}, t_\varepsilon(C_0)\rangle$$

Theorem 1. C_0 *is said to be SAT if and only if, the projector*

$$P_{n+\mu,1} = 1_{n+\mu-1} \otimes |1\rangle\langle 1| \quad ,$$

on the subspace of $\mathcal{H}$ spanned by the vectors

$$|\varepsilon_n, \varepsilon_{\mu-1}, 1\rangle$$

satisfies

$$P_{n+\mu,1}\psi_0 \neq 0 \qquad (*)$$

Proof. The condition (*) means that exists $t \in \mathcal{T}$ that $t(C_0) = 1$. On the contrary if such a t exists, $t \in \mathcal{T}$ such that $t(C_0) = 1$, then (*) holds. $\square$

Now, in the above notations, let

$$\rho = |\psi_o\rangle\langle\psi_o|$$

After a measurement of the observable with spectral projections $P_{n+\mu,0}, P_{n+\mu,1}$ the state becomes

$$\rho \to P_{n+\mu,1}\rho P_{n+\mu,1} + P_{n+\mu,0}\rho P_{n+\mu,0} =: \rho'$$

so that the probability of $P_{n+\mu,1}$ is

$$Tr\rho' P_{n+\mu,1} = \|P_{n+\mu,1}\psi\|^2 = \frac{|T(C_0)|}{2^n}$$

where $|T(C_0)|$ is the cardinality of the set $T(C_0)$, of all the truth functions t such that

$$t(C_0) = 1 \quad .$$

This shows that the Ohya–Masuda algorithm can in fact do more than simply verify satisfiability: in fact it allows to introduce a kind of *absolute measure of truth* for different set of clauses, measured in terms of the cardinality of the set of truth functions which give value 1 to the given set of clauses. We can introduce a total order in the family of these sets by saying that a family $\mathcal{C}$ is *more true* than another one $\mathcal{C}'$ if

$$|T(\mathcal{C})| \geq |T(\mathcal{C}')| \ .$$

To give a combinatorial estimate of this cardinality is quite difficult (in fact much more difficult than the SAT problem itself); however the Ohya–Masuda algorithm shows that one can get a probabilistic estimate by using a quantum computer.

4 Locality of the Ohya–Masuda Unitary Operator

To understand the Ohya–Masuda unitary operators structure, let us start to consider the one-clause C and let (I, I') a couple of subsets of $\{1, \ldots, n\}$, related to C, as described in Section 1. Then, given a truth function

$$t_\varepsilon \equiv (\varepsilon_1, \ldots, \varepsilon_n)$$

we have

$$t_\varepsilon(C) = \left(\bigvee_{i \in I} \varepsilon_i \right) \vee \left(\bigvee_{j \in I'} (1 - \varepsilon_j) \right)$$

Therefore, if U_C is an OM unitary operator for the SAT problem (C), we must have

$$U_C |\varepsilon, 0_\mu\rangle = |\varepsilon, x_\mu^\varepsilon, \left(\bigvee_{i \in I} \varepsilon_i \right) \vee \left(\bigvee_{j \in I'} (1 - \varepsilon_j) \right) \rangle$$

To implement U_C in a local way such an unitary operator we need *three kinds of gates*: (I.) Flip (or NOT) gates, (II.) localized OR gates, and (III.) localized AND gates (Toffoli gates).

5 Flip (or NOT) Gates

A flip, or NOT gate, is a completely localized gate, localized in a point h, and characterized by the property

$$N_h = \left(\bigotimes_{j \neq h} 1_j \right) \otimes ((|0\rangle\langle 1| + |1\rangle\langle +0|)) = (\otimes_{j \neq h} 1_j) \otimes N_h = (\otimes_{j \neq h} 1_j) \otimes U_{\text{NOT}} \ .$$

Its action on a binary string $|\varepsilon\rangle$ is

$$N_h |\varepsilon\rangle = |\varepsilon_1, \ldots, \varepsilon_{h-1}, 1 - \varepsilon_h, \varepsilon_{h+1}, \ldots\rangle$$

Because $h \neq k$, N_h and N_k act on different q-bits hence they commute and can, therefore, be parallelized. Given the clause $C \equiv (I, I')$, we define

$$N_{I'} = \otimes_{h \in I'} N_h$$

where N_h is as in (1). Therefore
$$N_{I'}|\varepsilon, 0_\mu\rangle = |(\varepsilon_i)_{i \notin I'}, (1 - \varepsilon_j)_{j \in I'}, 0_\mu\rangle =: |\varepsilon_{I'}, 0_\mu\rangle$$
where $\varepsilon_{I'}$, is obtained from ε by changing in $1 - \varepsilon_j$ all ε_j with $j \in I'$.

6 Localized OR, AND Gates

Lemma 3. *Let S be a set. Every function $f : S \to \{0, 1\}$ is unitarily implemented on $\mathbf{C}^{|S|} \otimes \mathbf{C}^2$, by the operator*
$$U_f = |x, f(x)\rangle\langle 0, x| + |x, 1 - f(x)\rangle\langle 1, x|$$
□

Example 1. The Toffoli gate corresponds to the choices
$$S = \{0, 1\}^{2n}$$
$$f : (\varepsilon_1, \varepsilon_2) \in \{0, 1\}^{2n} \to f(\varepsilon_1, \varepsilon_2) = \varepsilon_1 \wedge \varepsilon_2$$
where the $\wedge$–operation is meant componentwise. Explicitly
$$T = U_T = U_{AND} = |\varepsilon_1, \varepsilon_2, \varepsilon_1 \wedge \varepsilon_2\rangle\langle 0, \varepsilon_1, \varepsilon_2| + |\varepsilon_1, \varepsilon_2, 1 - (\varepsilon_1 \wedge \varepsilon_2)\rangle\langle 1, \varepsilon_1, \varepsilon_2|$$
The OR gate corresponds to the choice:
$$O = U_{OR} = |\varepsilon_1, \varepsilon_2, \varepsilon_1 \vee \varepsilon_2\rangle\langle 0, \varepsilon_1, \varepsilon_2| + |\varepsilon_1, \varepsilon_2, 1 - (\varepsilon_1 \vee \varepsilon_2)\rangle\langle 1, \varepsilon_1, \varepsilon_2|$$
When an AND (OR) is implemented a dust-q-bit is needed.

Definition 4. *A NOT (or Toffoli, or localized OR) gate is an operator of the form that:*
$$N_h = (\otimes_{j \neq h} 1_j) \otimes N \tag{1}$$
$$T_{i,j,k} = \left(\bigotimes_{h \neq i,j,k} 1_h\right) \otimes T \tag{2}$$
$$O_{i,j,k} = \left(\bigotimes_{h \neq i,j,k} 1_h\right) \otimes O \tag{3}$$
where (1) means that N acts on the h–th factor of $\otimes^N \mathbf{C}^2$ and (2), (3) that T and O act on the factor product i, j, k.

In the following it will be convenient to consider also a modified localized OR operator:
$$\overline{O}_{i,j,k} = O_{i,j,k} N_i \tag{4}$$
This new operator first changes the i–th 0 into 1 then acts with the ordinary $O_{i,j,k}$. Explicitly:
$$\overline{O}_{i,j,k}|\varepsilon_i, \varepsilon_j, 0_k\rangle = O_{i,j,k}|1 - \varepsilon_i, \varepsilon_j, 0_k\rangle$$

7 The Case of Only One Clause

In the notations of Section 1, a clause $C \equiv (I, I')$ and a truth function $t_\varepsilon \equiv \varepsilon \in \{0,1\}^{2n}$. Assuming that

$$I = \{1 \le j_1 < j_2 < \ldots < j_i \le n\}$$
$$I' = \{1 \le j'_1 < j'_2 < \ldots < j'_{i'} \le n\}$$

define the vector

$$\psi_\varepsilon^0 = |\varepsilon, 0_{2n}\rangle \in (\mathbf{C}^2)^{3n}$$

Then we have:

$$O_{j_1,j_2,n+1} \psi_\varepsilon^0 = |\varepsilon, \varepsilon_{j_1} \vee \varepsilon_{j_2}, 0_{2n-1}\rangle =: \psi_\varepsilon^1$$
$$O_{j_3,n+1,n+2} \psi_\varepsilon^1 = |\varepsilon, \varepsilon_{j_1} \vee \varepsilon_{j_2}, \varepsilon_{j_1} \vee \varepsilon_{j_2} \vee \varepsilon_{j_3}, 0_{2n-2}\rangle =: \psi_\varepsilon^2 \ .$$

Generally speaking, for $1 \le k \le i \le n$ we have

$$O_{j_k,n+(k-2),n+(k-1)} \psi_\varepsilon^{k-1} = |\varepsilon, \varepsilon_{j_1} \vee \varepsilon_{j_2}, \varepsilon_{j_1} \vee \varepsilon_{j_2} \vee \varepsilon_{j_3}, \ldots, \varepsilon_{j_1} \vee \ldots \vee \varepsilon_{j_k}, 0_{2n-k}\rangle = \psi_\varepsilon^k \ .$$

After i steps we arrive to:

$$\left(\prod_{k=3,\ldots,i}^{\leftarrow} O_{j_k,n+(k-2),n+(k-1)} \right) O_{j_1,j_2,n+1} \psi_\varepsilon^0 =$$
$$= |\varepsilon, \varepsilon_{j_1} \vee \varepsilon_{j_2}, \ldots, \varepsilon_{j_1} \vee \ldots \vee \varepsilon_{j_i}, 0_{2n-i}\rangle =: \psi_\varepsilon^i$$

where the symbol $\prod^{\leftarrow}$ means that the product is ordered from right to left, namely

$$\ldots O_{j_k,n+(k-2),n+(k-1)} \cdots O_{j_3,n+1,n+2} O_{j_1,j_2,n+1}$$

At this point the operation is repeated with the overlined OR $\overline{O}_{j_k,n+(k-2),n+(k-1)}$ (cf. (4)). With the usual notation $\varepsilon' = 1 - \varepsilon$ we get

$$\left(\prod_{k=3,\ldots,i'}^{\leftarrow} \overline{O}_{j'_k,n+(k-2),n+(k-1)} \right) \overline{O}_{j'_1,j'_2,n+1} \psi_\varepsilon^i =$$

$$|\varepsilon, \varepsilon_{j_1} \vee \varepsilon_{j_2}, \ldots, \varepsilon_{j_1} \vee \ldots \vee \varepsilon_{j_i}, \varepsilon'_{j'_1} \vee \varepsilon'_{j'_2}, \ldots, \varepsilon'_{j'_1} \vee \ldots \vee \varepsilon'_{j'_{i'}}, 0_{2n-i-i'}\rangle =: \psi_\varepsilon^{i+i'}$$

Then the unitary operator associated to a clause C will be

$$U_C = \left(\prod_{k=3,\ldots,i'}^{\leftarrow} \overline{O}_{j'_k,n+(k-2),n+(k-1)} \right) \overline{O}_{j'_1,j'_2,n+1}$$
$$\left(\prod_{k=3,\ldots,i}^{\leftarrow} O_{j_k,n+(k-2),n+(k-1)} \right) O_{j_1,j_2,n+1}$$

8 More Clauses: The General Case

Let
$$\mathcal{C}_0 = \{C_1, \ldots, C_m\}$$
be an (ordered) set of clauses and for every $j = 1, \ldots, n$, let
$$C_j \equiv (I_j, I'_j) \ .$$

Assume that
$$I_k = \{1 \le j_{k1} < j_{k2} < \ldots < j_{ki_k} \le n\}$$
$$I'_k = \{1 \le j'_{k1} < j'_{k2} < \ldots < j'_{ki'_k} \le n\}$$

For every clause we need at most $2n$ dust q-bits. Furthermore to realize AND between the clauses truth values, we need at most m new dust q-bits. Eventually we need at most $2mn + m$ dust q-bits and therefore the final space will be
$$(\mathbf{C}^2)^{2mn+m+n} = \mathcal{H}_\varepsilon \otimes \mathcal{H}_D^1 \otimes [\mathcal{H}_D^2 \otimes \mathbf{C}^2] \otimes \ldots \otimes [\mathcal{H}_D^m \otimes \mathbf{C}^2]$$
where $\mathcal{H}_\varepsilon \sim (\mathbf{C}^2)^n$ codes the truth function ε, $\mathcal{H}_D^k \sim (\mathbf{C}^2)^{2n}$ is the dust space of the k-th clause due to the OR operations and the additional factors $\mathbf{C}^2$ are needed for the AND operations. The unitary operator for the k-the clause C_k will then be
$$\hat{U}_{C_k} = (U_{C_k})_{\varepsilon, D_k} \otimes 1_{others} \ ; \quad k = 1, 2, \ldots, m$$
where U_{C_k} is defined as in the previous section and the index (ε, D_k), in the right hand side means that the operator acts on $\mathcal{H}_\varepsilon \otimes \mathcal{H}_D^k$. The unitary operator for the AND operation will be
$$T_{\mathcal{C}_0} = T_{c_{m-1},m,c_m} \ldots T_{c_3,4,c_5} T_{c_2,3,c_3} T_{1,2,c_2}$$
where the indices $1, 2, \ldots m$ refer to the q-bits encoding the truth value $t_\varepsilon(C_1), \ldots, t_\varepsilon(C_m)$ and the indices c_k to the $\mathbf{C}^2$ dust factor of the k-th clause

The unitary operator for the set $\mathcal{C}_0$ of clauses will be be an (ordered) set of clauses and
$$U = T_{\mathcal{C}_0} \prod_{k=1,\ldots,m} \hat{U}_{C_k}$$
Notice that no order is required in the above product because the unitary operators $\hat{U}_{C_k}$ commute.

References

1. M. Garey and D. Johnson, *Computers and Intractability – a Guide to the Theory of NP–Completeness*, Freeman, 1979.
2. T. Toffoli, Computation and construction universality of reversible cellular automata, *Journal of Computer and System Sciences* 15, (1977), 213.

3. M. Ohya, *Mathematical Foundation of Quantum Computer*, Maruzen Publ. Company, 1998.
4. M. Ohya and N. Masuda, *NP problem in Quantum Algorithm*, quant–ph/9809075.
5. M. Ohya, I.V. Volovich, Quantum computing, NP–complete problems and chaotic dynamics, *Preprint Vito Volterra* N426, 2000, Università degli Studi di Roma "Tor Vergata".
6. M. Ohya, N. Watanabe, On Mathematical treatment of Fredkin-Toffoli-Milburn gate, *Physica D*, 120 (1998), 206-213
7. I.V. Volovich, Quantum Computers and Neural Networks, Invited talk at the *International Conference on Quantum Information*, Meijo University, 4-8 Nov. 1997.
8. I.V. Volovich, Mathematical models of quantum computers and quantum decoherence problem, in the volume dedicated to V.A. Sadovnichij, Moscow State University, 1999, to be published.
9. I.V. Volovich, Quantum Kolmogorov Machine, Invited talk at the *International Conference on Quantum Information*, Meijo University, 4-8 Nov. 1997.
10. I.V. Volovich, Atomic Quantum Computer, quant–ph/9911062; *Preprint Vito Volterra* N403, 1999, Università degli Studi di Roma "Tor Vergata".
11. I.V. Volovich, Models of quantum computers and decoherence problem, *Preprint Vito Volterra* N358, 1999, Università degli Studi di Roma "Tor Vergata".

Computational Methods and Tools for Modeling and Analysis of Complex Processes

Ioannis Antoniou[1,2] and Viktor V. Ivanov[1,3]

[1] International Solvay Institutes for Physics and Chemistry, Brussels, Belgium
[2] Theoretische Natuurkunde, Free University of Brussels, Brussels, Belgium
[3] Joint Institute for Nuclear Research, Dubna, Russia

Abstract. This review is devoted to the computational methods and tools for modeling and analysis of various complex processes in physics, medicine, social dynamics and nature. We consider: 1) the multivariate data analysis based on Ω_n^k-criteria and artificial neural networks (ANN), 2) the applications of neural networks for the function approximation and for the reconstruction and prediction of chaotic time series, and 3) the use of cellular automata (CA) in pattern recognition and in modeling of complex dynamical systems.

Introduction

The modeling and statistical analysis of complex processes is a challenge to conventional methods. The probabilistic approach has been a well-known strategy for discussing complex systems as the underlying complexity gives rise to statistical simplicity [1–7].

We review here the results of statistical methods useful for the computation and prediction of complex systems, namely:

1. Multivariate data analysis,
2. Artificial neural networks for the identification of system dynamics,
3. Cellular automata for pattern recognition and modeling complex systems.

1 Multivariate Data Analysis Based on Ω_n^k-Criteria and ANN

The primary goal of experimental data processing consists in identification of relevant events among all events obtained in the experiment. By an event we mean a sample of feature variables characterizing the analyzed pattern. The classification of events in the one-dimensional case is carried out with the help of a simple cut on a feature variable. When an event is characterized by more then one variable, the procedure for constructing a multivariate classifier is not a trivial one.

Different effective methods were developed for multidimensional data analysis, like Fisher's discriminator [8], clustering [9], principal components

[10]. Recently the use of artificial neural networks in the analysis of experimental data, and, particularly in classifying multivariate events, has become widespread.

However, all methods developed till nowadays are directed to the classification of events belonging to different classes with comparable contribution into the distribution to be analyzed. At the same time, the maximal interest in physics, in biology, in medicine, and in other fields is now concerned with rare events. But, there does not exist an adequate systematic mathematical apparatus for the identification and analysis of rare events.

In paper [11] we have suggested and investigated a class of new non-parametric Ω_n^k-statistics

$$\Omega_n^k = -\frac{n^{\frac{k}{2}}}{k+1} \sum_{i=1}^{n} \left\{ \left[\frac{i-1}{n} - F(x_i) \right]^{k+1} - \left[\frac{i}{n} - F(x_i) \right]^{k+1} \right\}, \quad (1)$$

where $F(x)$ is the theoretical distribution function of the variable x and $x_1 < x_2 < \ldots < x_n$ is the ordered sample of size n. On the basis of (1) goodness-of-fit criteria have been constructed, which are usually applied for testing the correspondence of each sample (event) to the distribution known *apriori*.

The following procedure for identifying multidimensional feature events was developed in refs. [12, 14] on the basis of the Ω_n^k criteria:

1. *the spectra to be analyzed are transformed ("normalized"), so that the contributions of different dominant distributions (in most cases these are distributions concerning the background events) are described by a sole distribution function $F_b(x)$;*
2. *each sample, composed of values pertaining to the different transformed spectra, is tested with the aid of the Ω_n^k goodness-of-fit criterion for correspondence to the $F_b(x)$ hypothesis; in this process the signal events, which do not comply with the null-hypothesis, correspond to large absolute values of the Ω_n^k-statistic, resulting in their clustering in the critical region;*
3. *events that happen to be in the critical region are further subjected to a second test in accordance with items a) and b), only with the difference that now it is precisely the signal events that are collected in the admissible region (using the corresponding distribution function $F_s(x)$), and those events that fall into the critical region are rejected; this results in additional suppression of background events in the spectra being studied.*

The procedure for data handling described above was applied in analyzing the information obtained in several particle physics experiments [12, 14, 15].

In [12], the classification procedure was realized on the basis of the Ω_n^2 (Smirnov–Cramer–von Mises test) goodness-of-fit criterion. It was applied for the identification of the charged particles detected by the MASPIK spectrometer (see details in [13]) at the angle of 106 Mrad to the incident beam axis in the collisions of 9 GeV/c deuterons with targets from CD^2, CH^2 and C. The momentum spectrum of secondary particles in the interval 3.5 to 5.5 GeV/c was measured; the main contribution to the spectrum was given by protons, the admixture of deuterons did not exceed 1%. Distributions analyzed were presented as simultaneous time-of-flight measurements performed by two independent systems. The analysis of two independent mass spectra of secondary particles made it possible successful extraction of rare events due to secondary deuteron production.

A statistical method based on the Ω_n^3 goodness-of-fit criterion for the identification of relativistic charged particles using the measurements of ionization losses in several scintillation counters of the MASPIK spectrometer was suggested in [14]. The method was applied for identifying secondary particles (p, d, tritium, ^{3}He and ^{4}He nuclei) produced in collisions of ^{4}He nuclei (4.5 GeV/c per nucleon) with target nuclei at the angle 140 Mrad. It permitted reliable extraction of rare events due to the production of singly- and doubly-charged particles, the admixture of which did not exceed 0.1%.

Monte-Carlo simulation has been performed for the experimental studies of sub-threshold K^+-meson production processes to be carried out at the COSY accelerator (Juelich, Germany). Reliable identification of rare K^+-meson events in conditions of dominant background of π^+ (by estimation the ratio N_{K^+}/N_{π^+} may amount to 10^{-5}) was shown to be possible applying the traditional statistical method together with a procedure based on the Ω_n^3 goodness-of-fit criterion [15].

Recently, the use of artificial neural networks in multi-dimensional data analysis has became widespread [16–19]. One such problem consists in classifying individual events represented by empirical samples of finite volumes pertaining to one of the different partial distributions composing the distribution to be analyzed. A layered *feed-forward* network – multilayer perceptron (MLP) – is a convenient tool for constructing multivariate classifiers, although its learning speed and power of recognition critically depends on the choice of input data.

Thus, in [20] it has been shown that transformation of the data, input to a MLP, into the variational series leads to a significant acceleration of the network training and, also, to the improvement of the quality of recognition. Moreover, the representation of data in such a form permits to reduce the number of neurons in the hidden layer without loss of precision in the classification.

A model for a two-level trigger was developed for suppression of background and for the effective selection of feature events involving short-lived Λ-, Σ- and ϕ-particles in the experiment DISTO [21]. The first-level trigger is intended for selection of events by their multiplicity: only four-prong events are selected. Events accepted by the first-level trigger are then examined with the help of the second-level trigger, which is to be applied for the track recognition, in searching for a secondary vertex, and for identifying the secondary particles. It is based: 1) on the recognition of straight tracks applying a specialized cellular automaton (see details in the next section), 2) on the momentum variables permitting effective selection of events containing a secondary vertex, and 3) on the identification of secondary charged particles applying MLP [22].

A simple and efficient algorithm for identifying events with a secondary vertex making use of MLP was developed by Bonushkina et al [23]. The differences R_x, R_y (in the XOZ and YOZ projections respectively) between the largest and the smallest impact parameters[1] $\{D_i\}$ ($i = 1, 2, ..., n$; n is the number of tracks in the event) of all tracks belonging to each of the events analyzed were used in establishing the identification criteria for signal and background events. An effective method for identifying the tracks associated with a particular secondary vertex in an event was also developed. The method is based on the differences between the asymmetries exhibited by the sets $\{D_i\}$ of individual signal events and of background events.

A comparative study of multidimensional classifiers based on the goodness-of-fit criteria Ω_n^k and MLP has been carried out in [24, 25]. We show that, when *only the parameters of the dominant distribution are known*, the goodness-of-fit criteria Ω_n^k serve as a suitable tool for the recognition of events corresponding to various distributions. The repeated application of the Ω_n^k-criteria permits to extract the contributions of any number of partial distributions from the resultant spectrum observed in an experiment. This makes it possible, for instance, upon estimation of the parameters of the constituent distributions, to additionally make use, if necessary, of a neural network.

It is useful to note that the Ω_n^k-criteria usage is substantiated quantitatively, while the results yielded by ANN are only qualitative.

A procedure for the recognition of the electrocardiogram (ECG) features of one heart beat and from a single channel using MLP was developed in [26]. The main idea of the method was to present to the network not raw ECG, but the transformed data. We believe that a system of orthogonal polynomials on a set of uniformly spaced points is the adequate formalism for the analysis of ECGs, since measurements are taken in equal time intervals, and all points can be denoted $0, 1, 2, \ldots n$. The above mentioned polynomials $P_{k,n}(x)$, $k = 0, 1, 2, ..., m \leq n$ are related by the following recurrent equation (see, for instance, [27]):

$$\left(x - \frac{n}{2}\right) P_{m,n}(x) + \frac{(m+1)(n-m)}{2(2m+1)} P_{m+1,n}(x) + \frac{m(n+m+1)}{2(2m+1)} P_{m-1,n}(x) = 0,$$

$$1 \leq m \leq n, \qquad P_{0,n}(x) = 1, \quad P_{1,n}(x) = 1 - \frac{2x}{n}.$$

The polynomial $P_m(x)$ approximating the function $f(x)$ in this case is

$$P_m(x) = c_0 P_{0,n}(x) + c_1 P_{1,n}(x) + \ldots + c_m P_{m,n}(x),$$

where

$$c_i = \frac{(2i+1)n^{(i)}}{(i+n+1)^{(i+1)}} \sum_{k=0}^{n} f(k) P_{i,n}(k), \quad i = 0, 1, 2, \ldots m,$$

---

[1] The impact parameter of a track in the plane passing through the center of the target and perpendicular to the beam.

and

$$n^{(i)} = n(n-1)\ldots(n-i+1), \quad (i+n+1)^{(i+1)} = (i+n+1)(i+n)\ldots 1.$$

The proposed transformation results in a significantly simpler data structure, stability to noise and to other accidental factors. The method was tested on data generalizing the features of normal and modified ECGs and provided a high level of recognition for revealing barely noticeable pathologies.

A by-product of the method are the compression of raw data and the reduction of its amount; the compression coefficient has a value of $5 \div 10$ and can be improved. This is close to the result obtained in [28] where the Chebyshev polynomials were applied for the compression of cardiological data. Their procedure envisage the division of ECG on small subintervals where only three expansion coefficients provide the mean-squared error $\leq 1\%$ and the compression factor $6 \div 12$. But, in practical applications, when a function is given in fixed number of points, the problem of evaluation of expansion coefficients arises (see below).

2 ANN: Function Approximation and Identification of the Dynamics Underlying Time Series

The procedure adopted for the parameterization of functions defined on a finite set of argument values plays an essential role in the problem of experimental data processing. Diverse methods have been developed and are widely applied in constructing approximating functions in the form of algebraic or trigonometric polynomials. Among such polynomials, the orthogonal Chebyshev polynomials [27, 28] occupy an important place, since, in the case of a broad class of functions, expansions in Chebyshev polynomials converge more rapidly than expansions in any other set of polynomials [29]. A non-traditional approach to the interpolation of one-dimensional functions has been presented in [30].

Consider a function $f(x)$ defined on a finite set of values of x: $f(x_0)$, $f(x_1)$, ..., $f(x_n)$. If the values of $f(x)$ at intermediate x values are required for solution of the problem, then it is convenient to construct a function $\varphi(x)$, which is easy to calculate, and which assumes the values $f(x_0)$, $f(x_1)$, ..., $f(x_n)$ at the points $x_0, x_1, \ldots, x_n$, and which approximates $f(x)$ with a certain degree of accuracy in the rest of its domain. Such a function is called an interpolating function.

The function $\varphi(x)$ can be sought in the following form of an expansion in orthogonal Chebyshev polynomials of the I-st kind:

$$\varphi_n(x) = a_0 T_0(x) + a_1 T_1(x) + a_2 T_2(x) + \ldots + a_n T_n(x), \tag{2}$$

where $T_n(x) = \cos(n \arccos x), \quad |x| \leq 1$.

In order to avoid the difficulty in calculation of the coefficients a_i, proceeding from an arbitrary set of experimental points, we implement the

Chebyshev neural network (CNN) with the architecture presented in Fig. 1.

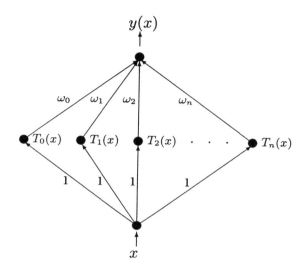

Fig. 1. Architecture of a feed-forward Chebyshev neural network.

The network has one input neuron, to which the argument x is supplied, a layer of hidden neurons and one output neuron, from which the calculated value of function $y(x)$ is taken.

The argument x is sent from the input neuron to the neurons of the hidden layer. Each i-th neuron of the hidden layer transforms the received signal in accordance with the transfer function $g_i(x)$ in the form of a Chebyshev polynomial $g_i(x) = T_i(x)$, where $i = 0, \ldots, n$. Then, the sum of weighted signals from the neurons of the hidden layer $a = \sum_{i=0}^{n} w_i \cdot T_i(x)$ is supplied to the output neuron, which transforms it in accordance with the linear function $g(a) = a$.

When the weights of the connections between the input neuron and the neurons in the hidden layer are all set to 1, the weights w_i will play the role of the expansion coefficients a_i. The number of neurons in the hidden layer coincides with the number of terms in Eq. (1) and determines the accuracy of approximation of the function being analyzed.

The weights w_i are calculated during the neural network "training" applying the *back-propagation* algorithm [31]. The correction of the weights $w_{i,k}$ at each k-th step is given by the following expression

$$\Delta w_{i,k} = -\eta \Delta E_{i,k} + \alpha \Delta w_{i,k-1}, \qquad (3)$$

where
$$\Delta E_{i,k} = \sum_p [y(x_p) - f(x_p)] T_{i-1}(x_p). \tag{4}$$

In Eqs. (3) and (4) η is the parameter controlling the learning speed, $\alpha \Delta \omega_{i,k-1}$ are the moments, which suppress the oscillations at the network output, $p = 1, 2, ..., N_{train}$ is the number of training patterns, $f(x_p)$ is the target value of the output signal $y(x_p)$ [31]. The neural network training goes on until an acceptable correspondence between output signals and target values is reached.

This approach permits to calculate the expansion coefficients during the network training process, for which arbitrary points (for instance, measured in experiments) from the function's domain are used. The neural network provides an accuracy of function approximation practically coinciding with the accuracy, that can be achieved within the traditional approach, when the values of the function at the nodal points are known.

Lapedes and Farber [32] studied the ability of layered neural networks to reconstruct and predict chaotic time series, which is an important problem in economics, meteorology and many other areas. They demonstrated that ANN with sigmoidal neurons in the hidden layer can be used for predicting points of highly chaotic time series with an increase by orders of magnitude in accuracy over conventional methods. In paper [33] we compared the performance of MLP with the CNN network in the reconstruction of a chaotic time series corresponding to the Logistic map [6, 34, 35].

Figure 2 shows the behavior of the identification level with respect to each epoch for MLP and for CNN during training of networks. By identification level we mean the fraction of patterns for which the relative error – the absolute deviation of the predicted value $y(x)$ from the actual value $f(x)$ divided by the actual value – does not exceed 0.05. One can see that the CNN network very quickly (approximately after $10 \div 15$ epochs) reaches a high level in reconstruction of the learning time series and then slowly improves it, reaching 95% after 5000 epochs. At the same time, the MLP network very slowly improves its reconstruction ability and after 5000 epochs reaches the maximal level, which equals to 69%. Both networks show similar behavior during their testing: 97% for CNN and 69% for MLP after the last training epoch.

More detailed analysis has shown (see [33]) that, compared with the conventional MLP, the CNN network provides a 50-fold improvement in the approximation. As a new approach provides better approximation of the time series analyzed, we have new possibilities for long-term prediction. Figure 3 shows the behavior of the deviation of the predicted value from the actual value when the MLP network (a) and the CNN network (b) are used for long-term prediction. We clearly see that MLP is reliable for 3 iterations while CNN can go up to 9 iterations.

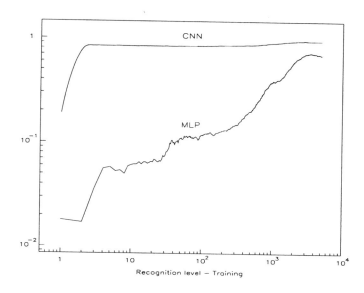

Fig. 2. The behavior of the recognition level vs. the number of epoch for MLP and CNN during training of networks.

3 CA: Pattern Recognition and Complex System Modeling

Cellular automata arose from numerous attempts to create a simple mathematical model describing complex biological structures and processes [36]. A cellular automaton is a most simple discrete dynamical system, the behavior of which is totally dependent upon the local interconnections between its elementary parts [37]. This model turned out to be very productive and has been widely and successfully applied in describing various complex structures and processes in physics, biology, chemistry and etc.

A model of the cellular automaton for the straight tracks recognition in the experiment DISTO [21] has been developed in [38]. In this case, a cell was identified with the straight-line segment connecting two hits in neighboring coordinate detectors: scintillation fibre chambers and proportional chambers. To take into account the inefficiency of the detectors one must consider, also, the segments connecting hits skipping one detector. Clearly, only such segments can be considered neighbors which have a common point serving as the end of one segment and the beginning of the second. At each step a cell can assume one of two possible states: 1, if the segment can be considered a part of the track, and 0 otherwise.

As the criterion in assigning segments to a track the angle between two adjacent segments was taken. Owing to the coordinate detectors having a

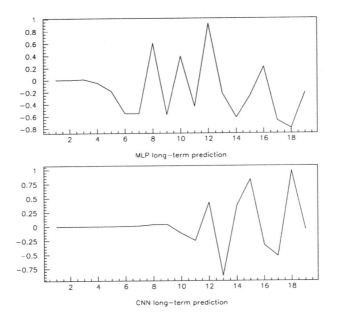

Fig. 3. The behavior of the deviation of the predicted value from the actual value when the MLP network and the CNN network are used for the long-term prediction.

discrete structure and to multiple scattering in the material of the experimental apparatus, the angles between track segments in the real experiment are not zero, but an upper limit can be imposed. Upon completion of the cellular automaton evolution, additional testing of the quality of reconstructed tracks (for instance, for the presence of at least two hits belonging only to each individual track) is carried out. This permits rejecting "phantom" tracks, which were accidentally constructed from hits belonging to different tracks.

Figure 4 shows the initial configuration of CA for a typical Monte-Carlo event in the spectrometer DISTO, and figure 5 presents the resultant configuration of CA for the same event.

In paper [39] the implementation of Probabilistic Cellular Automata (PCA) in the study of multi-species agent groups is investigated. PCA evolves in time according to two rules of communication: one describes interactions between agents and the other concerns the interactions of the agents with an additional agent which evolves in time in a periodic and stable manner. The application of these rules depends on a prescribed probability distribution. The

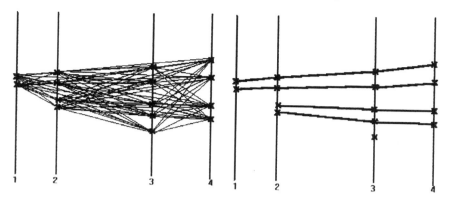

Fig. 4. Initial configuration of the cellular automaton for a typical Monte-Carlo event in the spectrometer DISTO.

Fig. 5. Resultant configuration of the cellular automaton for the event presented in the previous figure.

speed and the efficiency (coordination) with which collective patterns are constructed in time as a function of the parameters of the system is analyzed.

The ideas of self-organized criticality (SOC) and the observation of SOC-like behavior of simple dynamical systems, like cellular automata (CA), have motivated the application of CA to earthquake simulation. A number of papers on the construction of CA counterpart of the well-known Burridge-Knopoff (BK) model of earthquake faults [40] have been published [41, 42]. Both, the P. Bak & C. Tang model [41] of the complete stress release and the H. Nakanishi's model [42] with more elaborated stress relaxation function show the power law behavior which closely resembles the Gutenberg-Richter power law [43]. Both of them can not recover the main aspects of the BK system of differential equations for the spring-block model.

What can be done to make the existing CA models closer to the BK model? In the paper [44] a modification of the existing CA, which partially answers this question, is presented. One of the basic principles of the construction of cellular automata for physical applications consists in conserving the symmetry of the original physical system as precisely as possible.

Let us compare the CA models [41, 42] and the BK spring-block model in this aspect. The BK system is a system of N blocks of mass $m_i, i = \overline{1, N}$ resting on a rough surface and connected to each other by harmonic springs of stiffness k_c; each block is attached by a leaf spring of stiffness k_p to the moving upper line, see Fig. 1.

Initially (at $t = 0$) the system is at rest, and the elastic energy accumulated in "horizontal" springs is only due to randomly generated small initial displacements of the blocks from their neutral positions. The moving upper

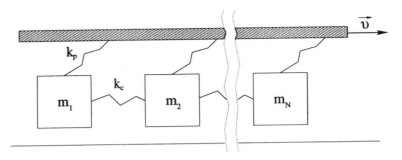

Fig. 6. The geometry of the BK model: the system is composed of N identical blocks of mass m_i, k_c is the stiffness of the "horizontal" springs, k_p is the stiffness of the pulling springs, v is the constant velocity of the pulling line.

line, which simulates the movement of an external driving plate, exerts a force $f_n = -k_p(x_n - vt)$ on each n-th block. The nonlinear friction is defined in such a way that it holds each block at rest until the sum of all the forces applied to this block exceeds a certain critical value F_0. Then the block makes a slip inhibited by nonlinear friction to a new position. A pause between two slips is believed to account for a pause between earthquakes.

The CA of P.Bak and C.Tang for artificial earthquake simulation [41] is a simple sandpile-like system which obeys the following rules:

- *The array of N values ("blocks") is initiated by certain randomly generated values $f_i^0, i = \overline{1, N}$ (the upper indices are used for discrete time).*
- *If the sum of the forces on i-th block exceeds a certain threshold value (usually $f_{Th} = 1$, without loss of generality), then "the accumulated stress" is shared with the nearest neighbors according to the rule*

$$f_{i\pm1} \to f_{i\pm1} + \frac{k_c}{2k_c + k_p}\delta f_i, \qquad (5)$$

where $\delta f_i = f_i - f_i'$ is the stress drop of the over-threshold i-th block, which is evaluated according to the law

$$f_i' = \phi(f_i - f_{Th}). \qquad (6)$$

The relaxation function ϕ is model dependent: it may be taken to be zero as in the P. Bak & C. Tang model, or it may be some decreasing function as in the Nakanishi cellular automaton.
- *When the evolution is completed the forces applied to all the blocks are incremented by the tectonic force*

$$f_i' \to f_i' + k_p v \Delta t, \quad i = \overline{1, N} \qquad (7)$$

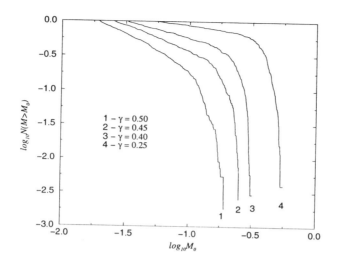

Fig. 7. Cumulative event distribution $N(M > M')$ for one-dimensional 35 cell CA calculated for different values of the asymmetry parameter $\gamma = 0.25, 0.4, 0.45, 0.5$ ($\gamma = 0.5$ corresponds to the symmetric case) plotted in logarithmic coordinates vs. the magnitude $m = \lg M$ with the event size understood to be the total stress relaxation $M = \sum_i (f_i - f_i')$. The significance level was taken as 0.1 of the maximal magnitude for each run. The relaxation function ϕ was taken as in the above cited Nakanishi's paper.

Now let us turn to the symmetries of the BK and CA models. Due to the presence of the tectonic driving force $k_p vt$, which implies the existence of a preferable space direction, namely, the direction of the moving plate velocity v, the BK system is apparently not invariant under inversion $x \to -x$. The stress redistribution law (5) of the CA, in contrast, is completely symmetric with respect to this inversion ($f_{i+1} \leftrightarrow f_{i-1}$). Therefore, the above mentioned CA has more symmetries than the parent BK model. What can be done about this? The answer is evident. An asymmetry should be introduced in the stress redistribution law (5) as follows:

$$f_{i+1} \to f_{i+1} + (1-\gamma)\alpha\delta f_i, \quad f_{i-1} \to f_{i-1} + \gamma\alpha\delta f_i. \tag{8}$$

The value of the asymmetry parameter $\gamma = 1/2$ corresponds to equal sharing (5), $\gamma = 0$ leads to completely asymmetric sharing. The factor $\alpha = \frac{2k_c}{2k_c + k_p}$ is chosen to comply with [42, 45].

We have performed computer simulations with one-dimensional 35 blocks CA for different values of the asymmetry parameter $\gamma = 0.25, 0.4, 0.45, 0.5$. The cumulative event distribution, i.e. the number of events of magnitude not less than a given value, which is often expected to have the form of the Gutenberg-Richter law, is presented in Fig. 7.

As it can be seen from the picture, the asymmetry parameter γ, significantly affects the slope of the curve (the logarithm of the cumulative event number vs. the magnitude). We also observed the amplitude of events to increase with asymmetry increasing. This means that γ is a control parameter of the system, the appropriate choice of which can tune the system close or far from the realistic value of $b \approx 1$ in the Gutenberg-Richter law [43]:

$$\lg N(M > M') = a - mb, \quad m = \lg M = \lg \sum_i (f_i - f_i'),$$

where a is the empirical constant.

We also developed an analytic method which gives rise to a new integral characteristic of the CA-generated event size [46]. By this method a functional quadratic in stress release, which can be regarded as an analog of the seismic event energy, is constructed. The distribution of seismic events with respect to this "energy" shows rather realistic behavior, even in two dimensions [46].

Acknowledgements

We acknowledge the European Commission for partial support within the framework of the EU–RUSSIA Collaboration under the ESPRIT project CTIAC-21042. We are grateful to Profs. I. Prigogine and G. Piragino for fruitful discussions and support.

These methods and tools reported in this paper were developed at the International Solvay Institutes for Physics and Chemistry (Brussels, Belgium), the Laboratory of Computing Techniques and Automation of the Joint Institute for Nuclear Research (Dubna, Russia), and INFN (Torino, Italy).

References

1. Prigogine I., "From Being to Becoming", Freeman, San Francisco (1980).
2. I. Prigogine, "The End of Certainty: Time, Chaos and Laws of Nature", The Free Press, New York (1997).
3. I. Antoniou and S. Tasaki, International J. of Quantum Chemistry 46, 425-474, (1993).
4. I. Antoniou and Z. Suchanecki, Nonlinear Analysis: Theory, Methods and Applications, 30, 1939-1958 (1997).
5. I. Antoniou and Z. Suchanecki, Found. Phys. **27**, 333-362 (1997).
6. A. Lasota and M. Mackey, Chaos, Fractals and Noise, Springer Verlag, 1994.
7. O. Barndorf-Nielsen, J. Jensen, W. Kendalf, "Networks and Chaos – Statistical and Probabilistic Aspects", Chapman & Hall, London (1993).
8. R.A. Fisher: Ann. Eugen.7 (1936), p.179.
9. Th. Nauman and H. Schiller: "Multidimensional Data Analysis", in "Formulae and Methods in Experimental Data Evaluation", vol.3 (European Physical Society, Geneve, 1983).
10. R. Brun et al. LINTRA, CERN Program Library E 5002.

11. V.V. Ivanov and P.V. Zrelov, Int. J. Comput. & Math. with Appl., vol. **34**, No. 7/8, (1997)703-726; JINR Communication P10-92-461, 1992 (in Russian).
12. P.V.Zrelov and V.V.Ivanov, In: Collection of Scientific Papers in Collaboration with Joint Institute for Nuclear Research, Dubna, USSR and Central Research Institute for Physics, Budapest, Hungary. "Algorithms and Programs for Solution of Some Problems in Physics". Sixth Volume. Preprint KFKI-1989-62/M, 1989, p.127-142.
13. L.S. Azhgirey et al, "Russian Nucl. Phys.", **46**, issue 4(10), (1986)1134-1141.
14. P.V. Zrelov and V.V. Ivanov, Nucl. Instr. and Meth. in Phys. Res., A**310** (1991)623-630.
15. P.V. Zrelov et al, JINR Preprint, P10-92-369, Dubna, 1992; Mathematical Modeling, vol. **4**, N%11, (1993)56-74 (in Russian).
16. S. Haykin, "Neural Networks: A Comprehensive Foundation", Prentice-Hall, Inc., 1999.
17. C. Peterson and Th. Rögnvaldsson, in Proc. II Int. Workshop on "Software Engineering, Artificial Intelligence and Expert Systems in High Energy Physics". New Comp. Tech. in Phys. Res. II, edited by D. Perret-Gallix, World Scientific, (1992)113.
18. B. Denby, in Proc. II Int. Workshop on "Software Engineering, Artificial Intelligence and Expert Systems in High Energy Physics". New Comp. Tech. in Phys. Res. II, edited by D. Perret-Gallix, World Scientific, (1992)287.
19. D.T. Pham and L. Xing, "Neural Networks for Identification, Prediction and Control", Springer-Verlag Berlin (1995).
20. A.Yu. Bonushkina et al, in Proc. IV Int. Workshop on Software Engineering, Artificial Intelligence and Expert Systems for High Energy and Nuclear Physics, April 3-8, 1995, Pisa, Italy; "New Computing Techniques in Physics Research IV", edited by B. Denby & D. Perret-Galix, "World Scientific", (1995)751.
21. F. Balestra et al, Nucl. Instr. and Meth. A **426** (1999) 385-404.
22. M.P. Bussa et al, Nuovo Cimento, vol. **109**A, (1996)327-339.
23. A.Yu. Bonushkina et al, JINR Rapid Communications 5(79)-96, (1996)5-14.
24. V.V. Ivanov, in Proc. IV Int. Workshop on Software Engineering, Artificial Intelligence and Expert Systems for High Energy and Nuclear Physics, April 3-8, 1995, Pisa, Italy; "New Computing Techniques in Physics Research IV", edited by B. Denby & D. Perret-Galix, "World Scientific", (1995)765.
25. A.Yu. Bonushkina et al, Int.J.Comput. & Math. with Appl., vol. **34**, No. 7/8, (1997)677-685.
26. A. Babloyantz and V.V. Ivanov, in ICANN'95, Paris, France, 1995.
27. I.S. Berezin and N.P. Zhydkov, in Computing Methods, vol. **1**, Moscow, 1959 (in Russian).
28. A.F. Nikiforov and S.K. Suslov: "Classical orthogonal polynomials", "Znanie", series "Mathematics, cibernatics", 1985/12, 27 (in Russian).
29. R.W. Hamming, in Numerical Methods for Scientists and Engineers, McGraw-Hiil, 1962.
30. V. Basios et al, Int. J. Comput. & Math. with Appl., vol. **34**, No. 7/8, (1997)687-693.
31. D.E. Rumelhart, G.E. Hinton and R.J. Williams, in Parallel Distributed Processing: Explorations in the Microstructure of Cognition, Ed. D.E. Rumelhart and J.L.McClelland, **1**, MIT Press, 1986.
32. A. Lapedes and R. Farber, Los Alamos Report LA-UR 87-2662 (1987).
33. P. Akritas et al, Chaos, Solitons and Fractals **11** (2000) 337-344.
34. E.A. Jackson, Perspectives of Nonlinear Dynamics, Cambridge University Press, 1989.
35. S. Rabinovich, G. Berkolaiko and S. Havlin, J. Math. Phys., 1996, **37**, 5828–5836.
36. S. Wolfram (ed.), Theory and Applications of Cellular Automata, World Scientific, 1986.

37. T. Toffoli and N. Margolus, Cellular Automata Machines: A New Environment for Modeling, MIT Press, Cambridge, Mass., 1987.
38. M.P. Bussa M.P. et al, Int. J. Comput. & Math. with Appl., vol. **34**, No. 7/8, (1997)695-701.
39. I. Antoniou et al, in "Discrete Dynamics in Nature and Society", v. **3**, No. 1 (1999)15-24.
40. R. Burridge and L. Knopoff, "Bulletin of the Seismological Society of America", **57**, No. 3, (1967)341–371.
41. P. Bak and C. Tang, "Journal of Geophysical Research", **94**, No. B11, (1989)15635-15637.
42. H. Nakanishi, "Phys. Rev. A", **43**, No. 12, (1991)6613–6621.
43. B. Gutenberg and C.F. Richter, "Bulletin of the Seismological Society of America", **34**, (1944)185–188.
44. P.G. Akishin P.G. et al, in Proc. IFAC LSS'98, Large Scale Systems: Theory and Applications, Rio Patras, Greece, July 15-17, (1998)907-913.
45. Z. Olami, H.S. Feder, and K. Christensen, "Phys.Rev.Lett.", **68**, No. 8, (1992)1244–1246.
46. P.G. Akishin et al, in "Discrete Dynamics in Nature and Society", **2**, (1998)267-279.

Quantum Recognizable Tree Functions

Symeon Bozapalidis

Department of Mathematics, Aristotle University of Thessaloniki, Thessaloniki, Greece

Abstract. We study functions computed by quantum tree automata and quantum algebras. A cutpoint theorem is established.

1 Introduction

To better understand computation in a quantum system, very recently C. Moore and J. Crutchfield have extended many concepts from Mathematical Language Theory (such as finite state and pushdown automata, regular and context free grammars) to the quantum case and they found analogs of several classical results (cf. [MC]).

The present paper can be considered as a continuation of the above work. We propose quantum versions of recognizable series on trees (cf. [BR2],[BL],[Bo2], [BA]) which constitute the simplest, from complexity point of view, class of tree functions.

To do this we define quantum bottom up tree automata and quantum algebras; the functions they compute assign probabilities to trees and have nice closure properties. However, traditional and quantum models are computationally distinct.

The paper is divided into four sections. Section 1 is devoted to an exhibition of preliminary facts. *Quantum tree automata* (QTA) are introduced in section 4 such a machine $\mathcal{M}$ is a linearized tree automaton whose transitions are $k^n \times k$ complex matrices having a certain unitarity property. Its behavior $|\mathcal{M}|$ is computed as follows: having fed $\mathcal{M}$ with an input tree t, it will move to a quantum state $\mu_\mathcal{M}(t) \in \mathbb{C}^n$; then $|\mathcal{M}|(t)$ is the norm of the projection of $\mu_\mathcal{M}(t)$ into an acceptance subspace.
Basic properties of functions coming in this way (and called *quantum recognizable* (QR)) are investigated. More precisely it is shown that QR functions form a convex set closed under product, complement and right derivatives.

Moreover, in this setup, a pumping lemma is stated. QTA are the extension to trees of the finite quantum automata of [MC].

In section 3 we introduce and study *Hermitian Γ-algebras* (Γ ranked alphabet); Such an algebra is an ordinary Γ-algebra $\mathcal{A} = (A, \alpha)$ whose carrier set A is a Hermitian space and Γ-operations

$$\alpha_f : A^n \to A, f \in \Gamma_n, n \geq 0$$

preserve Hermitian product:

$$(\alpha_f(x_1,...,x_n)/\alpha_f(x'_1,...,x'_n)) = (x_1/x'_1)...(x_n/x'_n), x_i, x'_i \in A (i=1,...,n).$$

A Hermitian Γ-algebra endowed with a projection $p: A \to A$ ($p^2 = p$) constitutes what we call a quantum Γ-algebra (QA). It is an algebraic structure equivalent to QTA from computational power point of view. The word analog of a QA is just the real time automaton in [MC]. Using quantum algebras we can prove that inverse linear non deleting tree homomorphisms preserve QR functions.

In order to achieve the construction of the reachable part of a given QA $\mathcal{A} = (A, \alpha, p)$ it is necessary to replace the projection $p: A \to A$ by a weaker output system which is a finite list of a linear form $\varphi_i : A \to \mathbb{C} (1 \leqslant i \leqslant k)$ fulfilling the normalization condition. For $x \in A$,

$$|x| \leqslant 1 \Rightarrow \sum_{i=1}^{k} |\varphi_i(x)|^2 \leqslant 1.$$

Quantum recognizable functions are properly included into the class of recognizable functions; this implies that the equality problem is solvable on QR.

In the last section we achieve an elegant result: cutting quantum recognizable functions to range isolated positions we get recognizable tree languages. Notice that an analogous statement holds for stochastic functions [Pa].

2 Preliminaries

Basic notation and facts concerning trees and Hermitian space follow.

2.1 Trees

Consider a finite ranked alphabet $\Gamma = (\Gamma_n)_n \geqslant 0$ and a set of variables $\Xi_k = \{\xi_1, ..., \xi_k\}, k \geqslant 0$ The set $T_\Gamma(\Xi_k)$ of Γ-trees indexed by Ξ_k is the smallest set verifying
- $\Gamma_0 \cup \Xi_k \subseteq (\Xi_k)$
- if $f \in \Gamma_n$ and $t_1,...,t_n \in T_\Gamma(\Xi_k)$ then $f(t_1,...,t_n) \in T_\Gamma(\xi_k)$

Denote by P_Γ the subset of $T_\Gamma(\xi)$ consisting of all trees with just one occurrence of the variable ξ. For $\tau, \pi \in P_\Gamma$, $\tau\pi$ is the result of substituting τ at the place of ξ inside π. This multiplication law structures P_Γ into a monoid which actually is free, that is each $\tau \in P_\Gamma$ is uniquely written as a product of trees of the form

$$f(t_1,\ldots,t_{i-1},\xi,t_{i+1},\ldots,t_n) \ f \in \Gamma_n, \ n \geqslant 1, t_j \in T_\Gamma$$

P_Γ acts on T_Γ :

$$T_\Gamma \times P_\Gamma \to T_\Gamma \ (t,\tau) \mapsto t\tau$$

$t\tau$ coming by substitution of t at the unique ξ occurring in t.
Subsets $L \subseteq T_\Gamma$ are called *tree languages*. We are interested in such languages recognized by finite state machines.
A (bottom up) *tree automaton* is a 4-tuple $M = (\Gamma, Q, m, T)$ where
- Γ is a finite ranked alphabet of input symbols
- Q is a finite set of states
- $T \subseteq Q$ is the set of final states and
- for all $f \in \Gamma_n$, $m_f : Q^n \to Q$ describes the moves of M.

The *behavior* of M is the tree language

$$L(M) = \{t \in T_\Gamma / \mu_M(t) \in T\}$$

where $\mu_\mathcal{M} : T_\Gamma \to Q$ is inductively defined by

$$\mu_M(f(t_1, \ldots, t_n)) = m_f(\mu_M(t_1), \ldots, mu_M(t_1), \ldots, \mu_M(t_n))$$

$L \subseteq T_\Gamma$ is *recognizable* whenever $L = L(M)$ for some tree automaton M.

A syntax characterization of recognizability will be needed later on. The *context* of a tree $\tau \in P_\Gamma$ with respect to $L \subseteq T_\Gamma$ is

$$C_L(\tau) = (t, \pi)/t \in T_\Gamma, \pi \in P_\Gamma \text{ and } t\tau\pi \in L.$$

Equality of contexts define a congruence $\equiv_L$ on the monoid P_Γ the quotient $M_L = P_\Gamma / \equiv_L$ is the *syntactical monoid* of L. In [BI] has been shown that

Theorem 1. $L \subseteq T_\Gamma$ *is recognizable if and only if* M_L *is finite.*

Tree language recognizability through left derivatives is also needed:

Theorem 2. *Next conditions are equivalent:*
i) $L \subseteq T_\Gamma$ *is recognizable*
ii) $\text{card} t^{-1}L/t \in T_\Gamma < \infty$ *where* $t^{-1}L = \{\tau \in P_\Gamma / t\tau \in L\}$.

Some words for alphabet change. Consider finite ranked alphabets Γ and Δ.
A *tree homomorphism* from T_Γ to T_Δ is just a sequence of functions

$$h_n : \Gamma_n \to T_\Delta(\xi_1, \ldots, \xi_n) \; n \geq 0$$

They are organized into a single function

$$h : T_\Gamma \to T_\Delta$$

as follows:
- $h(c) = h_0(c)$, $c \in \Gamma_0$
- $h(f(t_1, \ldots, t_n)) = h_n(f)[h(t_1), \ldots, h(t_n)]$

where the right hand side member of the above equality is the result of substituting the tree $h(t_i)$ at all occurrences of the variables ξ_i inside $h_n(f)$.
h is said to be *linear* (resp. *non deleting*) whenever for all $f \in \Gamma_n$, the variables $\xi_1, \ldots, \xi_n$ occur in the tree $h_n(f)$ at most (resp. at least) once.

2.2 Some Hermitian Calculus

A Hermitian space is a $\mathbb{C}$-space A equipped with a Hermitian product, i.e. a function

$$A \times A \to \mathbb{C} \qquad (x,y) \mapsto (x/y)$$

which is linear on the left

$$(\lambda_1 x_1 + \lambda_2 x_2 / y) = \lambda_1(x_1/y) + \lambda_2(x_2/y)$$

and satisfies the identity

$$(x/y) = \overline{(y/x)} \quad \text{(complex conjugate)}$$

and the condition $(x/x) \geqslant 0$ for all $x \in A$ with the equality $(x/x) = 0$ to hold just for $x = 0$.

The norm of $x \in A$ is $|x|^2 = (x/x)$.
$\mathbb{C}$-linear functions $h : A \to A$ respecting norm $|h(x)| = |x|$, $\forall x \in A$ or equivalently Hermitian products

$$(h(x)/h(y)) = (x/y), \ \forall x, y \in A$$

are called isometries. Their set Isom(A) is a monoid under function composition. Any isometry $h : A \to A$ is invertible (A finite dimensional) and holds $h^{-1} = h^*$, h^* denoting the adjoint of h which is the $\mathbb{C}$-linear function $h^* : A \to A$ defined by the formula

$$(h(x)/y) = (x/h^*(y)), \ \forall x, y \in A$$

A list of vectors $e_1, \ldots, e_k$ of A is orthonormal if $(e_i/e_j) = \delta_{ij}$, δ_{ij} being the Kronecker's delta: $\delta_{ij} = 1$, if $i = j$ and 0 else
If $e_1, \ldots, e_n$ is an orthonormal basis of A then for each $x \in A$ we have

$$x = (x/e_1)e_1 + \ldots + (x/e_n)e_n$$

and therefore

$$(x/y) = \sum_{i=1}^{n}(x/e_i)\overline{(y/e_i)} \quad \text{and} \quad |x|^2 = \sum_{i=1}^{n}|(x/e_i)|^2$$

The matrix of an isometry $h : A \to A$ with respect to an orthonormal basis $e_1, \ldots, e_n$ of A, is an $n \times n$ complex matrix $U = (u_{ij})$ satisfying

$$U^{-1} = U^*$$

U^* denoting the transpose-conjugate of U. Such matrices are called unitary they are characterized by the relation

$$\sum_{k=1}^{n} u_{ik} u_{jk} = \delta_{ij} \quad \forall i, j$$

expressing that the rows of U form an orthonormal basis of the Hermitian space $\mathbb{C}^n$ endowed with the standard product

$$(x/y) = x_1 \bar{y}_1 + \ldots + x_n \bar{y}_n$$

for $x = (x_1, \ldots, x_n)$, $y = (y_1, \ldots, y_n)$.
Notice that row unitarity is equivalent to column unitarity.

3 Quantum Tree Automata

A *bottom up quantum tree automaton* is a 4-tuple

$$\mathcal{M} = (\Gamma, Q, \mu, e)$$

consisting of a finite ranked alphabet Γ of inputs, a finite set Q of states, a Γ-indexed family of functions

$$\mu_f : Q^n \to \mathbb{C}^Q, \gamma \in \Gamma$$

describing the moves of $\mathcal{M}$ and a finite list $e = (e_1, \ldots, e_k)$ of perpendicular unit vectors $e_1, \ldots, e_k$ of the Hermitian space $\mathbb{C}^Q$. The transition function must obey the next unitarity condition: for all $q_1, \ldots, q_n, q'_1, \ldots, q'_n \in Q$ and all $f \in \Gamma_n$

$$\sum_{q \in Q} \mu_f(q_1, \ldots, q_n)(q) \overline{\mu_f(q'_1 \ldots, q'_n)(q)} = \delta_{q_1 \ldots q'_1 \ldots q'_n}$$

where $\delta_{q_1} \ldots, q_1' \ldots q_n'$ is equal to 1 whenever for all i $q_i = q'_i$ and 0 else. In particular, for $c \in \Gamma_0$ the above condition collapses to

$$\sum_{q \in Q} \mu_c(q) \overline{\mu_c(q)} = 1$$

expressing that μ_c is a unit vector of $\mathbb{C}^Q$.
μ_f is multilinearly extended into a function (still denoted by the same symbol)

$$\mu_f : (\mathbb{C}^Q)^n \to \mathbb{C}^Q$$

via

$$\mu_f(x_1,\ldots,x_n) = \sum_{q_1,\ldots,q_n \in Q} x_1(q_1)\ldots x_n(q_n)\mu_f(q_1,\ldots,q_n)$$

for all $x_1,\ldots,x_n \in \mathbb{C}^Q$.

The *reachability map* of $\mathcal{M}$ is the function $\mu_\mathcal{M} : T_\Gamma \to \mathbb{C}^Q$ inductively defined by
- $\mu_\mathcal{M}(c) = \mu_c$, $c \in \Gamma_0$
- $\mu_\mathcal{M}(f(t_1,\ldots,t_n)) = \mu_f(\mu_\mathcal{M}(t_1),\ldots,\mu_\mathcal{M}(t_n))$, for all $f \in \Gamma_n$, $t_j \in T_\Gamma (1 \leqslant j \leqslant n)$.

Finally the *behavior* of $\mathcal{M}$ is the function $|\mathcal{M}| : T_\Gamma \to \mathbb{R}_+$ given by

$$|\mathcal{M}|(t) = \sum_{i=1}^{k} |(\mu_\mathcal{M}(t)/e_i)|^2, \forall t \in T_\Gamma$$

where $(\mu_\mathcal{M}(t)/e_i)$ means the Hermitian product of the vectors $h_\mathcal{M}(t)$ and e_i

Call a function $F : T_\Gamma \to \mathbb{R}_+$ *quantum recognizable* (QR) if $F = |\mathcal{M}|$, for some quantum tree automaton $\mathcal{M}$.

Proposition 1. *For each $t \in T_\Gamma$, the vector $\mu_\mathcal{M}(t)$ is unit.*

Proof. This is true for $t = c \in \Gamma_0$. Assume that $\mu_\mathcal{M}(t)$ is a unit vector for all trees t having height $< n$ and let s be such that $height(s) = n$. Then $s = f(t_1,\ldots,t_n)$, $f \in \Gamma_n$ and $t_i \in T_\Gamma$ with $height(t_i) < n$, for all i. Further

$$|\mu_\mathcal{M}(t)|^2 = (\mu_\mathcal{M}(t)/\mu_\mathcal{M}(t))$$
$$= \sum_{q_i,q'_i \in Q} \mu_\mathcal{M}(t_1)(q_1)\ldots\mu_\mathcal{M}(t_n)(q_n) \cdot \overline{\mu_\mathcal{M}(t_1)(q'_1)}\ldots\overline{\mu_\mathcal{M}(t_n)(q'_n)} \left(\frac{\mu_f(q_1,\ldots,q_n)}{\mu_f(q'_1,\ldots,q'_n)}\right)$$
$$= \sum_{q_i,q'_i \in Q} \mu_\mathcal{M}(t_1)(q_1)\overline{\mu_\mathcal{M}(t_1)(q'_1)}\cdots\mu_\mathcal{M}(t_n)(q'_n)\overline{\mu_\mathcal{M}(t_n)(q'_n)} \cdot \delta_{q_1,\ldots,q_n,q'_1,\ldots,q'_n}$$
$$= [\sum_{q_1} \mu_\mathcal{M}(t_1)(q_1) \cdot \overline{\mu_\mathcal{M}(t_1)(q_1)}]\cdots[\sum_{q_n} \mu_\mathcal{M}(t_n)(q_n) \cdot \overline{\mu_\mathcal{M}(t_n)(q_n)}]$$
$$= |\mu_\mathcal{M}(t_1)|^2 \cdots |\mu_\mathcal{M}(t_n)|^2$$
$$= 1$$

Corollary 1. *The behavior $|\mathcal{M}|$ of a quantum tree automaton $\mathcal{M}$, ranges over the unit interval $[0,1]$.*

Proof. Complete $e_1,\ldots,e_k$ into an orthonormal basis of $\mathbb{C}^Q$ by adding vectors $e_{k+1},\ldots,e_{cardQ}$. The components of $\mu_\mathcal{M}(t)$ along this basis are $(\mu_\mathcal{M}(t)/e_i)$, $1 \leqslant i \leqslant cardQ$. Therefore

$$|\mathcal{M}|(t) \leqslant \sum_{i=1}^{cardQ} |(\mu_\mathcal{M}(t)/e_i)|^2 = |\mu_\mathcal{M}(t)|^2 = 1$$

hence the result.

In the sequel we list some properties of quantum recognizable functions.

Proposition 2. *If the function $F : T_\Gamma \to [0,1]$ is quantum recognizable, then so is the complement $\overline{F} : T_\Gamma \to [0,1]$, $\overline{F}(t) = 1 - F(t)$ for all $t \in T_\Gamma$.*

Proof. Let $\mathcal{M} = (\Gamma, Q, \mu, e = (e_1, \ldots, e_k))$ be a quantum automaton such that $|\mathcal{M}| = F$. Then $\overline{\mathcal{M}} = (\Gamma, Q, \mu, \overline{e})$ with $\overline{e} = (e_{k+1}, \ldots, e_{cardQ})$ being the list completing $|e_1, \ldots, e_k|$ into an orthonormal basis of $\mathbb{C}^{cardQ}$ has as behavior $\overline{F}$.

Proposition 3. *The class $QR(T)$ of all quantum recognizable functions $T_\Gamma \to [0,1]$ is a closed under product convex subset of the real space of all functions $T_\Gamma \to \mathbb{R}$.*

Proof. Let, indeed

$$\mathcal{M}_i = (\Gamma, Q^{(i)}, \mu^{(i)}, e^{(i)} = (e_1^{(i)}, \ldots, e_{k_i}^{(i)})) \quad i = 1, 2$$

be two quantum tree automata. We construct

$$\mathcal{M} = (\Gamma, Q^{(1)} \times Q^{(2)}, \mu, e = (e_{j_1 j_2})_{1 \leqslant j_1 \leqslant k_1, 1 \leqslant j_2 \leqslant k_2})$$

1) by setting for all $f \in \Gamma_n$, $n \geqslant 0$

$$\mu_f[(q_1^{(1)}, q_1^{(2)}), \ldots, ((q_n^{(1)}, q_n^{(2)})](q_1, q_2) = \mu_f^{(1)}(q_1^{(1)}, \ldots, q_n^{(1)})(q_1) \cdot \mu_f^{(2)}(q_1^{(2)}, \ldots, q_n^{(2)})(q_2)$$

with $q_j^{(1)} \in Q^{(1)}$, $q_j^{(2)}$.
2) for all j_1, j_2

$$e_{j_1 j_2}(q_1, q_2) = e_{j_1}^{(1)}(q_1) \cdot e_{j_2}^{(2)}(q_2).$$

An induction argument shows that for all $t \in T_\Gamma$ $q_1 \in Q^{(1)}$ and $q_2 \in Q^{(2)}$

$$\mu_\mathcal{M}(t)(q_1, q_2) = \mu_{\mathcal{M}_\infty}(t)(q_1) \cdot \mu_{\mathcal{M}_\epsilon}(t)(q_2)$$

Therefore

$$|\mathcal{M}| = \sum_{j_1, j_2} |(\mu_\mathcal{M}(t)/e_{j_1 j_2})|^2$$

$$= \sum_{j_1, j_2, q_1, q_2} |\mu_\mathcal{M}(t)(q_1, q_2) \cdot \overline{e_{j_1 j_2}(q_1, q_2)}|^2$$

$$= \sum_{j_1, q_1} |\mu_{\mathcal{M}_1}(t)(q_1) \cdot \overline{e_{j_1}(q_1)}|^2 \cdot \sum_{j_2, q_2} |\mu_{\mathcal{M}_2}(t)(q_2) \cdot \overline{e_{j_2}(q_2)}|^2$$

$$= |\mathcal{M}_1|(t) \cdot |\mathcal{M}_2|(t)$$

$$= (|\mathcal{M}_1| \cdot |\mathcal{M}_2|)(t)$$

Next, we shall show that $QR(\Gamma)$ is closed under convex combinations; for this let $F_1, F_2 \in QR(\Gamma)$ and $\lambda_1, \lambda_2 \in [0,1]$ such that $\lambda_1 + \lambda_2 = 1$.
Consider the quantum tree automata $\mathcal{M}_i$ ($i = 1, 2$) as before and construct

$$\mathcal{M}' = (\Gamma, Q^{(1)} \times Q^{(2)}, \mu, e' = (e'_1, \ldots, e'_{k_1 + k_2}))$$

with μ to be the same as in $\mathcal{M}$, whereas e'_i are defined by
- if $1 \leqslant i \leqslant k_1$, then $e'_i(q_1, q_2) = \sqrt{\lambda_i} \cdot e_i^{(1)}(q_1)$
- if $k_1 < i \leqslant k_1 + k_2$, $e'_i(q_1, q_2) = \sqrt{\lambda_i} \cdot e_{i-k_1}(q_2)$ for all $q_i \in Q^{(i)}$, $i = 1, 2$. It is easy to constate that e' is an orthonormal system of vectors of $\mathbb{C}^{Q^{(1)} \times Q^{(2)}}$ and that

$$|\mathcal{M}'| = \lambda_1|\mathcal{M}_1| + |\mathcal{M}_2|.$$

Example 1. (Word quantum automata)
Let X be an ordinary alphabet and Γ the ranked alphabet associated with it:

$$\Gamma_1 = X, \; \Gamma_0 = \{\hat{\perp}\}, \; \Gamma_n = \varnothing \; \text{for } n > 1$$

where $\hat{\perp}$ is a fixed 0-ranked symbol.
For such an alphabet a bottom up quantum tree automaton $\mathcal{M}$ consists of the next data:
- a finite state set Q.
- a unit vector $\mu_{\hat{\perp}}$ of $\mathbb{C}^Q$
- a finite list $e_1, \ldots, e_k$ of orthonormal vectors of $\mathbb{C}^Q$
- a function $\mu : X \to U(Q \times Q)$, $U(Q \times Q)$ denoting the monoid of all unitary $Q \times Q$ matrices that is functions $A : Q \times Q \to \mathbb{C}$ verifying:

$$\sum_{q \in Q} A(p, q)A(p', q) = \delta_{pp'}, \qquad \forall p, p' \in Q.$$

Such a machine, is nothing but a finite state quantum automaton in the terminology of [MC]. Its *behavior* is the function $|M| : X^* \to [0, 1]$.

$$(|\mathcal{M}|(w) = \sum_{i=1}^{k} |(\mu^*(w)(\mu_{\hat{\perp}})/e_i)|^2 \; w \in X^*$$

with μ^* denoting the canonical extension of μ on the free monoid X^*.

Proposition 4. *(cf. [MC]). The set $QR(X)$ of all behaviors of quantum word automata, is a convex set closed under product and complement.*

To examine the relationship between tree languages and quantum functions we need some additional notational matter.

Let $\mathcal{M} = (\Gamma, Q, \mu, (e_1, \ldots, e_k))$ be a quantum tree automaton; we are going to define an action of the free monoid P_Γ on the $\mathbb{C}$-space $\mathbb{C}^Q$:

$$\mathbb{C}^Q \times P_\Gamma \to \mathbb{C}^Q \qquad (X, \tau) \mapsto X\tau$$

- if $\tau \in P_\Gamma$ has the generic form.
(g) $\qquad \tau = f(t_1, \ldots, t_{i-1}, x, t_{i+1}, \ldots, t_n) \; f \in \Gamma_n, t_j \in T_\Gamma$
then for each $q \in Q$ we set

$$q\tau = \mu_f(\mu_\mathcal{M}(t_1),\ldots,\mu_\mathcal{M}(t_{i-1}),q,\mu_\mathcal{M}(t_{i+1}),\ldots,\mu_\mathcal{M}(t_n))$$

and for each $X \in \mathbb{C}^Q$ we set

$$X\tau = \sum_{q \in X} X(q)(q\tau)$$

- if τ is a product of trees of the form (g)

$$\tau = \tau_1 \ldots \tau_k$$

then

$$q\tau = (\ldots(q\tau_1)\tau_2\ldots)\tau_k$$

<u>Fact</u> For all $\tau \in P_\Gamma$, the $Q \times Q$-matrix $U_\tau = (u_{qp})$ whose entries are given by

$$u_{qp} = (q\tau)(p) \qquad p,q \in Q$$

is unitary.

Further, for all $t \in T_\Gamma$ and all $\tau \in P_\Gamma$ it holds

$$\mu_\mathcal{M}(t\tau) = \mu_\mathcal{M}(t)U_\tau.$$

Proposition 5. *If the function $F : T_\Gamma \to [0,1]$ is QR, then so is its right derivative at $\tau \in P_\Gamma$*

$$F\tau^{-1} : T_\Gamma \to [0,1], \ (F\tau^{-1})(t) = F(t\tau) \ \forall t \in T_\Gamma$$

Proof. Let the quantum free automaton

$$\mathcal{M} = (\Gamma, Q, \mu, (e_1,\ldots,e_k))$$

compute F; then $F\tau^{-1}$ is computed by

$$\mathcal{M}\tau^{-1} = (\Gamma, Q, \mu, (e_1\tau^{-1},\ldots,e_k\tau^{-1}))$$

where for all i $(1 \leqslant i \leqslant k)$

$$e_i\tau^{-1} = e_i U_\tau^{-1}$$

(Recall that $U_\tau^{-1} = U_\tau^*$ the transpose-conjugate of U_τ.) Indeed, for all $t \in T_\Gamma$ we have:

$$|\mathcal{M}\tau^{-1}|(t) = \sum_{i=1}^{k} |(\mu_{\mathcal{M}\tau^{-1}}(t)/e_i\tau^{-1})|^2$$

$$= \sum_{i=1}^{k} |(\mu_\mathcal{M}(t)/e_i U_\tau^{-1})|^2$$

$$= \sum_{i=1}^{k} |(\mu_\mathcal{M}(t)/e_i U_\tau^*)|^2$$

$$= \sum_{i=1}^{k} |(\mu_\mathcal{M}(t) U_\tau/e_i)|^2$$

$$= \sum_{i=1}^{k} |(\mu_\mathcal{M}(t\tau)/e_i)|^2$$

$$= |\mathcal{M}|(t\tau)$$

$$= (|\mathcal{M}|\tau^{-1})(t)$$

In other words $|\mathcal{M}\tau^{-1}| = |\mathcal{M}|\tau^{-1}$ and the proof is completed

Corollary 2. *If the word function $f : X^* \to [0,1]$ is quantum recognizable, then so is any right derivative*

$$fw^{-1} : X^* \to [0,1] \quad , \quad (fw^{-1})(u) = f(uw), \; \forall u \in X^*$$

Proposition 6. *Let $F : T_\Gamma \to [0,1]$ be a QR function; for every $\tau \in P_\Gamma$ and every $\varepsilon > 0$, there is a natural number $k \geqslant 1$ such that the inequality*

$$|F(t\tau^k \pi) - F(t\pi)| < \varepsilon$$

holds for all trees $t \in T_\Gamma$ and $\pi \in P_\Gamma$.

Proof. Adapt the proof of theorem 5 in [MC] by working with the matrix U_τ instead of U_w.

As a consequence we get

Proposition 7. *If the characteristic function of a tree language $L \subseteq T_\Gamma$ is QR, then the syntactical monoid M_L of L will be a group.*

Proof. Indeed, for all $\tau \in P_\Gamma$, choosing $\varepsilon < 1$ in the above proposition, we get

$$|\mathcal{X}_L(t\tau^k \pi) - \mathcal{X}_\mathcal{L}(t\pi)| < 1$$

for some $k \geqslant 1$ and all $t \in T_\Gamma$, $\pi \in P_\Gamma$; thus

$$\mathcal{X}_L(t\tau^k \pi) = \mathcal{X}_L(t\pi) \text{ or } \tau^k \equiv \xi (\text{mod } \equiv_L)$$

where ξ is the unit element of the monoid P_Γ.
So, the equivalence class $[\tau]$ of τ modulo $\equiv_L$ is an invertible element of the monoid $M_L = P_\Gamma / \equiv_L : [\tau]^{-1} = [\tau]^{k-1}$.

Very simple recognizable tree languages fall to the above situations; for instance

$$L_f = \{t \in T_\Gamma / |t| = 1\}$$

consisting of all trees with just one occurrence of the symbol $f \in \Gamma$. Its syntactical monoid is

$$\begin{array}{c|ccc} \cdot & 1 & \alpha & 0 \\ \hline 1 & 1 & \alpha & 0 \\ \alpha & \alpha & \alpha & 0 \\ 0 & 0 & 0 & 0 \end{array}$$

So, it has non QR characteristic function.

4 Quantum Algebras

We introduce an algebraic structure equivalent to quantum tree automata from computational point of view.

Let Γ be a finite ranked alphabet.
A $\mathbb{C}$-*multilinear* Γ-*algebra* is a pair $\mathcal{A} = (A, a)$ consisting of a $\mathbb{C}$-space A together with a Γ-indexed family of functions (called Γ-*operations* of $\mathcal{A}$)

$$\alpha_f : A^n \to A, \ f \in \Gamma_n, \ n \leqslant 0.$$

which are $\mathbb{C}$-linear at any argument i.e. for all k $(1 \leqslant k \leqslant n)$

$$\alpha_f(\ldots, \underbrace{\sum_{i=1}^p \lambda_i x_i^{(k)}}_{k-th\,place}, \ldots) = \sum_{i=1}^p \lambda_i \alpha_f(\ldots, x_i^{(k)}, \ldots)$$

for all $\lambda_j \in \mathbb{C}$ and $x_j^k \in A$.
Especially the elements α_c ($c \in \Gamma_0$) are named *constants* of A .
The *reachability map* $H_\mathcal{A} : T_\Gamma \to A$ of A is inductively defined as follows:
- for all $c \in \Gamma_0$; $H_\mathcal{A}(c) = \alpha_c$
- for all $t = f(t_1, ..., t_n)$, $f \in \Gamma_n$, $t_j \in T_\Gamma$

$$H_\mathcal{A}(f(t_1, ..., t_n)) = \alpha_f(H_\mathcal{A}(t_1), \ldots, H_\mathcal{A}(t_n)).$$

The monoid P_Γ acts on any multilinear Γ-algebra $\mathcal{A} = (A, \alpha)$

$$A \times P_\Gamma \to A \ (x, \tau) \mapsto x\tau \ (x \in A, \ \tau \in P_\Gamma)$$

in a way similar to that described in section 2.
<u>Fact</u> For all $t \in T_\Gamma$ and $\tau \in P_\Gamma$ it holds

$$H_{\mathcal{A}}(t\tau) = H_{\mathcal{A}}(t)\tau.$$

Now, a $\mathbb{C}$-multilinear Γ-algebra $\mathcal{A} = (A, \alpha)$ is said to be *Hermitian Γ-algebra* whenever A is a Hermitian space and all the functions α_f preserve the Hermitian product in the sense that

$$(\alpha_f(x_1, \ldots, x_n)/\alpha_f(x'_1, \ldots, x'_n)) = (x_1/x'_1) \cdots (x_n/x'_n)$$

for every $x_i, x'_i \in A$, $(i = 1, \ldots, n)$.
For the constants α_i ($i \in \Gamma_n$) we impose to be unit vectors of A : $|\alpha_c| = 1$.
<u>Fact</u> In any Hermitian Γ-algebra $\mathcal{A}=(A, \alpha)$ it holds:
i) $|\alpha_f(x_1, \ldots, x_n)| = |x_1| \cdots |x_n|$ for all $f \in \Gamma_n$ ($n \leqslant 1$) and $x_i \in A$ ($i = 1, \ldots, n$)
and
ii)$|H_{\mathcal{A}}(t) = 1$ for all $t \in T_\Gamma$.

A *quantum Γ-algebra* is a triple $\mathcal{A} = (A, \alpha, p)$, where (A, α) is a Hermitian Γ-algebra and $p : A \to A$ is a projection, that is a $\mathbb{C}$-linear function satisfying $p^2 = p$.
The *function compute* by such a structure $F_{\mathcal{A}} : T_\Gamma \to [0, 1]$ is given by

$$F_{\mathcal{A}}(t) = |p(H_{\mathcal{A}}(t))|^2 \text{ for all } t \in T_\Gamma.$$

Assume quantum Γ-algebras

$$\mathcal{A} = (A, \alpha, p) \text{ and } \mathcal{A}' = (A', \alpha', p')$$

are given; a $\mathbb{C}$-linear function $h : A \to A'$ commuting with Γ-operations i.e.

$$h(\alpha_f(x_1, \ldots, x_n)) = \alpha'_f(h(x_1), \ldots, h(x_n))$$

for all $f \in \Gamma_n$ ($n \geqslant 0$), $x_j \in A$ and such that

$$|p(x)| = |p'(h(x))| \text{ for all } x \in A$$

is called a *morphism* of quantum Γ-algebras.

Proposition 8. *Assume quantum Γ-algebras*

$$\mathcal{A} = (A, \alpha, p) \text{ and } \mathcal{A}' = (A', \alpha', p')$$

are given. If there is a morphism $h : A \to A'$, then $\mathcal{A}$ and $\mathcal{A}'$ compute the same function: $F_{\mathcal{A}} = F_{\mathcal{A}'}$.

Proof. Indeed, for all $t \in T_\Gamma$ we have

$$H_{\mathcal{A}'}(t) = h(H_{\mathcal{A}}(t))$$

and therefore

$$\begin{aligned} F_{\mathcal{A}'}(t) &= |p'(H_{\mathcal{A}'}(t))|^2 \\ &= |p'(h(H_{\mathcal{A}}(t)))|^2 \\ &= |p(H_{\mathcal{A}}(t))|^2 \\ &= F_{\mathcal{A}}(t) \end{aligned}$$

as stated.

The construction of the generic object in the category of Hermitian algebra, is our next scope.
Consider the $\mathbb{C}$-space $\mathbb{C} < T_\Gamma >$ of all finite formal sums:

$$\mathbf{P} = \sum_i \lambda_i t_i \quad \lambda_i \in \mathbb{C},\ t_i \in T_\Gamma$$

that we call *tree polynomials* with complex coefficients.
In order to structure $\mathbb{C} < T_\Gamma >$ in a Hermitian space we need a (natural) Hermitian product. For this given:

$$\mathbf{P} = \sum_i \lambda_i t_i \quad \mathbf{Q} = \sum_i \mu_i t_i \quad \lambda_i, \mu_i \in \mathbb{C}$$

we put

$$(\mathbf{P}/\mathbf{Q}) = \sum_i \lambda_i \bar{\mu}_i$$

Let further $f \in \Gamma_n$ and take tree polynomials

$$\mathbf{P}_k = \sum_i \lambda_i^{(k)} t_i \quad k = 1, \ldots, n$$

Their *f-catenation* $f(\mathbf{P}_1, \ldots, \mathbf{P}_n)$ is by definition the tree polynomial

$$f(\mathbf{P}_1, \ldots, \mathbf{P}_n) = \sum_{i_1, \ldots, i_n} \lambda_{i_1}^{(1)} \cdots \lambda_{i_n}^{(n)} f(t_{i_1}, \cdots, t_{i_n})$$

It holds

$$(f(\mathbf{P}_1, \ldots, \mathbf{P}_n)/f(\mathbf{Q}_1, \ldots, \mathbf{Q}_n)) = (\mathbf{P}_1/\mathbf{Q}_1) \cdots (\mathbf{P}_n/\mathbf{Q}_n)$$

Indeed, suppose

$$\mathbf{Q}_k = \sum_i \mu_i^{(k)} t_i$$

we have

$$(f(\mathbf{P}_1,\ldots,mathbf{P}_n)/f(\mathbf{Q}_1,\ldots,mathbf{Q}_n)) = \sum_{i_1,\ldots,i_n} \lambda_{i_1}^{(1)}\cdots\lambda_{i_n}^{(n)} \bar{\mu}_{i_1}^{(1)}\cdots\bar{\mu}_{i_n}^{(n)}$$
$$= (\sum_{i_1} \lambda_{i_1}^{(1)} \bar{\mu}_{i_1}^{(1)})\cdots(\sum_{i_n} \lambda_{i_n}^{(n)} \bar{\mu}_{i_n}^{(n)})$$
$$= (\mathbf{P}_1/\mathbf{Q}_1)\cdots(\mathbf{P}_n/\mathbf{Q}_n)$$

as stated

Proposition 9. $\mathbb{C}<T_\Gamma>$ *is the initial object in the category whose objects are the Hermitian Γ-algebras and whose arrows are the linear functions preserving Γ-operations.*

Proof. Given a Hermitian Γ-algebra (A,α), the unique arrow from $\mathbb{C}<T_\Gamma>$ to A is obtained by taking the $\mathbb{C}$-linear extension of $H_A : T_\Gamma \to A$

$$\tilde{H}_A : \mathbb{C}<T_\Gamma> \to A , \quad \tilde{H}_A(\mathbf{P}) = \sum_{i=1}^{k} \lambda_i H_A(t_i)$$

where

$$\mathbf{P} = \sum_{i=1}^{k} \lambda_i t_i$$

Quantum tree automata and quantum algebras can be identified from machine point of view.
Precisely any quantum tree automaton

$$\mathcal{M} = (\Gamma, Q, \mu, e = (e_1,\ldots,e_n))$$

gives rise to a quantum Γ-algebra

$$\mathcal{A}_\mathcal{M} = (\mathbb{C}^Q, \alpha, p)$$

where $a_f : (\mathbb{C}^Q)^n \to \mathbb{C}^Q$ ($f \in \Gamma_n$) is the multilinear extension of μ_f whereas $p : \mathbb{C}^Q \to \mathbb{C}^Q$ is the projection on to the subspace of $\mathbb{C}^Q$ generated by the vectors $e_1,\ldots,e_k$. Of course the behavior of $\mathcal{M}$ coincides with the function computed by $\mathcal{A}_\mathcal{M}$:

$$|\mathcal{M}| = \mathcal{F}_{\mathcal{A}_\mathcal{M}}$$

Conversely to a given finite dimensional quantum $\Gamma - algebra \mathcal{A} = (A,\alpha,p)$, corresponds a quantum tree automaton:

$$\mathcal{M}_\mathcal{A} = (\Gamma, Q, \mu, e = (e_1,\ldots,e_k))$$

by taking $Q = \{1,2,\ldots,n\}$, ($n = dim A$) as state set choose an orthonormal basis $e_1,\ldots,e_k$ of the subspace $Im(p) = \{p(\alpha)/\alpha \in A\}$ and complete it into an orthonormal basis $E = \{e_1,\ldots,e_k, e_{k+1},\ldots,e_n\}$ of the whole space A. The move function of $\mathcal{M}_A$

$$\mu_f : \{1, 2, \ldots, n\}^k \to \mathbb{C}^{\{1,2,\ldots,n\}} \quad f \in \Gamma_k, \, k \geq 0$$

is given by

$$\mu_f(i_1, \ldots, i_k)(i) = (\alpha_f(e_{i_1}, \ldots, e_{i_k})/e_i) \text{ for all } i \in \{1, \ldots, n\}$$

i.e., it is the i-th component of the vector $\alpha | f(e_{i_1}, \ldots, e_{i_k})$ along E.

The unitary condition holds because $\mathcal{A}$ is a Hermitian algebra. Indeed, for every $f \in \Gamma_k$ we have

$$(\mu_f(i_1, \ldots, i_k)/\mu_f(j_1, \ldots, j_k)) = \sum_{i=1}^{n} \mu_f(i_1, \ldots, i_k)(i) \cdot \overline{\mu_f(j_1, \ldots, j_k)(i)}$$

$$= \sum_{i=1}^{n} (\alpha_f(e_{i_1}, \ldots, e_{i_k})/e_i) \cdot \overline{\alpha_f(e_{j_1}, \ldots, e_{j_k})/e_i}$$

$$= (\alpha_f(e_{i_1}, \ldots, e_{i_k})/\alpha_f(e_{j_1}, \ldots, e_{j_k}))$$

$$= (e_{i_1}/e_{j_1}) \cdots (e_{i_k}/e_{j_k})$$

$$= \delta_{i_1 j_1} \cdots \delta_{i_n j_n}$$

$$= \delta_{i_1 \ldots i_n, j_1 \ldots j_n}$$

Once again, the behavior of $\mathcal{M}_\mathcal{A}$ is the same as the function realized by $\mathcal{A}$:

$$F_\mathcal{A} = |\mathcal{M}_\mathcal{A}|.$$

We summarize

Proposition 10. *A function $F : T_\Gamma \to [0, 1]$ is quantum recognizable if and only if it can be realized by a finite dimensional quantum Γ-algebra.*

Example 2. We shall determine all QR functions $T_\Gamma \to [0, 1]$ computed by 1-dimensional quantum Γ-algebras: $\mathcal{A} = (A, \alpha, p)$, $\dim A = 1$.
Let $e \in A$, $|e| = 1$; for each $f \in \Gamma_n$, we set

$$k_f = \alpha_f(e, \ldots, e), \, k_f \in \mathbb{C}.$$

Clearly $|k_f| = 1$. Then

$$H_\mathcal{A}(t) = \prod k_f^{|t|_f}$$

$|t|_f$ denoting the number $f \in \Gamma$ occurs in t.
The possible projection $p : A \to A$ are either the identity or the zero functions.

Consequently, the corresponding computed functions are either the $F_1 : T_\Gamma \to [0, 1]$, $F_1(t) = 1$ for all $t \in T$ or the zero function F_0. Their convex combination

$$\lambda \cdot F_1 + (1 - \lambda) \cdot F_0 \quad \lambda \in [0, 1]$$

yields the function $F_\lambda : T_\Gamma \to [0, 1]$, $F_\lambda(t) = \lambda$, for all $t \in T_\Gamma$.

Example 3 (Quantum modules). Assume Γ is the ranked (monadic) alphabet associated with an ordinary finite alphabet X, i.e.

$$\Gamma_0 = \{\hat{1}\}, \Gamma_1 = X, \Gamma_k = \emptyset \quad \text{for } k > 1$$

In this case a quantum Γ-algebra consists of a Hermitian space A, a distinguished vector $\alpha_{\hat{1}} \in A$, $|\alpha_{\hat{1}}| = 1$, a function $\alpha : X \to Isom(A)$ sending each letter $x \in X$ to an isometry $\alpha(x) : A \to A$ and a projection $p : A \to A$. We call such a structure a *quantum X-module*. The function $F : X* \to [0,1]$ computed by such a module is given by

$$F(w) = |p(\alpha^*(w)(\alpha_{\hat{1}}))|^2 \quad w \in X^*$$

$\alpha^* : X^* \to Isom(A)$ denoting the monoid morphism extending α.

Proposition 11. *A function $F : X^* \to [0,1]$ is quantum recognizable if and only if it can be computed by a finite dimensional quantum X-module.*

Remark Quantum modules are essentially identical with the real time quantum automata of [MC].

Alphabet change is investigated below

Theorem 3. *Let $h : T_\Gamma \to t_\Delta$ be a linear non deleting tree homomorphism and $F : T_\Delta \to [0,1]$ a quantum recognizable function. Then the composition*

$$T_\Gamma \xrightarrow{h} T_\Delta \xrightarrow{F} [0,1]$$

is quantum recognizable, as well.

Proof. Let us consider a finite dimensional quantum Δ-algebra $\mathcal{A} = (A, \alpha, p)$ computing F. For each tree $t \in T_\Gamma(\xi_1, ..., \xi_n)$ we define the function $\alpha_t : A^n \to A$ inductively as follows:
- if $t = \xi_i$ $(1 \leq i \leq n)$ then α_t is the i-th projection $pr_i : A^n \to A$, $pr_i(x_1, ..., x_n) = x_i$
- if $t = f(t_1, ..., t_p)$, then for all $x_1, ..., x_n \in A$

$$\alpha_t(x_1, ..., x_n) = \alpha_f(\alpha_{t_1}(x_1, ..., x_n), ..., \alpha_{t_p}(x_1, ..., x_n))$$

Claim: Let $t \in T_\Gamma(\xi_1, ..., \xi_n)$ and assume that the variable ξ_i occurs in t exactly k_i times, i.e. $k_i = |t|_{\xi_i}$. Then for all $\xi_1, ..., \xi_n, \xi'_1, ..., \xi'_n \in A$ it holds

$$(\alpha_t(x_1, ..., x_n)/\alpha_t(x'_1, ..., x'_n)) = (x_1/x'_1)^{|t|_{\xi_1}} ... (x_n/x'_n)^{|t|_{\xi_n}}$$

By induction. If $t = \xi_i$ then

$$(pr_i(x_1, ..., x_n)/pr_i(x'_1, ..., x'_n)) = (x_i/x'_i)$$

Suppose our assertion true for all trees with height $< m$ and let $t \in T_\Gamma(\xi_1, ..., \xi_n)$ be such that $height(t) = m$.

Then $t = f(t_1, ..., t_p)$, $f \in \Gamma_k$, $t_j \in T_\Gamma(\xi_1, ..., \xi_n)$ and

$(\alpha_t(x_1, ..., x_n)/\alpha_t(x'_1, ..., x'_n))$
$= (\alpha_f(\alpha_{t_1}(x_1, ..., x_n), ..., \alpha_{t_p}(x_1, ..., x_n))/\alpha_f(\alpha_{t_1}(x'_1, ..., x'_n), ..., \alpha_{t_p}(x'_1, ..., x'_n)))$
$= (\alpha_{t_1}(x_1, ..., x_n)/\alpha_{t_1}(x'_1, ..., x'_n)) \cdots (\alpha_{t_k}(x_1, ..., x_n)/\alpha_t(x'_1, ..., x'_n))$
$= (x_1/x'_1)^{|t_1|\xi_1} \cdots (x_n/x'_n)^{|t_1|\xi_n} \cdots (x_1/x'_1)^{|t_k|\xi_1} \cdots (x_n/x'_n)^{|t_k|\xi_n}$
$= (x_1/x'_1)^{|t_1|\xi_1 + \cdots + |t_k|\xi_n} \cdots (x_n/x'_n)^{|t_1|\xi_n + \cdots + |t_k|\xi_n}$
$= (x_1/x'_1)^{|f(t_1,...,t_k)|\xi_1} \cdots (x_n/x'_n)^{|f(t_1,...,t_k)|\xi_n}$
$= (x_1/x'_1)^{|t|\xi_1} \cdots (x_n/x'_n)^{|t|\xi_n}$

Our claim is therefore established.

Now, consider the quantum Γ-algebra $\mathcal{B} = (A, b, p)$ by defining its operation via the formula
- $b_c = a_{h_0(c)}$, $c \in \Gamma_0$
- $b_f(x_1, ..., x_n) = \alpha_{h_n(f)}(x_1, ..., x_n)$ for all $f \in \Gamma_n$, $x_i \in A$

The unitarity condition holds because of the linear non deleting character of h:

$(b_f(x_1, ..., x_n)/b_f(x'_1, ..., x'_n)) = (\alpha_{h_n(f)}(x_1, ..., x_n)/\alpha_{h_n(f)}(x'_1, ..., x'_n))$
$= (x_1/x'_1)^{|h_n(f)|\xi_1} \cdots (x_n/x'_n)^{|h_n(f)|\xi_n}$
$= (x_1/x'_1) \cdots (x_n/x'_n)$

It remains to show that $F_\mathcal{B}$ equals $F_\mathcal{A} \circ h$. For all $t \in \Gamma$ we have

$$H_\mathcal{B}(t) = H_\mathcal{A}(h(t))$$

and thus

$$F_\mathcal{B}(t) = |p(H_\mathcal{B}(t))|^2 = |p(H_\mathcal{A}(h(t)))|^2 = F_\mathcal{A}(h(t)) = (F_\mathcal{A} \circ h)(t)$$

as asserted.

When both Γ and Δ are the monadic alphabets associated with the ordinary alphabets X and Y, then T_Γ and T_Δ are isomorphic to X^* and Y^* respectively and the ordinary monoid morphisms $X^* \to Y^*$ can be viewed as linear non deleting tree homomorphisms $T_\Gamma \to T_\Delta$. Thus, the previous result collapses to the known one of [MC].

Corollary 3. *Given a monoid morphism $h : X^* \to Y^*$ and a quantum recognizable function $F : Y^* \to [0, 1]$. The composition $F \circ h : X^* \to [0, 1]$ is again quantum recognizable.*

A $\mathbb{C}$-multilinear Γ-algebra $\mathcal{A} = (A, \alpha)$ is said to be *reachable* if the extension of its reachability map $\widetilde{H}_\mathcal{A} : \mathbb{C} < T_\Gamma > \to A$ is a surjective function.
Notice that

$$Im(\widetilde{H}_\mathcal{A}) = \{\widetilde{H}_\mathcal{A}(P)/P \in \mathbb{C} < T_\Gamma >\}$$

is a $\mathbb{C}$-multilinear sub-Γ-algebra of A.

However, in the case $\mathcal{A} = (A, \alpha, p)$ is a quantum Γ-algebra of A, $Im(\widetilde{H}_A)$ is far from being closed under the projection $p : A \to A$, and consequently $Im(\widetilde{H}_A)$ has no longer the property of being quantum.

We can overlap this obstacle by replacing $p : A \to A$ with a weaker accepting system. That is by a finite list of $\mathbb{C}$-linear form $\varphi_i : A \to \mathbb{C}$ $(1 \leqslant i \leqslant k)$ verifying the normalization condition: for all $x \in A$

$$(nor) \quad \sum_{i=1}^{k} |\varphi_i(x)|^2 \leqslant 1 \quad \text{whenever } |x| \leqslant 1$$

Indeed in the case $\mathcal{A} = (A, \alpha, p)$ is a finite dimensional quantum Γ-algebra, then choosing an orthonormal basis $e_1, ..., e_k$ for the subspace $Im(\widetilde{H}_A)$ we can define $\phi_i : A \to \mathbb{C}$ $(1 \leqslant i \leqslant k)$ by setting

$$\varphi_i(x) = (x/e_i) \quad x \in A,\ 1 \leqslant i \leqslant k$$

It holds, for all $x \in A$

$$|p(x)|^2 = \sum_{i=1}^{k} |(x/e_i)|^2 = \sum_{i=1}^{k} |\varphi_i(x)|^2$$

In other words the structure

$$(2) \quad (A, \alpha, (\varphi_1, ..., \varphi_k))$$

computes the same function with $\mathcal{A}$.

A triple (2) is termed an *extended quantum Γ-algebra (EQA)* if (A, α) is a Hermitian Γ-algebra and $\varphi_1, ..., \varphi_k$ are $\mathbb{C}$-linear forms satisfying conditions (nor) Functions computed by finite dimensional such algebras are called *extended quantum recognizable functions*; their set, (EQR) obviously contains QR *extended quantum tree automaton* come from quantum tree automata

$$\mathcal{M} = (\Gamma, Q, \mu, (e_1, ..., e_k))$$

by just taking $e_1, ..., e_k$ to be any family of vectors of $\mathbb{C}^Q$ verifying the condition: for all $X \in \mathbb{C}^Q$ $|X| \leqslant 1$ implies

$$\sum_{i=1}^{k} |(X/e_i)|^2 \leqslant 1.$$

Then using the techniques of section 2 we can show that EQR is a convex chosed under product,etc.

Our aim now is to compare quantum with ordinary function recognizability. First some necessary matter.

Consider a field K and a ranked alphabet Γ.
A tree function $S : T_\Gamma \to K$ is termed *recognizable* if it is the behavior of a k-linearized (bottom up) tree automaton. Such functions can be characterized in a more algebraic manner as follows:

Theorem 4. *(cf. [BR],[BA])$S : T_\Gamma \to K$ is recognizable if and only if it can be computed by a finite dimensional realization i.e. a triple ($\mathcal{A} = (A, \alpha), \varphi$) where $\mathcal{A}$ is a K-multilinear Γ-algebra and $\varphi : A \to K$ a k-linear form such that*

$$S(t) = \varphi(H_A(t)), \text{ for all } t \in T_\Gamma.$$

In [BL] the minimal realization (V_S, U_S, φ_S) of a given function $S' : T_\Gamma \to K$ is constructed. V_S is the K-subspace of K^{P_Γ} generated by all the left derivatives of S

$$t^{-1}S : P_\Gamma \to K \quad , \quad (t^{-1}S)(\tau) = S(t\tau), \ \tau \in P_\Gamma$$

i.e.

$$V_S = < t^{-1}S/t \in T_\Gamma >$$

The Γ-operations $(v_S)_f : V_S^n \to V_S$ $(f \in \Gamma_n)$ are given by

$$(v_S)_f(t_1^{-1}S, ..., t_1^{-1}S) = (f(t_1, ..., t_n))^{-1}S$$

whereas $\varphi_S : V_S \to K$ is

$$\varphi_S(t^{-1}S) = S(t) \quad , \quad \forall t \in T_\Gamma$$

Theorem 5. *(cf. [BL])$S : T_\Gamma \to K$ is recognizable if and only if $dim V_S < \infty$.*

Example 4. Take a recognizable function $S : T_\Gamma \to \mathbb{C}$. Then its conjugate

$$\overline{S} : T_\Gamma \to \mathbb{C} \quad , \quad \overline{S}(t) = \overline{S(t)} \text{ for all } t \in T_\Gamma$$

is recognizable, too. Indeed, the function

$$t^{-1}\overline{S} \mapsto t^{-1}S \quad t \in T_\Gamma$$

is an isomorphism of $V_{\overline{S}}$ onto V_S so that $dim V_{\overline{S}} = dim V_S < \infty$.

Proposition 12. *The class EQR of extended quantum recognizable functions is included into the class R of recognizable functions.*

Proof. Consider an EQR

$$\mathcal{A} = (A, \alpha, (\varphi_1, ..., \varphi_k))$$

According to theorem 4 the function $S_i : T_\Gamma \to \mathbb{C}$ computed by the realization (A, α, φ_i) is recognizable. $(1 \leqslant i \leqslant k)$:

$$S_i(t) = \varphi_i(H_A(t)), \quad t \in T_\Gamma$$

thus so is its conjugate $\overline{S}_i : T_\Gamma \to \mathbb{C}(1 \leqslant i \leqslant k)$. Since

$$F_A(t) = \sum_{i=1}^{k}(S_i \cdot \overline{S}_i)(t) = \sum_{i=1}^{k}|S_i(t)|^2$$

the result comes by observing that the product and sum of recognizable function is again a recognizable function (cf. [BR]).

To complete information about SQR functions we state

Proposition 13. *A function* $S : T_\Gamma \to \mathbb{R}$ *is $\mathbb{R}$-recognizable iff it is $\mathbb{C}$-recognizable. Consequently, any SQR function is $\mathbb{R}$-recognizable.*

Proof. Denote by $(V_S)_\mathbb{R}$ and $(V_S)_\mathbb{C}$ the $\mathbb{R}$ and $\mathbb{C}$-spaces respectively generated by the left derivatives of S. Then

$$dim(V_S)_\mathbb{R} < \infty \iff dim(V_S)_\mathbb{C} < \infty$$

Here is an interesting consequence of proposition 12.

Corollary 4. *The equality problem is decidable for (semi)quantum recognizable functions ; i.e. for $F, F' \in QR(\Gamma)$ we can decide whether $F = F'$ or not.*

In particular, for $F \in QR(\Gamma)$ we can decide whether $F = 0$ or not.

Proof. Since equality is decidable for recognizable functions (cf. [Bo1]) , the result follows from proposition 14.

Closure questions for SQR functions are investigated below.

5 The Cutpoint Theorem

A set of trees of the form

$$[F > \lambda] = \{t \in T_\Gamma / F(t) > \lambda\}$$

where $F : T_\Gamma \to [0, 1]$ is a QR function and $\lambda \in [0, 1]$ is called *quantum tree language (QTL)*.
The main question here is whether QTL's are more powerful than ordinary recognizable tree languages. The positive answer is in the next statement

Theorem 6 (Cutpoint Theorem). *Let $F : T_\Gamma \to [0, 1]$ be quantum recognizable and $\lambda \in [0, 1]$ an isolated cutpoint of it (i.e. in some neighborhood of λ there is no value of F). Then $[F > \lambda]$ is recognizable.*

We need some auxiliary facts

Lemma 1. *Let $p : A \to A$ be a projection of the finite dimensional Hermitian space A ; then*

$$|p(x)| \leqslant |x|, \text{ for all } x \in A$$

Lemma 2. *Let $\mathcal{A} = (A, \alpha)$ be a Hermitian Γ-algebra. Then for all $\tau \in P_\Gamma$ and all $x \in A$,*

$$|x\tau| = |x|$$

Proof. Assume first e of the form

$$(\mu) \quad \tau = f(t_1, \ldots, t_{i-1}, \xi, t_{i+1}, \ldots, t_n),\, A \in \Gamma_n, t_j \in T_\Gamma$$

Then

$$\begin{aligned}|x\tau| &= |\alpha_f(H_\mathcal{A}(t_1), \ldots, H_\mathcal{A}(t_{i-1}), x, H_\mathcal{A}(t_{i+1}), \ldots, H_\mathcal{A}(t_n))| \\ &= |H_\mathcal{A}(t_1)| \cdots |H_\mathcal{A}(t_{i-1})| \cdot |x| \cdot |H_\mathcal{A}(t_{i+1})| \cdots |H_\mathcal{A}(t_n)| \\ &= |x|.\end{aligned}$$

In the case $\tau = \tau_1 \cdots \tau_k$, with all τ_i of the form (μ), we have

$$|x\tau| = |(\cdots((x\tau_1)\tau_2)\cdots\tau_k| = \cdots = |x\tau_1| = |x|.$$

Lemma 3. *Let $\varepsilon > 0$ and consider the set U_ε of all vectors $x \in \mathbb{C}^n$, $|x| = 1$, such that*

$$x, y \in U_\varepsilon \Rightarrow |x - y| \geqslant \varepsilon$$

Then U_ε is finite.

Proof (Proof of cutpoint theorem). Let $\mathcal{A} = (A, \alpha, p)$ be a quantum Γ-algebra of finite dimensions such that $F_\mathcal{A} = F$.
Claim: For all $t, s \in T_\Gamma$ it holds

$$H_\mathcal{A}(t) = H_\mathcal{A}(s) \Rightarrow t^{-1}[F > \lambda] = s^{-1}[F > \lambda]$$

Indeed, we have
$\tau \in t^{-1}[F > \lambda]$ iff $t\tau \in [F > \lambda]$ iff $F(t\tau) > \lambda$ iff $|p(H_\mathcal{A}(t\tau))|^2 > \lambda$ iff $|p(H_\mathcal{A}(t)\tau)|^2 > \lambda$ iff $|p(H_\mathcal{A}(s)\tau)|^2 > \lambda$ iff $\tau \in s^{-1}[F > \lambda]$
as claimed. Now suppose

$$t^{-1}[F > \lambda] \neq s^{-1}[F > \lambda]$$

that is for some $\tau \in P_\Gamma$

$$\tau \in t^{-1}[F > \lambda] \text{ and } \tau \notin s^{-1}[F > \lambda]$$

or vice versa, i.e.

$$F(t\tau) > \lambda \text{ and } \tau \notin s^{-1}[F > \lambda]$$

or vice versa, i.e.

$$F(t\tau) > \lambda \text{ and } F(s\tau) \leqslant \lambda$$

or vice versa, i.e.

$$|p(H_\mathcal{A}(t\tau))|^2 > \lambda \text{ and } |p(H_\mathcal{A}(s\tau))|^2 \leqslant \lambda$$

or vice versa. By hypothesis λ is an isolated cutpoint of F, so that for some $\varepsilon > 0$

$$|F(t) - \lambda| \geqslant \varepsilon \quad \text{for all } t \in T_\Gamma$$

Hence

$$(1) \quad |p(H_A(t\tau))|^2 - |p(H_A(s\tau))|^2 \geqslant 2\varepsilon$$

Since

$$|p(H_A(w))|^2 \leqslant 1 \text{ for all } w \in T_\Gamma$$

from the previous inequality (1) we get

$$|p(H_A(t\tau))| - |p(H_A(s\tau))| \geqslant \varepsilon$$

But

$$\begin{aligned}
|p(H_A(t\tau))| - |p(H_A(s\tau))| &\leqslant |p(H_A(t\tau)) - p(H_A(s\tau))| \\
&= |p(H_A(t\tau) - H_A(s\tau))| \\
&\stackrel{lemma\,1}{\leqslant} |H_A(t\tau) - H_A(s\tau)| \\
&= |H_A(t)\tau - H_A(s)\tau| \\
&= |(H_A(t) - H_A(s)\tau| \\
&\stackrel{lemma\,2}{=} |H_A(t) - H_A(s)|
\end{aligned}$$

In other words

$$|H_A(t) - H_A(s)| \geqslant \varepsilon$$

meaning that

$$H_A(t), H_A(s) \in U_\varepsilon$$

since by lemma 3 U_ε is finite, the set $\{t^{-1}[F > \lambda]/t \in T_\Gamma\}$ is finite because as we just have proven, the function

$$t^{-1}[F > \lambda] \mapsto H_A(t)$$

is injective. According to proposition 2 the tree language $[F > \lambda]$ is recognizable

Remarks:

1. Using a similar argument we can show that, under the assumption of cutpoint theorem, the tree language $[F < \lambda]$ is recognizable, as well.

2. We should point out that a Cutpoint Theorem already holds for stochastic functions (cf. [Pt]).

Corollary 5. *Assume λ is an isolated cutpoint of the quantum function $f : X^* \to [0, 1]$ then the language $\{w \in X^*/f(w) > \lambda\}$ is rational.*

Further

Corollary 6. Let $F : T_\Gamma \to [0, 1]$ be a quantum recognizable function with finite range. Then the tree language:

$$[F = \lambda] = \{t \in T_\Gamma / F(t) = \lambda\}$$

are recognizable.

Proof. Suppose $A = \{\alpha_1, ..., \alpha_n\}$ is the range of F $\alpha_1 < ... < \alpha_n$. For each $i(1 \leqslant i \leqslant n)$ choose constants λ, μ as follows

$$\alpha_{i-1} < \lambda < \alpha_i < \mu < \alpha_{i+1}.$$

Then the tree language

$$[F = \alpha_i] = T_\Gamma - ([F > \mu] \cup [F < \lambda])$$

is recognizable.

References

[BA] S. Bozapalidis and A. Alexandrakis. Representations matricielles des series d'arbres Reconnaissables, *RAIRO Inf. Theor.* 23(4) (1988) pp.449-459.
[BI] S. Bozapalidis and S. Ioulidis. Actions, Finite state sets and applications to trees, *Int. J. Comp. Math.* 31 (1989) pp. 11-17.
[BI] S. Bozapalidis and O. Louskou-Bozapalidou. The rank of a formal tree power series, *Theoret. Comput. Sci.*, 27 (1983) pp. 211-215.
[Bo1] S. Bozapalidis. Effective construction of the syntactic algebra of a recognizable series on trees, *Acta Informatica* 28 (1991) pp. 351-363.
[Bo2] S. Bozapalidis *Theory of Polypodes I: Recognizability*, TR. Aristotle University Thessaloniki, (1999).
[BR] J. Berstel and C. Reutenauer. Recognizable formal power series on trees, *Theoret. Comput. Sci.*, 18 (1982) pp. 115-142
[CS] F. Cesceg and M. Steiby. *Tree Automata*, Akademiai Kiado, (1974).
[En] J. Engelfriet. *Tree Automata and Tree Grammars*, Daimi., (1977).
[MC] C. Moore and J. Crutchfield. Quantum automata and quantum grammars, *Theoret. Comput. Sci.*, 237 (2000) pp. 275-306.
[Pa] A.Paz. *Probabilistic Automata*, Academic Press, (1971).

An Unconventional Computational Linear Algebra: Operator Trigonometry

Karl Gustafson[1,2]

[1] University of Colorado, Department of Mathematics, Boulder, U.S.A.
[2] International Solvay Institute for Physics and Chemistry, Brussels, Belgium

Abstract. Conventional computational linear algebra is chiefly concerned with the linear solver problem: $Ax = b$. Many fine algorithms for many situations have been found and these are widely utilized in scientific computing. The theory and practice of general linear solvers is intimately linked to the eigenvalues and eigenvectors of the matrix A. Here I will present a theory of antieigenvalues and antieigenvectors for matrices or operators A and show how that theory is also intimately linked to computational linear algebra. I will also present new applications of this theory, which I call operator trigonometry, to Statistics and Quantum Mechanics. Also I will give some hitherto unpublished insights into the operator trigonometry, its origins, meaning, and future potential. Finally, I will present a fundamental new view of $\sin \phi(A)$.

1 Operator Trigonometry: An Overview

In this paper, viz. this section, I want to be quite brief about the older history of what I now call operator trigonometry, so that I can get quickly to the more recent results and issues. Therefore I immediately refer the reader to the two recent books [1,2] which describe the operator trigonometry from its inception in 1967 to 1995, and its history. In this paper I will go beyond [1,2] historically by interspersing a few lesser-known motivating insights as I describe the basic elements of the operator trigonometry and its applications to date. Sections 2, 3, 4 and 5 will stress the very latest developments and issues as I see them.

The problem that induced the operator trigonometry in 1967 was an abstract multiplicative perturbation question in the Hille–Yosida theory of evolution semigroups. Given a contraction semigroup $U(t) = e^{-tA}$ on a Banach space X, the question was: when does a multiplicatively perturbed generator BA still generate a contraction semigroup? Let me move immediately to the simpler case of X a Hilbert space and A and B bounded operators on X. It was known that any generator A must satisfy a spectral condition and also be accretive: $\operatorname{Re} \langle Ax, x \rangle \geq 0$ for all x. Let B be strongly accretive: $\operatorname{Re} \langle Bx, x \rangle \geq m_B \langle x, x \rangle$, $m_B > 0$. Then I showed that BA is a generator of a contraction semigroup $W(t) = e^{-tBA}$ iff BA is accretive: $\operatorname{Re} \langle BAx, x \rangle \geq 0$. It turned out that the spectral conditions needed for BA to be a generator in the

Hille–Yosida theory were much more straight-forward to deal with than was this product accretivity question.

Let me intersperse here a first side-comment of perhaps some interest. In the 1960's, that strange decade which was also the heyday of pure mathematics in the United States, all of us young guys were encouraged to send our new results even if incomplete to the *Notices of the American Mathematical Society* to ascertain feedback from others who might be doing similar things. So I did: you will find my operator trigonometry beginnings scattered in the 1967 *Notices* [3]. As I thus pondered the product accretivity question above, I found that if I looked more closely at strongly accretive A and B, for BA to be accretive it was sufficient that

$$\inf_{-\infty < \epsilon < \infty} \|\epsilon B - I\| \leq \inf_{x \neq 0} \frac{\operatorname{Re} \langle Ax, x \rangle}{\|Ax\| \|x\|}. \tag{1.1}$$

The right side of this expression I called: $\cos \phi(A)$, surely a natural enough definition, to connote an angle $\phi(A)$ which represented the limit of A's turning capability when applied to all vectors x. I called $\phi(A)$ the angle of the operator A. I thought it would be very nice if somehow the left side of the expression was also in some way trigonometric. It is, it was not so easy to prove, I announced it also in the *Notices* in [4], the result is: $\inf_\epsilon \|\epsilon B - I\| = \sin \phi(B)$, where $\sin \phi(B)$ satisfies $\sin \phi(B) = (1 - \cos^2 \phi(B))^{1/2}$. So the operator product of two strongly accretive operators A and B will itself be accretive when: $\sin \phi(B) \leq \cos \phi(A)$.

Although I maintained a side interest in the operator trigonometry, among other things, after 1968 for the most part I moved on to other research, e.g. scattering theory, then computational fluid dynamics, then neural and optical computing. For one account of some of the latter activities see for example [5]. Because I had always suspected that the operator trigonometry could have useful applications to computational linear algebra, ten years ago I returned to the operator trigonometry to bring it into the computational contexts. I will discuss the main results in the following Section 2. Then in Section 3 I will present some very recent connections and uses of the operator trigonometry for statistics and econometrics. These are only being submitted for publication this year. In Section 4 I will show beginning relations of the operator trigonometry to quantum mechanics and quantum computing. I had hoped to have more of the latter for this conference, but the best I can do here is describe some of the possibilities and issues. In Section 5 I will present a completely new perspective on the meaning of $\sin \phi(A)$. I expect this to open up a number of exciting new opportunities for the operator trigonometry.

The basic elements of the operator trigonometry are $\sin \phi(B)$ and $\cos \phi(A)$ as defined by the two sides of (1.1), as just discussed. Also I introduced notions of antieigenvalue and antieigenvector which play key roles in the operator trigonometry. I called $\cos \phi(A)$, i.e. right side of (1.1), the first antieigenvalue of A. A vector x turned that maximum amount by A was called a first antieigenvector. For simplicity in this paper we will for the most part describe the operator trigonometry as it is known for A a symmetric positive definite $n \times n$ matrix A. Then

$$\sin \phi(A) = \frac{\lambda_n - \lambda_1}{\lambda_n + \lambda_1}, \qquad \cos \phi(A) = \frac{2\sqrt{\lambda_1 \lambda_n}}{\lambda_n + \lambda_1} \qquad (1.2)$$

where A's eigenvalues are $0 < \lambda_1 \leq \cdots \leq \lambda_n$ ordered from smallest to largest. The operator maximum turning angle $\phi(A)$ is attained by the antieigenvector pair

$$x_{\pm}^1 = \pm \left(\frac{\lambda_n}{\lambda_1 + \lambda_n} \right)^{1/2} x_1 + \left(\frac{\lambda_1}{\lambda_1 + \lambda_n} \right)^{1/2} x_n \qquad (1.3)$$

where x_1 and x_n are eigenvectors corresponding to λ_1 and λ_n, respectively. I have normalized all of x_1, x_n and $x_{\pm}^1$ to norm one in (1.3). Multiplicity is no problem: in the case of repeated eigenvalues, the x_1 and x_n may be taken from anywhere in the unit sphere in their respective eigenspaces. Using intermediate eigenvalues λ_j and λ_k, e.g., $0 < \lambda_1 < \lambda_j < \lambda_k < \lambda_n$, one can define intermediate operator critical turning angles according to (1.2) and (1.3). Because there was always some confusion about these "higher" antieigenvalues and antieigenvectors, recently [6] I have clarified and so extended the theory. Also in [6] I have extended the operator trigonometry to arbitrary matrices A. I refer the interested reader to that paper for further details.

There is one important caveat to bear in mind concerning the operator trigonometry when viewed as an alternate spectral theory of principal turning vectors (i.e., the antieigenvectors) rather than the conventional spectral theory of principal stretching vectors (i.e. the eigenvectors). The antieigenvectors $x_{\pm}^1$ of (1.3) do not "span an antieigenspace". Only those two particular weighted combinations of the eigenvectors x_1 and x_n will be most turned by A.

One will find all the References at the end of this paper, to be to my work on the operator trigonometry. But from those you may find adequate literature about the application areas of computational linear algebra, statistics and econometrics, quantum mechanics and quantum computing, that literature being too large and diverse to attempt to give in this paper.

2 Computational Linear Algebra

In the 1980's my research emphasis had turned to computational fluid dynamics. There one quickly learns that a major percentage (e.g., 70% of the computer time for applications in two space dimensions, 90% for those in three space dimensions) of the computational effort is consumed in the linear solvers for the large linear systems $Ax = b$ that result from discretizing the partial differential operators describing the fluid flows. The same statement applies to the discretized partial differential equations underlying semiconductor design and other physical modeling in two and three space dimensions. Linear solvers may be roughly broken down into direct methods (e.g.,

Gauss, LU, or FFT-based schemes) or indirect methods (e.g., iterative Gradient and Splitting schemes). Being invited to speak at the 1990 conference [7], I decided to take that opportunity to examine the possible connections of the operator trigonometry to computational linear algebra. I had suspected such connections 30 years ago, I even mentioned it in one of the 1968 papers [8]. Very quickly the following result was obtained.

One way to solve $Ax = b$ is by steepest descent. Although this algorithm converges very slowly, it is the conceptual basis of many much faster gradient based schemes. One of the fundamental numerical analysis results (1948) of the past century is that of Kantorovich: for A a symmetric positive definite matrix with eigenvalues $0 < \lambda_1 \leq \cdots \leq \lambda_n$, the steepest descent worst case convergence error bound is

$$E_A(x_{k+1}) \leq \left(1 - \frac{4\lambda_1\lambda_n}{(\lambda_1 + \lambda_n)^2}\right) E_A(x_k) \tag{2.1}$$

From (1.2) juxtaposed with (2.1) I immediately had the geometric meaning of (2.1)

$$E_A(x_{k+1}) \leq \sin^2 \phi(A) E_A(x_k). \tag{2.2}$$

In (2.1) and (2.2) E_A denotes the energy inner product

$$E_A(x) = \langle (x - x^*), A(x - x^*) \rangle \tag{2.3}$$

where x^* is the true solution to the system: $Ax^* = b$.

I insert here an observation. Kantorovich's bound (2.1) had stood since 1948, over 40 years before I published [7]. Kantorovich received the Nobel Prize in Economics in 1975 for his mathematical contributions thereto. Moreover there is a very large related literature of inequalities related to the socalled Kantorovich inequalities. But until I put the operator trigonometry into the picture, no one had seen the geometry of those several theories.

It turns out that some had tried. There is a socalled Kantorovich–Wielandt angle θ defined in terms of A's condition number $\kappa = \lambda_n/\lambda_1$. But a preoccupation by those working in numerical linear algebra with condition number κ makes θ not a natural angle, whereas the operator trigonometry operator turning angle $\phi(A)$ is *the* natural angle for A. For more on this matter see [9].

A similar bound to (2.1) holds for the much better conjugate gradient scheme and also I have recently given new trigonometric understandings for a number of related schemes. These include Preconditioned Conjugate Gradient, Jacobi, Gauss–Seidel, SOR, SSOR, Uzawa, AMLI, ADI, Multigrid, Domain Decomposition, and related iterative solution methods for $Ax = b$. More details may be found in [10,11,12], other papers cited there, and work in progress. I don't really want to go into great detail here. However, I would like to complete this section with the following. First, important empirical observations about the relative occurrences and roles of $\sin \phi(A)$ and $\cos \phi(A)$ as they enter computational linear algebra. Second, my very recent extension of the operator trigonometry from A symmetric positive definite to arbitrary invertible matrices A. Third, a related new result concerning linear solvers based upon the QR factorization of A. Fourth, a few closing comments/insights concerning the culture of the computational linear algebra community.

First an empirical observation: I have found in these ten years of work relating the operator trigonometry to the many competing linear solver schemes, that it is more often $\sin\phi(A)$ which is lurking inherently under the convergence behavior of these schemes. To illustrate this point, let me consider the very basic Richardson scheme (not a gradient scheme)

$$x_{k+1} = x_k + \alpha(b - Ax_k) \tag{2.4}$$

This scheme when much refined underlies a number of important iterative solvers. The scheme starts with an arbitrary initial guess x_0 and proceeds with iteration matrix $G_\alpha = I - \alpha A$. One wants to choose the parameter α to produce an optimal convergence rate. I showed in [10] that in Richardson iterative solution of $Ax = b$ for symmetric positive definite A, the optimal parameter α is

$$\alpha = \epsilon_m = \frac{\langle Ax_\pm^1, x_\pm^1\rangle}{\|Ax_\pm^1\|^2} \tag{2.5}$$

where ϵ_m is the minimizing parameter for $\sin\phi(A)$ in (1.1) and where $x_\pm^1$ are A's first antieigenvectors (1.3). The optimal convergence rate of the Richardson scheme is $\sin\phi(A)$. Let me just assert that the implicitness of $\sin\phi(A)$ in related but more complicated linear solver schemes where one is free to try to choose some optimal parameters, occurs essentially in the same way.

Less often it is $\cos\phi(A)$ which shows up in these schemes. A rather dramatic (in my opinion) instance of this is my finding it as the explanation of the exciting semi-iterative speedup of ADI (Alternating Direction Implicit scheme, also sometimes called the Peaceman–Rachford scheme), a very important splitting scheme which under other guises still plays an essential role in some of the latest NASA codes. I cannot resist presenting here the beautiful way in which a whole sequence of operator sub-angles are seen as the underlying reason for this important processing speedup.

The true power of ADI comes to the fore when it is employed semi-iteratively, i.e., the optimizing parameter is allowed to change with each iteration. Then the convergence factors $\rho(M^{(q)})$ are minimizd when

$$d(\alpha,\beta,q) = \max_{\alpha\leq x\leq\beta} \prod_{p=1}^{q}\left|\frac{1-\tau_p x}{1+\tau_p x}\right| \tag{2.6}$$

is minimized, see [12] for literature on ADI. For $q = 1, 2, 4, \ldots$ the convergence rate bounds (2.6) possess an interesting recursive relationship $d(\alpha,\beta,2q) = d\left(\sqrt{\alpha\beta}, \frac{\alpha+\beta}{2}, q\right)$. Taking $\tau_1 = \sqrt{\alpha\beta} = \sqrt{\lambda_{11}\lambda_{N-1,N-1}}$, we have

$$d(\alpha,\beta,1) = \frac{1-\sqrt{\alpha/\beta}}{1+\sqrt{\alpha/\beta}} = \frac{\lambda_{N-1,N-1}^{1/2} - \lambda_{1,1}^{1/2}}{\lambda_{N-1,N-1}^{1/2} + \lambda_{1,1}^{1/2}} = \sin(A_h^{1/2}). \tag{2.7}$$

In these expressions A_h is a discretized Laplacian operator with Dirichlet boundary conditions with eigenvalues λ_{ij}, $1 \leq i,j \leq N-1$ reflecting the finite differences grid size, see [12].

For $q = 2$ and 4 we then have

$$d(\alpha, \beta, 2) = \frac{1 - \left(\frac{\sqrt{\alpha\beta}}{\frac{\alpha+\beta}{2}}\right)^{1/2}}{1 + \left(\frac{\sqrt{\alpha\beta}}{\frac{\alpha+\beta}{2}}\right)} = \frac{1 - (\cos A_h)^{1/2}}{1 + (\cos A_h)^{1/2}} \qquad (2.8)$$

and

$$d(\alpha, \beta, 4) = \frac{1 - \left[\frac{2\sqrt{\sqrt{\alpha\beta}\cdot\frac{\alpha+\beta}{2}}}{\sqrt{\alpha\beta}\cdot\frac{\alpha+\beta}{2}}\right]^{1/2}}{1 + \left[\frac{2\sqrt{\sqrt{\alpha\beta}\cdot\frac{\alpha+\beta}{2}}}{\sqrt{\alpha\beta}+\frac{\alpha+\beta}{2}}\right]^{1/2}} = \frac{1 - \frac{(\cos A_h)^{1/4}}{\cos\left(\frac{\phi(A_h)}{2}\right)}}{1 + \frac{(\cos A_h)^{1/4}}{\cos\left(\frac{\phi(A_h)}{2}\right)}} \qquad (2.9)$$

The operator-trigonometric last expression in (2.9) follows from looking at the expression in the brackets as

$$\frac{2\sqrt{\sqrt{\alpha\beta}\cdot\frac{\alpha+\beta}{2}}}{\sqrt{\alpha\beta}+\frac{\alpha+\beta}{2}} = \frac{2^{3/2}(\alpha\beta)^{1/4}(\alpha+\beta)^{-1/2}}{\frac{2\sqrt{\alpha\beta}}{\alpha+\beta}+1}$$

$$= \frac{\left(\frac{2\sqrt{\alpha\beta}}{\alpha+\beta}\right)^{1/2}}{\left(\frac{2\sqrt{\alpha\beta}}{\alpha+\beta}+1\right)\cdot\frac{1}{2}} = \frac{(\cos A_h)^{1/2}}{\left(\frac{\cos A_h + 1}{2}\right)} \qquad (2.10)$$

For a coarse 3×3 mesh example these semi-iterative convergence rate numbers (2.7) to (2.9) are calculated to be $d_1 \cong 0.268$, $d_2 \cong 0.036$, $d_4 \cong 0.00065$, illustrating trigonometrically the inherent power of semi-iterative ADI as a linear solver.

Second, in computational linear algebra, important applications and hence interest has recently been turning to the case of A a general, *nonsymmetric*, perhaps sparse, perhaps very large, matrix, often $n \times n$, invertible, and perhaps with only real entries. Having these applications in the back of my mind, for the last couple of years I have thought about the most natural way to extend the operator trigonometry to arbitrary matrices, and my opinion became that one should use polar form. There are two strong contributing reasons for this, and I would like to expose them here. First, I have never been motivated in the operator trigonometry to think of uniformly turning operators A, e.g., those which rotate all vectors by a fixed angle. What interests us in the operator trigonometry is the relative turning of vectors, just as in the classical eigenvalue theory we are interested in the relative stretching of vectors. Thus polar form $A = U|A|$ efficiently removes the "uniform turning", e.g., in U, and we already have the operator angle theory for $|A|$. Second, for invertible operators A, polar form is better than singular value decomposition for our purposes of an extended operator trigonometry, because we can show that the essential min-max Theorem $\sin^2 \phi(A) + \cos^2 \phi(A) = 1$ extends to arbitrary A via $|A|$.

The key move is to change the key definition of the operator sine in (1.1) to:

$$\sin \phi(A) = \inf_\epsilon \|\epsilon A - U\|. \tag{2.11}$$

Of course in the A symmetric positive definite case, U is just the Identity. Then, considering first for example A to be an arbitrary $n \times n$ nonsingular matrix with singular values $\sigma_1 \geq \sigma_2 \geq \cdots \geq \sigma_n > 0$, we obtain from (2.11) and (1.2) applied to $|A|$, that

$$\sin \phi(A) = \min_{\epsilon > 0} \|\epsilon |A| - I\| = \frac{\sigma_1(A) - \sigma_n(A)}{\sigma_1(A) + \sigma_n(A)}. \tag{2.12}$$

One may check that the key min-max identity in its essential form $\sin^2 \phi(A) + \cos^2 \phi(A) = 1$ is then satisfied if one modifies the definition of the operator cosine in (1.1) to:

$$\cos \phi(A) = \inf_{x \neq 0} \frac{\langle |A|x, x \rangle}{\||A|x\| \|x\|}. \tag{2.13}$$

Then $\cos \phi(A)$ is given as in (1.2) with λ_1 and λ_n replaced by σ_n and σ_1, respectively.

Thus one obtains a full operator trigonometry of relative turning angles for arbitrary invertible matrices A from their $|A|$. See [6] for more details. This construction may be extended to noninvertible A. The price that one pays is that one has ignored U. Actually you have not totally ignored U. It is easy to check that equivalent to (2.13) is the definition

$$\cos \phi(A) = \inf_{x \neq 0} \frac{\langle Ax, Ux \rangle}{\|Ax\| \|Ux\|}. \tag{2.14}$$

So one is looking at the turning power of A relative to its U. Howeer, this choice of extended operator trigonometry which I have made here for the context of computational linear algebra will come back to haunt us in Section 4 when we turn to quantum mechanics.

Third, let me insert here a new result which could have important ramifications as a bridge between my extended operator trigonometry just described above and its eventual application to $Ax = b$ linear solvers for general A. Many of the general, more sophisticated solvers which attempt to be robust for general A, for example the widely used GMRES and its variants, are essentially based upon a Gram–Schmidt-like orthogonalization procedure, or equivalently, the well-known fundamental matrix factoriztion $A = QR$, where Q is an orthogonal matrix and R is upper triangular. However, note the following. From $A = QR$ combined with the polar factorization $A = U|A|$ we see immediately that

$$|A| = (A^*A)^{1/2} = (R^*Q^*QR)^{1/2} = |R| \tag{2.15}$$

from which we may conclude that the extended operator trigonometry angle $\phi_{ext}(A)$ defined via $|A|$ from either (2.11) or (2.13) is the same as the angle $\phi_{ext}(R)$ defined via $|R|$. Let me also note the potential of this observation to connect the extended operator trigonometry to the so-called "structured shift" fast linear solvers based upon regarding A as the transfer operator of a time varying system. Those algorithms depend essentially upon an inner-outer factorization of A which is essentially a QR procedure.

Fourth, a few comments/insights concerning the nature and culture of the computational linear algebra community. Extremely good algorithms, software, theory, are already in place. Thus when I have presented my operator trigonometry recently at numerical linear algebra conferences, usually I am generally well received but often I am also harshly rebuked by two criticisms from one or two members of the audience. One, well, can you treat arbitrary matrices A, rather than just symmetric positive definite ones? Two, well, can you use your operator trigonometry to create a "better" linear solver? Let me reply here. One, yes, now. Moreover, I find much of *your* theory still stuck at the symmetric positive definite case. Two, no, not yet. Moreover upon reflection I have found that none of *you* have produced a "better" linear solver. It is an extremely competitive field into which a lot of money has been spent over the last 30 years.

3 Statistics and Econometrics

Very recently I have connected the operator trigonometry to certain basic matrix methods in statistics and econometrics. Application of the operator trigonometry to provide new results there to date include statistical efficiency of parameter estimation and certain results in correlation theories. More details will be provided in [13]. Here I will be extremely brief and I will describe only the new geometrical meanings underlying the much used least squares estimators OLSE and BLUE.

We focus attention on the general linear model

$$y = X\beta + \epsilon \qquad (3.1)$$

where y is an n-vector composed of n random samplings of a random variable Y, X is an $n \times p$ matrix usually called the design or model matrix, β is a p vector composed of p unknown nonrandom parameters to be estimated, and e is an n-vector of random errors incurred in observing y. The elements x_{ij} of X may have different statistical meanings depending on the application. We assume for simplicity that

$$E(e) = 0, \quad \text{Cov}(e) = \sigma^2 V, \quad V > 0, \quad \text{rank } X = r \leq p. \qquad (3.2)$$

The assumption $E(e) = 0$ means that each error e_i is distributed with mean zero, the assumptions $\text{Cov}(e) = \sigma^2 V$ and $V > 0$ mean the covariance matrix has been assumed here (for simplicity) to be nonsingular, and moreover a naturally occurring variance factor σ^2 has been taken out so that all further discussion may center on V which is symmetric positive definite. Below we will just drop the σ^2 factor and speak in terms of V. Although there are statistical applications in which V or X may be unknown, here for simplicity we assume that V and X are known. Moreover for simplicity here we may take an often assumed simplifying assumption that $n \geq 2p$. Generally one thinks of X as composed of just a few (regressor) columns. In practice this is sometimes justified by statements such as "it is not easy to experimentally get good independent columns of X".

There is a large literature on the general linear model (3.1) and I will not attempt to describe it here: see [13] or any books. We may consider an ordinary least squares estimator (OLSE) $\hat{\beta}$ and the best linear unbiased estimator (BLUE) β^*. Then the (relative) efficiency of $\hat{\beta}$ is defined as

$$RE(\hat{\beta}) = \frac{|\text{Cov}(\beta^*)|}{|\text{Cov}(\hat{\beta})|} = \frac{1}{|X'VX||X'V^{-1}X|} \quad (3.3)$$

where $|T|$ denotes $\det(T)$. Here we have assumed without loss of generality that

$$X'X = I_p \quad (3.4)$$

A fundamental lower bound for efficiency is:

$$RE(\hat{\beta}) \geq \prod_{i=1}^{p} \frac{4\lambda_i \lambda_{n-i+1}}{(\lambda_i + \lambda_{n-i+1})^2}; \quad (3.5)$$

The following result [13] constitutes a new geometrical meaning of fundamental lower bound on efficiency and a new result for how it may be attained. For the general linear model (3.1) with SPD covariance matrix $V > 0$, for $p = 1$ the geometrical meaning of the relative efficiency (3.3) of an OLSE estimator $\hat{\beta}$ against BLUE β^* is

$$RE(\hat{\beta}) \geq \cos^2 \phi(V) \quad (3.6)$$

where $\phi(V)$ is the operator angle of V. For $p \leq n/2$ the geometrical meaning is

$$RE(\hat{\beta}) \geq \prod_{i=1}^{p} \cos^2 \phi_i(V) = \prod_{i=1}^{p} \mu_i^2(V) \quad (3.7)$$

where the $\phi_i(V)$ are the successive decreasing critical turning angles of V, i.e., corresponding to the higher antieigenvalues $\mu_i(V)$. The lower bound (3.5) as expressed geometrically in (3.6) is attained for $p = 1$ by either of the two first antieigenvectors of V. For $p \leq n/2$ the lower bound (3.5) as expressed geometrically in (3.7) is attained as

$$\prod_{i=1}^{p} \left[\frac{\langle Vx_{\pm}^i, x_{\pm}^i \rangle}{\|Vx_{\pm}^i\| \|x_{\pm}^i\|} \right]^2 \quad (3.8)$$

where $x_\pm^i$ denotes either of the ith higher antieigenvectors of V given by

$$x_\pm^i = \pm \left(\frac{\lambda_i}{\lambda_i + \lambda_{n-i+1}}\right)^{1/2} x_{n-i+1} + \left(\frac{\lambda_{n-i+1}}{\lambda_i + \lambda_{n-i+1}}\right)^{1/2} x_i. \tag{3.9}$$

In (3.9) x_i denotes the normalized ith eigenvector of V corresponding to the eigenvalue λ_i.

There are many other interesting new connections between the operator trigonometry and some Lagrange multiplier techniques used for optimization of regression in statistical methods, that are established in [13]. In particular I establish fundamental comparisons between what I call reduced Inefficiency equations

$$\frac{V^2 x_i}{x_i' V x_i} + \frac{x_i}{x_i' V^{-1} x_i} = 2V x_i, \qquad i = 1, \ldots, p. \tag{3.10}$$

that are key to the statistical theory, and the Euler equations

$$\frac{A^2 x}{\langle A^2 x, x\rangle} - \frac{2Ax}{\langle Ax, x\rangle} + x = 0$$

for the antieigenvectors in my operator trigonometry. Suffice it to say: this is a new geometry for statistics, heretofore lacking, with many more details to be worked out in the future.

4 Quantum Mechanics and Quantum Computing

In this section I will be brief, because the work is ongoing and more importantly, still quite incomplete. Recently [14,15] I have shown that certain important inequalities for quantum spin probabilities are special cases of the operator trigonometry. Certain arguments about existence or nonexistence of hidden variables are reducible to a general Hilbert space triangle inequality we discovered within the operator trigonometry. I will describe these recent results of the operator trigonometry as the first part of this section. More details and the extensive background literature in physics may be found from [14,15].

As to quantum computing, I will present some preliminary thoughts, as the second part of this section of this paper. I had hoped to have more for this conference. A preliminary conclusion is that the operator trigonometry needs to be tailored much more extensively to the specific operators used in quantum computing theory. Those are for the most part unitary operators. As I stated at the end of Section 2, my recent extension of the operator trigonometry from symmetric positive definite operators to arbitrary operators, ignored the unitary factors of those operators. I will also show here that my earlier notion of total operator cosines, although applicable to unitary operators, does not tell us much.

The background for my entering the operator trigonometry into quantum mechanics centers on the famous 1935 EPR paradox of Einstein, Podolski and

Rosen. This was an argument for the need for additional variables in quantum theory to restore causality and locality to that theory. Then Bell's 1964 inequalities which provided a necessary condition for such a classical probability hidden variable model were shown in 1982 by Aspect and others to be violated beyond any doubt by a number of quantum mechanical physical systems. The debates and experiments about locality and classical interpretation of quantum mechanics of course continue to this day.

What I was able to show was that such inequalities are a direct mathematical consequence of the operator trigonometry of Hilbert space. That is, no physical arguments or assumptions or consequences need be involved at all. The Bell's inequalities and other similar ones may be seen as special cases of the operator trigonometry. Let me explain how this comes about.

Within the operator trigonometry there occurs a basic triangle inequality for operator angles, namely

$$\phi(BA) \leq \phi(B) + \phi(A). \tag{4.1}$$

This follows from an individual vector triangle inequality as follows. Given any three unit vectors x, y, z in a Hilbert space, let $\langle x, y \rangle = a_1 + ib_1$, $\langle y, z \rangle = a_2 + ib_2$, $\langle x, z \rangle = a_3 + ib_3$, and define the angles $\phi_{xy}, \phi_{yz}, \phi_{xz}$ in $[0, \pi]$ by $\cos \phi_{xy} = a_1$, $\cos \phi_{yz} = a_2$, $\cos \phi_{xz} = a_3$. To show the vector triangle inequality

$$\phi_{xz} \leq \phi_{xy} + \phi_{yz}, \tag{4.2}$$

it suffices to show

$$\cos \phi_{xz} \geq \cos(\phi_{xy} + \phi_{yz}), \tag{4.3}$$

which by the sum formula for cosines is equivalent to

$$\sqrt{1 - a_1^2}\sqrt{1 - a_2^2} \geq a_1 a_2 - a_3. \tag{4.4}$$

The desired result (4.2) follows trivially when the right side of (4.4) is negative. In the other case we need

$$(1 - a_1^2)(1 - a_2^2) \geq (a_1 a_2 - a_3)^2, \tag{4.5}$$

But for unit vectors the Gram matrix

$$G = \begin{bmatrix} \langle x, x \rangle & \langle x, y \rangle & \langle x, z \rangle \\ \langle y, x \rangle & \langle y, y \rangle & \langle y, z \rangle \\ \langle z, x \rangle & \langle z, y \rangle & \langle z, z \rangle \end{bmatrix} \tag{4.6}$$

has determinant (using complex cancellations)

$$|G| = \begin{vmatrix} 1 & a_1 & a_3 \\ a_1 & a & a_2 \\ a_3 & a_2 & 1 \end{vmatrix} = 1 + 2a_1 a_2 a_3 - (a_1^2 + a_2^2 + a_3^2) \geq 0 \tag{4.7}$$

which gives (4.3) and hence also the desired individual vector triangle inequality (4.2).

To now show the connection of the operator trigonometry to the quantum mechanical hidden variable spin systems, I follow [14,15]. See those papers for citations to other work by Bell, Wigner, Accardi, others, which [14] and [15] themselves follow. Conditional probabilities

$$P(A = a_\alpha \mid B = b_\beta), P(B = b_\beta \mid C = c_\gamma), P(C = c_\gamma \mid A = a_\alpha) \quad (4.8)$$

are assumed to satisfy symmetry conditions $P(A = a_\alpha \mid B = b_\beta) = P(B = b_\beta \mid A = a_\alpha)$, etc. and are said to satisfy a complex Hilbert space probability model if there exists a complex Hilbert space $\mathcal{H}$ such that for each observable A, B, C there exists an orthonormal basis $(\phi_\alpha), (\psi_\beta), (\chi_\gamma)$ such that for each α, β, γ, $P(A = a_\alpha \mid B = b_\beta) = |\langle \phi_\alpha, \psi_\beta \rangle|^2$, etc. Limiting discussion to three observables taking only two values, the conditional probabilities (4.8) may be represented by the following transition probability matrices:

$$P = P(A \mid B) = \begin{bmatrix} p & 1-p \\ 1-p & p \end{bmatrix} = \begin{bmatrix} \cos^2(\alpha/2) & \sin^2(\alpha/2) \\ \sin^2(\alpha/2) & \cos^2(\alpha/2) \end{bmatrix}$$

$$Q = P(B \mid C) = \begin{bmatrix} q & 1-q \\ 1-q & q \end{bmatrix} = \begin{bmatrix} \cos^2(\beta/2) & \sin^2(\beta/2) \\ \sin^2(\beta/2) & \cos^2(\beta/2) \end{bmatrix} \quad (4.9)$$

$$R = P(C \mid A) = \begin{bmatrix} r & 1-r \\ 1-r & r \end{bmatrix} = \begin{bmatrix} \cos^2(\gamma/2) & \sin^2(\gamma/2) \\ \sin^2(\gamma/2) & \cos^2(\gamma/2) \end{bmatrix}.$$

For simplicity we may assume $0 < p, q, r < 1$, $0 < \alpha, \beta, \gamma < \pi$. For conditions for a quantum mechanical spin system to have a complex Hilbert space model existing, the Pauli matrices

$$\sigma_1 = \begin{bmatrix} 0 & 1 \\ 1 & 0 \end{bmatrix}, \sigma_2 = \begin{bmatrix} 0 & -i \\ i & 0 \end{bmatrix}, \sigma_3 = \begin{bmatrix} 1 & 0 \\ 0 & -1 \end{bmatrix} \quad (4.10)$$

and spin operators $\sigma \bullet a = \sigma_1 a_1 + \sigma_2 a_2 + \sigma_3 a_3$ for $a = (a_1, a_2, a_3)$ a real 3-vector of norm 1 are considered. A spin model for the transition probabilities (4.8) is said to exist if there exists three normalized 3 vectors a, b, c such that the orthonormal bases $\psi_\alpha(a), \psi_\beta(b), \psi_\gamma(c)$ realize the matrices P, Q, R of (4.9). In this way the question of the existence of a Hilbert space probability model is reduced to the question of the existence of three norm-1 vectors a, b, c such that

$$|\langle \psi_1(a), \psi_1(b) \rangle|^2 = \cos^2 \theta_{ab}/2,$$
$$|\langle \psi_1(a), \psi_2(b) \rangle|^2 = \sin^2 \theta_{ab}/2; \quad (4.11)$$

where $\cos \alpha = \cos \theta_{ab}$, $\cos \beta = \cos \theta_{bc}$, $\cos \gamma = \cos \theta_{ac}$ link the angles of P, Q, R in (4.9) to the vector directions a, b, c. These three vectors in the Bell picture represent hidden variable measuring directions of a second apparatus relative to a fixed first apparatus some distance away. One can then show that such vectors a, b, c exist if and only if

$$\cos^2 \alpha + \cos^2 \beta + \cos^2 \gamma - 1 \leq 2 \cos \alpha \cos \beta \cos \gamma. \quad (4.12)$$

But (4.12) is a special case of (4.7) which is equivalent to the vector triangle inequality (4.2).

It is also shown in [15] that certain fundamental eigenstates of lifetime and mass in particle physics are antieigenvectors in the operator trigonometry of the associated operators.

Turning now to quantum computing and the potential use of the operator trigonometry therein, the main idea is to develop a trigonometric theory of decoherence and also perhaps for the dynamics of certain quantum computing algorithms. However, this will require a further extension of the operator trigonometry, because thus far I have extended the theory from A symmetric positive definite to arbitrary $A = U|A|$ via polar form, at the expense of ignoring U. It is exactly unitary matrices we need to examine in quantum computing.

Let me illustrate the situation as follows. Early in the operator trigonometry I defined a "total" turning angle $\phi_{\text{total}}(A)$ for arbitrary operators A by

$$\cos \phi_{\text{total}}(A) = \inf_{Ax \neq 0} \frac{|\langle Ax, x \rangle|}{\|Ax\| \|x\|} \quad (4.13)$$

For invertible normal operators A it is known [1,2] that a general formula for total antieigenvalues of normal operators, namely,

$$\cos^2 \phi_{\text{tot}}(A) = |\mu_1|^2(A) = \frac{(\beta_i|\lambda_j| + \beta_j|\lambda_i|)^2 + (\delta_i|\lambda_j| + \delta_j|\lambda_i|)^2}{(|\lambda_i| + |\lambda_j|)^2 |\lambda_i||\lambda_j|}. \quad (4.14)$$

There $\lambda_i = \beta_i + i\delta_i$ and $\lambda_j = \beta_j + i\delta_j$ are the two eigenvalues of a normal operator A which contribute to the first total antieigenvectors of A. For the matrix example

$$A = \begin{bmatrix} \sqrt{3}/2 + 1/2 & 0 \\ 0 & 1/2 + i\sqrt{3}/2 \end{bmatrix} \quad (4.15)$$

this means that

$$\cos \phi_{\text{tot}}(A) = \left[\frac{(\sqrt{3}/2 + \tfrac{1}{2})^2 + (\tfrac{1}{2} + \sqrt{3}/2)^2}{(1+1)^2(1)(1)} \right]^{1/2} = \left(\frac{2 + \sqrt{3}}{4} \right)^{1/2} \approx 0.96592582. \quad (4.16)$$

We note that $\phi_{\text{tot}}(A) = 15°$ and that this is the angle of bisection of the angle between the eigenvalues λ_1 and λ_2. This simple unitary operator (4.15) indicates that we do not get much information from the total angle operator trigonometry.

I took this example from [6] where it is shown that generally the three angles $\phi(A)$, $\phi_{\text{tot}}(A)$, and $\phi_e(A)$, the latter meaning the angle of A given by $\phi(|A|)$ in the extended operator trigonometry I have described here in Section 2, may be different. Recall that the original $\phi(A)$ of the operator trigonometry was defined by (1.1) for use with accretive operators A and hence it is determined in terms of a "real" cosine. For A of (4.15), we have

$$\frac{1}{2} = \cos \phi(A) < 0.966 = \cos \phi_{\text{tot}}(A) < 1 = \cos \phi_e(A) \quad (4.17)$$

If fact it must be admitted: very few examples of the operator trigonometry have been worked out beyond A symmetric positive definite. Not even for A unitary.

Therefore, let us jump into this vacuum. And to gain some intuition toward an eventual operator trigonometry for quantum computing, why not look at some of the basic unitary operators which arise there as representations of elementary quantum computing operations?

Consider for example the quantum computing conditional phase shift operator

$$B(\phi) = \begin{bmatrix} 1 & 0 & 0 & 0 \\ 0 & 1 & 0 & 0 \\ 0 & 0 & 1 & 0 \\ 0 & 0 & 0 & e^{i\phi} \end{bmatrix} \quad (4.18)$$

The total operator trigonometry turning angle $\phi_{\text{total}}(B(\phi))$ will be determined according to (4.14) by the eigenvalues 1 and $e^{i\phi}$, from which we obtain

$$\cos^2 \phi_{\text{total}}(B(\phi)) = \frac{(\cos\phi + 1)^2 + (\sin\phi)^2}{4} = \frac{1 + \cos\phi}{2} \quad (4.19)$$

Thus for example for phase $\phi = 45°$ we find $\phi_{\text{total}}(B(\phi)) = 22.5°$. As a second example we may consider the quantum controlled-not operator

$$C_{XOR} = \begin{bmatrix} 1 & 0 & 0 & 0 \\ 0 & 1 & 0 & 0 \\ 0 & 0 & 0 & 1 \\ 0 & 0 & 1 & 0 \end{bmatrix} \quad (4.20)$$

The total antieigenvectors are determined by the eigenvalues $\lambda_3 = 1$ and $\lambda_4 = -1$ which according to (4.14) tell us that

$$\cos^2 \phi_{\text{total}}(C_{XOR}) = \frac{(1-1)^2 + (0+0)^2}{4} = 0 \quad (4.21)$$

and hence $\phi_{\text{total}}(C_{XOR}) = 90°$. Again all we have obtained is a total turning angle which is exactly the angle which bisects the extreme angles of the spectrum of the unitary operator on the complex unit circle.

There is another way to see this limitation on the total turning angle $\phi_{\text{tot}}(A)$, and let us note it here since technically the formula (4.14) was derived assuming that the normal operator A was already in diagonalized form. For unitary matrices $A = U$ from (4.13) we have equivalently

$$\cos \phi_{\text{total}}(U) = \min_{\|x\|=1} \frac{|\langle Ux, x\rangle|}{\|Ux\|\|x\|} = \min_{\|x\|^2=1} |\langle Ux, x\rangle| = \min_{\lambda \in W(U)} |\lambda| \quad (4.22)$$

where $W(U)$ is the numerical range of U, e.g. see [2]. Because for unitary operators U we know $W(U)$ is exactly the closed convex hull of the spectrum $\sigma(U)$, it follows that its total cosine $|\mu_1(U)|$ is the distance from the origin in the complex plane to the nearest point of $W(U)$. For the example (4.15) the numerical range $W(A)$ is exactly the chord between the two eigenvalues on the unit circle. Thus the total antieigenvectors are just those vectors whose numerical range values $\langle Ux_\pm, x_\pm \rangle$ in (4.22) attain the midpoint of that chord. When U has eigenvalues 180 degrees apart on the unit circle, as C_{XOR} does, $|\mu_1(U)|$ will always be 0.

To try to gain some intuition for a better operator trigonometry for unitary quantum computing operators, let us consider one more here. Further investigation will have to take place elsewhere. Let us consider the rotational portions of universal 2-bit quantum gate operators such as

$$U = \begin{bmatrix} i\cos\theta & \sin\theta \\ \sin\theta & i\cos\theta \end{bmatrix} \quad (4.23)$$

Instead of only looking at pure (real) state vectors $|\text{off}>$ and $|\text{on}>$, let us consider (4.23) applied to arbitrary (mixed, complex) state vectors $x = (x_1, x_2)$. Then one calculates

$$\langle Ux, x \rangle = 2\,\text{Re}\,(\bar{x}_1, x_2) + i(|x_1|^2 \cos\theta + |x_2|^2 \sin\theta) \quad (4.24)$$

U has eigenvalues $\lambda_1 = \sin\theta + i\cos\theta$ and $\lambda_2 = -\sin\theta + i\cos\theta$ and so the total cosine by (4.22) is attained by those x which make the imaginary part of (4.24) vanish. We also note that U's eigenvectors are $(1,1)$ and $(1,-1)$ and that these may be seen, in the sense of Section 3, as maximal inefficiency vectors. For more details about this, see [13].

Thus we admit that the operator trigonometry applied profitably to quantum computing operators needs more work and thought. The angle $\phi_{\text{total}}(U)$ seems too crude. The angles $\phi_{re}(U)$ and $\phi_{imag}(U)$ also seem insufficient.

Let me insert a few new ideas here.

First, just above, I mentioned that the eigenvectors of quantum gate operators seemed to have much in common with the maximally inefficient regressor vectors in statistical parameter estimation theory. In particular: maximal inefficiency vectors are always "maximally" mixed states (all 1's) whereas antieigenvectors are always "minimally" mixed states (only 2 1's), of course eigenvectors being "unmixed" states (only 1 1). What are the implications of this to decoherence in quantum computing?

Second, quantum gates like (4.23) or perhaps more complicated gates such as

$$\begin{bmatrix} 1 & 0 & 0 & 0 \\ 0 & 1 & 0 & 0 \\ 0 & 0 & e^{i\alpha}\cos\theta & -ie^{i(\alpha-\phi)}\sin\theta \\ 0 & 0 & -ie^{i(\alpha-\phi)}\sin\theta & e^{i\alpha}\cos\theta \end{bmatrix} \quad (4.25)$$

can be placed into simple quantum circuits to produce a "counterfactual" quantum computation which truly realizes the potential speedup of quantum computation. This concept utilizes the fact of "doing nothing" to achieve the speedup. In such quantum computation loops after N cycles the probability of no computation at all having taken place is $(\cos^2(\pi/2N))^N$. How can this theory be combined with that of the spin models discussed at the beginning of this section, by use of the operator trigonometry?

Third, possibly more interesting operator trigonometrically than just the unitary quantum computation basic operators such as (4.18), (4.20), (4.23), (4.24), would be their representation $U = e^{iA}$ in terms of their Hermitian logarithms A. We do have a full operator trigonometry for positive selfadjoint operators A and this is "untainted" by the complex phases implicit in U. Further, dynamical system perspectives $U^n = (e^{iA})^n$ or $U_t = e^{itA}$ for quantum computer processing could be approached in this way.

In the next section, from a completely independent motivation, I will present a completely new definiton of $\sin\phi(A)$. Possibly it will lead to better ways to open doors to quantum computing via the operator trigonometry.

5 A New Formulation of $\sin\phi(A)$

As the operator trigonometry has become more widely known and useful recently, sometimes I am somewhat skeptically asked: how can you get so much out of such a natural object such as $\cos\phi(A)$? My answer: $\sin\phi(A)$. Therefore I stress: you cannot have an operator trigonometry without both entities present, and moreover with a $\sin\phi(A)$ defined "independently" from $\cos\phi(A)$. In my original experience I was lucky: the Hille–Yosida abstract multiplicative perturbation question as I resolved it needed both sides of the inequality (1.1). The right side leads relatively naturally to a notion of $\cos\phi(A)$. However, I was required to see and then prove the MinMax theorem [4] before I could arrive at $\sin\phi(A)$:

$$\inf_{-\infty<\epsilon<\infty} \|\epsilon A - I\| \equiv \min_{\epsilon>0} \|\epsilon A - I\| \equiv \sin\phi(A) \tag{5.1}$$

For more on this theme that it is $\sin\phi(A)$ which really makes the operator trigonometry work, see my recent papers [6] and [13].

In this section I want to present a new formulation of $\sin\phi(A)$ which I believe is quite fundamental and which therefore could have interesting future implications in situations where an expression $\min_\epsilon \|\epsilon A - I\|$ is not central. What I present here is incomplete in terms of generality but for the case of A a symmetric positive definite matrix, for which we have a rather complete operator trigonometry now, it opens up a very interesting new view of $\sin\phi(A)$. I am not sure now in hindsight how I missed this. On the other hand, my original definition of $\sin\phi(A)$ according to (5.1) was exactly what was needed not only in the original abstract Hille–Yosida operator theory but also in many of the computational linear algebra applications of the operator trigonometry as we described in Section 2. Furthermore (5.1) is more general than the new formulation of $\sin\phi(A)$ which as I present it here will be only for finite matrices and not for the general accretive bounded operators on a Hilbert space to which (5.1) applies. However I have no doubt that one can extend my new formulation of $\sin\phi(A)$ to match the generality of (5.1).

The key to this new formulation of $\sin \phi(A)$ is to start with the Lagrange identity

$$\left(\sum_{i=1}^n a_i^2\right)\left(\sum_{i=1}^n b_i^2\right) - \left(\sum_{i=1}^n a_i b_i\right)^2 = \sum_{1 \le i < j \le n}(a_i b_j - a_j b_i)^2 \qquad (5.2)$$

Here $a = (a_1, \ldots, a_n)$ and $b = (b_1, \ldots, b_n)$ are vectors of real numbers. Let $b = Aa$ where A is a SPD matrix. Then (5.2) becomes

$$\|a\|^2 \|Aa\|^2 - \langle Aa, a\rangle^2 = \sum_{1 \le i < j \le n}(a_i b_j - a_j b_i)^2 \qquad (5.3)$$

or equivalently for $a \ne 0$,

$$1 - \frac{\langle Aa, a\rangle^2}{\|a\|^2 \|Aa\|^2} = \frac{\sum_{1 \le i < j \le n}(a_i b_j - a_j b_i)^2}{\|a\|^2 \|Aa\|^2} \qquad (5.4)$$

When a is one of the two (1.3) normalized first antieigenvectors $x_\pm^1$ of A, and let us choose x_+^1, namely

$$x_+^1 = (\lambda_n^{1/2} x_1 + \lambda_1^{1/2} x_n)(\lambda_n + \lambda_1)^{-1/2} \qquad (5.5)$$

where x_1 and x_n are the extreme eigenvectors for A, then

$$A x_+^1 = (\lambda_n^{1/2} \lambda_1 x_1 + \lambda_1^{1/2} \lambda_n x_n)(\lambda_n + \lambda_1)^{-1/2} \qquad (5.6)$$

and $\|A x_+^1\| = \lambda_n^{1/2} \lambda_1^{1/2}$. Thus (5.4) becomes

$$1 - \cos^2 \phi(A) = \frac{\sum_{1 \le i < j \le n}(a_i b_j - a_j b_i)^2}{\lambda_n \lambda_1} \qquad (5.7)$$

In this way we have arrived at a new formulation of $\sin^2 \phi(A)$: the right side of (5.7).

By way of illustration and verification, suppose A was already diagonalized so that $x_1 = (1, 0, \ldots, 0)$ and $x_n = (0, \ldots, 0, 1)$ are the two extreme eigenvectors corresponding respectively to λ_1 and λ_n. Then

$$x_+^1 = ((\lambda_n/(\lambda_1 + \lambda_n))^{1/2}, 0, \ldots, 0, (\lambda_1/(\lambda_1 + \lambda_n))^{1/2})$$

and

$$A x_+^1 = (\lambda_1 (\lambda_n/(\lambda_1 + \lambda_n))^{1/2}, 0, \ldots, 0, \lambda_n (\lambda_1/(\lambda_1 + \lambda_n))^{1/2})$$

so that the a_i and b_i in the right side sum of (5.7) all vanish except for a_1, a_n, b_1, b_n. Hence in that simplified case this sum collapses to

$$\sum_{1 \le i < j \le n}(a_i b_j - a_j b_i)^2 = (a_1 b_n - a_n b_1)^2 = \left(\left(\tfrac{\lambda_n}{\lambda_1+\lambda_n}\right)^{1/2} \lambda_n \left(\tfrac{\lambda_1}{\lambda_1+\lambda_n}\right)^{1/2}\right.$$
$$\left. - \left(\tfrac{\lambda_1}{\lambda_1+\lambda_n}\right)^{1/2} \lambda_1 \left(\tfrac{\lambda_n}{\lambda_1+\lambda_n}\right)^{1/2}\right)^2$$
$$= \left(\tfrac{\lambda_n - \lambda_1}{\lambda_n + \lambda_1}\right)^2 \cdot \lambda_n \lambda_1 \qquad (5.8)$$

Notice now that with (5.8) inserted there, the right side of (5.7) is exactly $\sin^2 \phi(A)$ in accordance with (1.2).

To conclude: we have arrived at a new formulation of $\sin \phi(A)$ for A any symmetric positive definite matrix:

$$\sin \phi(A) = \max_{x \neq 0} \frac{\left(\sum_{1 \leq i < j \leq n} (x_i y_j - y_j x_i)^2\right)^{1/2}}{\|x\| \|y\|} \tag{5.9}$$

where $y = Ax$.

Let me make a few preliminary comments about this new result. More understanding will have to come later after further investigation.

First, for $n = 3$, this result may be understood in terms of the well-known vector cross product:

$$\|a \times Aa\| = \|a\| \|Aa\| \sin \phi(A) \tag{5.10}$$

when $a = x_{\pm}^1$ is an antieigenvector of A. Thus in 3 dimensions we have a way to see $\sin \phi(A)$ with exactly the same geometrical content as that for $\cos \phi(A)$: $\langle Aa, a \rangle = \|a\| \|Aa\| \cos \phi(A)$, for $a = x_{\pm}^1$ an antieigenvector of A.

Second, as is well-known, vector cross products only exist for the special vector space dimensions $n = 1, 3$, and 7. However, we really don't need to be in a cross product dimension. For arbitrary dimensions one may go back to the old theory of "compounds" in the theory of minor expansions of determinants. I am in the process of doing this as time permits.

Third, for $n = 2$ we may obtain a "symplectic" representation for $\sin \phi(A)$. For, from (5.9) one may verify that

$$\sin \phi(A) = \max_{x \neq 0} \frac{\langle Ax, J^*x \rangle}{\|Ax\| \|J^*x\|} \tag{5.11}$$

where $J = \begin{bmatrix} 0 & 1 \\ -1 & 0 \end{bmatrix}$. For $n = 3$ in a similar way one can express the sum in (5.9) in an analogous way. That is, consider the sum $\sum(a_i b_j - a_j b_i)^2$, $1 \leq i < j \leq 3$, $x = (a_1, a_2, a_3)$, $Ax = (b_1, b_2, b_3)$. Then

$$\sum_{1 \leq i < j \leq 3} (a_i b_j - a_j b_i)^2 = \langle Ax, J_1 x \rangle^2 + \langle Ax, J_2 x \rangle^2 + \langle Ax, J_3 x \rangle^2 \tag{5.12}$$

where

$$J_1 = \begin{bmatrix} 0 & -1 & 0 \\ 1 & 0 & 0 \\ 0 & 0 & 0 \end{bmatrix}, J_2 = \begin{bmatrix} 0 & 0 & -1 \\ 0 & 0 & 0 \\ 1 & 0 & 0 \end{bmatrix}, J_3 = \begin{bmatrix} 0 & 0 & 0 \\ 0 & 0 & -1 \\ 0 & 1 & 0 \end{bmatrix} \tag{5.13}$$

Since symplectic theory holds for even dimensions one could work out a representation more like (5.11) for $n = 4, 6, \ldots$. This general situation for higher dimensions will be worked out as time permits.

Fourth, similar formulations hold for complex vector spaces. Quaternions, Clifford algebras, spin theories wait to be investigated, and, hopefully, applied to the quantum mechanical and quantum computing problems discussed in Section 4.

6 Conclusions

Let me state capsule summaries as follows, restricted to single sentences one-to-one with each of the sections of this paper.

1. Operator trigonometry is an unconventional new chapter in linear algebra.
2. For computational linear algebra, it appears most often in a form not heretofore recognized as meaning: $\sin \phi(A)$.
3. For statistics and econometrics it appears most often in a form not heretofore recognized as meaning: $\cos \phi(A)$.
4. For quantum mechanics it appears most often in spin systems over the complex field.
5. $\sin \phi(A)$ has in this paper been formulated in a new and more geometrical way relative to $\cos \phi(A)$.
6. The operator trigonometry has many avenues open for further development and application, both in theory and for application and computation: some of these (e.g., control theory, wavelets, other) may be found in [16–20].

References

1. K. Gustafson, *Lectures on Computational Fluid Dynamics, Mathematical Physics, and Linear Algebra*, World Scientific, Singapore (1997).
2. K. Gustafson and D. Rao, *Numerical Range*, Springer (1997).
3. K. Gustafson, *Notices American Mathematical Society* 14 (1967), 520, 717, 824, 943.
4. K. Gustafson, *Notices American Mathematical Society* 15 (1968), 699.
5. K. Gustafson, Ergodic learning algorithms, *Unconventional Models of Computation* (eds: C. Calude, J. Casti, M. Dinneen), Springer (1998), 228–242.
6. K. Gustafson, An extended operator trigonometry, *Linear Algebra and Applications* (2000), to appear.
7. K. Gustafson, Antieigenvalues in analysis, *Proceedings Fourth International Workshop in Analysis and its Applications* (eds: C. Stanojevic, O. Hadzic), Dubrovnik, 1990; Novi Sad, Yugoslavia (1991), 57–69.
8. K. Gustafson, Positive (noncommuting) operator products and semigroups, *Math. Z.* 105 (1968), 160–172.
9. K. Gustafson, The geometrical meaning of the Kantorovich–Wielandt inequalities, *Linear Algebra and its Applications* 296 (1999), 143–151.
10. K. Gustafson, Operator trigonometry of iterative methods, *Num. Lin. Alg. with Applic.* 4 (1997), 333–347.
11. K. Gustafson, Domain decomposition, operator trigonometry, Robin condition, *Contemporary Mathematics* 218 (1998), 455–460.
12. K. Gustafson, Operator trigonometry of the model problem, *Num. Lin. Alg. with Applic.* 5 (1998), 377–399.
13. K. Gustafson, Operator trigonometry of statistics and econometrics, (2000), to appear.
14. K. Gustafson, The geometry of quantum probabilities, *On Quanta, Mind, and Matter* (eds: H. Atmanspacher, A. Amann, U. Müller-Herold), Kluwer, Dordrecht (1999), 151–164.

15. K. Gustafson, Quantum trigonometry, *Infinite Dimensional Analysis, Quantum Probability, and Related Topics* 3 (2000), 33–52.
16. K. Gustafson, The trigonometry of quantum probabilities, *Trends in Contemporary Infinite Dimensional Analysis and Quantum Probability*, (eds: L. Accardi, H-H Kuo, N. Obata, K. Saito, S. Si, L. Streit) Italian Institute of Culture, Kyoto (2000), 159–173.
17. K. Gustafson, Operator trigonometry of linear systems, *Proc. 8th IFAC Symposium on Large Scale Systems* (eds: N. Koussoulas, P. Groumpos), Pergamon Press (1999), 950–955.
18. K. Gustafson, Operator trigonometry of wavelet frames, *Iterative Methods in Scientific Computation* (eds: J. Wang, M. Allen, B. Chen, T. Mathew), IMACS Series in Computational and Applied Mathematics 4, New Brunswick, NJ (1998), 161–166.
19. K. Gustafson, A computational trigonometry and related contributions by Russians Kantorovich, Krein, Kaporin, *Computational Technologies* 4 (No. 3), Novosibirsk (1999), 73–93.
20. K. Gustafson, Parallel computing forty years ago, *Mathematics and Computers in Simulation* 51 (1999), 47–62.

Splicing Systems, Aqueous Computing, and Beyond

Tom Head

Department of Mathematical Sciences, Binghamton University, Binghamton, U.S.A.

Abstract. The origin of the splicing system concept is reviewed and the original motivation for the concept is given. The concept of an *aqueous computing* architecture is sketched in a manner independent of specific implementations. Wet lab computations made using biomolecular implementations are reported. Hopes for future non-biomolecular realizations are confided.

1 Introduction

My involvement with unconventional methods of computing began with my initiation of the theory of splicing systems. This theory has been extensively developed by a wide community of researchers interested in the theories of formal languages and computation. The concerns of the splicing community have favored augmented splicing systems powerful enough to generate all recursively enumerable languages. Since the original splicing concept modeled the cut and paste activities carried out on double stranded DNA molecules, computing by controlled splicing has been viewed as abstract DNA computing. It is especially gratifying that much of this research has been included in the later chapters of [PRS98] as well as in a special chapter of [RS97]. With these advanced forms of computing by splicing in expert hands, my own splicing research has taken the opposite direction. I have studied the generation of subclasses of the class of regular languages by highly restricted forms of splicing. For DNA computing itself, I prefer a wet lab to an abstract theory. On the other hand, I am not attracted by ad hoc wet lab solutions. Study of the operations that were originally modeled by splicing theory has allowed wet lab implementations of a computational architecture that is suitable for a broad class of algorithmic problems. I like to call this computation aqueous computing since it is quite distinct from the other forms of biomolecular computing currently being explored. Thus splicing theory blends naturally into aqueous computing. Although the concept of the aqueous architecture arose from the study of molecular biology, only the water is a permanent feature of the architecture. Beyond the current biomolecular implementations may lie implementations in which molecules in aqueous solution are modified by electromagnetic radiation.

Section 2 describes the earliest phase of the development of the splicing system concept. Section 3 explains the stimulus and motivation for testing the splicing system concept in a wet lab. Section 4 is a sketch of the aqueous computing concept at a level of generality that is above any specific biomolecular implementation. Section 5 reports briefly three aqueous computations that have been successfully completed. Section 6 provides examples of (1) a computation in progress, (2) a planned computation that will test an alternate biochemical form of writing, and (3) a planned feasibility test of a technology that might liberate aqueous computing from biochemical processing altogether. All sections emphasize the critical importance of contributions made by many colleagues and friends to the developments described here. As a background reference for each algorithmic problem discussed here we suggest [GJ79].

2 The Origin of the Splicing System Concept

In 1975 Meera Blattner, Stuart Zimmerman and I met (in Houston, Texas) to discuss possible joint research in what was then called biologically motivated automata theory. Each of us was acquainted with, and inspired by, the theory of L-systems. This is the theory that Aristid Lindenmayer initiated to model developmental processes in biology [Lin68]. Meera had recently given an important related undecidability result. Stuart believed that important results could be obtained if we would think through the basic processes of molecular biology (DNA → RNA → Proteins) and express as much as possible using the concepts of automata theory and L-systems. Our paths diverged as two of us left the Houston area, but I continued to consider this project that we had discussed. Since Lindenmayer's string theoretic encoding of the development of algae and plants stimulated such sweeping developments in formal language theory [HR75], mathematics [RS80], and computer graphics [PL90], the dynamic string theoretic world of the biological macromolecules was sure, I thought, to provide material for similarly extensive formal developments.

In the spring of 1984 I audited a course offered by Jerold Shields based on B. Lewin's exciting new (at that time) book "Genes" [Le83]. Virtually all the phenomena of molecular biology were interesting to view as string behaviors. However, my inclination was to avoid probabilistic or fuzzy phenomena, at least initially. This required considering only biomolecular processes that had been described with almost perfect precision and which were nearly error free. Only gene splicing technology seemed to meet these requirements adequately at that time. Consequently the splicing model was created to represent the cutting of double stranded DNA molecules with specific restriction enzymes and the re-ligation of the resulting fragments with a ligase enzyme.

Although we now allow ourselves to view any form of biomolecular interaction as a form of computing, the splicing system concept was not created as a model of computation. The intent was to take one tiny step in the direction of providing formal representations of the generative potential of specific enzymatic systems. Paraphrasing Darwin and Oparin, the question was: 'What can develop in a warm little pond containing a specified set of molecules?' Specifically, splicing modeled enzymatic systems that consist of a finite collection of restriction enzymes and a ligase. Perhaps the original purpose of the splicing model can best be made clear by observing a later success: From any splicing system, a regular expression can be constructed for the language the system generates [CH91] [Pi96] [Pi00]. The interpretation of this theorem is that, for any finite list of restriction enzymes and any finite list of DNA sequences, a regular expression E, over the four symbol alphabet $\{[A/T], [C/G], [G/C], [T/A]\}$ of double stranded DNA, can be constructed for which we may assert: A DNA molecule can be constructed by applying enzymes in the given list and a ligase to sufficiently many DNA molecules having sequences in the given list if and only if the sequence of the DNA molecule is represented in the regular expression E.

My former colleague in Alaska, Ronald Gatterdam, in joint work with K. Denninghoff, was the first researcher to modify the splicing concept in such a way as to alter the generative power. In [DG89] they demonstrated that each Turing machine can be simulated by a splicing system if one employs the concept of a multiset appropriately. The splicing system concept was developed in the context of Gatterdam's interest [Ga89] [Ga92] and encouragement. Here in Binghamton, Dennis Pixton has thoroughly interwoven the splicing concept with the fundamental themes of formal language theory by proving that every full abstract family of languages is closed under (iterated) splicing [Pi00].

The first work on splicing theory, done by people I had not yet met, was by T. Yokomori in Japan, by Gh. Paun in Romania, and Rudolf Freund in Austria. Yokomori related splicing to algorithmic learning and biomolecular sequence questions. He discussed splicing with others in Japan and with visitors from India, Rani Siromoney [SSD92], and Italy, Claudio Ferretti [YKF97]. Paun immediately began a wide ranging foundational investigation of the splicing concept and characterized the effect of many forms of control structure on splicing systems. His extensive joint research activities introduced the splicing concept to a wide European research community. Much of this work is included in [HPP97] and [PRS98]. With others, especially G. Rozenberg and A. Salomaa, splicing with control was observed to provide a theoretical basis for computing [PS96] [PRS96]. It was during this early period of interest in splicing theory that Leonard Adleman's paper [Ad94] introduced wet lab DNA computing. Those of us in the splicing community were primed and ready to learn about wet lab computation.

3 Wet Splicing Systems and Dynamical Questions

There were assumptions made in [He87] about how actual sets of enzymes and DNA molecules would behave together. Those of us in Binghamton wanted to confirm for ourselves that such assumed behavior actually does take place. This was especially important to us because other theoretical researchers were referring now to our work. We needed to confirm that a set consisting of at least two distinct restriction enzymes and a ligase could operate, simultaneously in the same test tube, on an initially given set of DNA molecules without mutually destructive interference. The restriction enzymes should cut; the ligase should re-ligate matching fragments; the enzymes should re-cut those re-ligated molecules which still contained restriction sites; the ligase should re-ligate matching fragments; etc., etc. (ideally, ad infinitum). Notice what we are describing is a highly dynamical process.

To provide an experiment that allowed a demonstration that such a coherent multifaceted activity actually does take place, we chose a pair of enzymes, Dra III and Bgl I, that act at different sites, but produce the same three base (therefore not dyadically symmetric, i.e., not palindromic) overhangs. After a close study of the DNA sequence of the bacteriophage lambda [HRSW83, Appendix II], we selected two segments from this sequence, one with a unique Dra III site and no Bgl I site, and one with a unique Bgl I site and no Dra III site. Let whw' and xhx' represent these two segments having the unique Dra III site and the unique Bgl I site, respectively. Here h has been used to represent the identical three base double stranded regions that provide the common overhangs. From bacteriophage lambda we copied (using PCR) the segments whw' and xhx' for use in our experiment. A common buffer (i.e., water containing the appropriate ion concentrations, etc.) was found in which all three enzymes could function efficiently. To this buffer were added Dra III, Bgl I, a ligase and the two varieties of DNA. The segments of lambda, whw' and xhx', had been chosen so that, when the recombinant molecules whx' and xhw' formed, the lengths would relate as follows: length xhw' > length whw' > length xhx' > length whx'. In fact these lengths, measured in kilobase pairs, were approximately: 2.7, 2.1, 1.6, 1.0. It is fundamental for this experiment that, because Dra III and Bgl I act at distinct sites, when the recombinants xhw' & whx' are formed by the ligase, they cannot be cut by either of the restriction enzymes. By contrast, when the original molecules whw' & xhx' are reconstructed by the ligase, they can be recut by the enzymes. At a series of time intervals, the content of the test tube was sampled and the DNA molecules separated by length electrophoretically on a gel. Unless the splicing system concept fails to represent actual biomolecular behavior, the time series of gel photos should show: two initial intense bands at 2.1 kb & 1.6 kb that fade through time and ultimately vanish, and two final intense bands at 2.7 kb & 1.0 kb that grew through time from initial absence. (In the intermediate stages there should also be bands representing wh', $h''w'$, xh', $h''x'$ which are the temporarily formed fragments with single stranded overhangs h' & h'' arising from the cut through the double stranded segment h.) This experiment was carried out by Elizabeth (Laun) Goode and K.J. Reddy in Reddy's lab and reported with the confirming gel photos in [LR99]. The wet lab time series obtained from the experiment was consistent with the splicing system model, the only disappointment being that some of the fragments with overhangs were not completely re-ligated by the end of the experiment. I regard this experiment as a confirmation that splicing theory has a basis in biomolecular reality.

Designing experiments can cause one to notice new natural concepts for theoretical investigation. The experiment reported in [LR99] suggests that $\{xhw', whx'\}$ be called the adult language of the full language

$\{xhx', whw', xhw', whx'\}$ generated by the formal model of the biomolecular context. The two adult strings are not subject to further development. We suggest below that formal definitions be developed for the concept of an adult language and also for the subtler concept of a limit language. These two concepts are illustrated with the following thought experiment:

The restriction enzymes Aci I and Hpa II cut at the sites $x = [C/G]\text{'}[C/G][G/C], [C/G]$ and $y = [C/G]\text{'}[C/G][G/C], [G/C]$ respectively, where ' indicates the cut made in the top strand and the comma (,) indicates where the cut is made in the bottom strand. These two enzymes produce the same dyadically symmetric (i.e., palindromic) two base overhangs, but note that Aci I acts at a site that does not have complete dyadic symmetry! We perform the following thought experiment: Construct a double stranded DNA molecule $w = uxv$ where length $u = 400$ base pairs, x is the site at which Aci I acts, length $v = 100$ base pairs, and there is no other occurrence of sites at which either Aci I or Hpa II act in w. To an appropriate buffer, we add Aci I, Hpa II, a ligase, ATP, and a vast number of the molecules $w = uxv$. Aci I cuts each w. Since x is dyadically symmetric, the ligase re-ligates the fragments of w at random. The re-ligation reconstructs not only the original molecule $w = uxv$, but also the two recombinant molecules $r = uyu'$ & $s = v'x''v$, where u' & v' are obtained from u & v by rotating them through 180 degrees, and $x'' = [G/C][C/G][G/C][C/G]$. Now w is cut again by Aci I and $r = uyu'$ is cut by Hpa II, but neither enzyme cuts $s = v'x''v$. Thus as long as there are either w molecules or r molecules they will be cut, but s remains permanently without being cut. In the limit, all the v segments ($v' = v$ since molecules in water are unchanged under rotation) are incorporated into the s molecules. Thus, in the limit, there are no w molecules. Consequently, in the limit, only the molecules $r = uyu'$ and the fragments produced when they are cut by Hpa II remain in a dynamic state. Ignoring the fragments with single stranded overhangs, we have initially in the tube only w and as time goes to infinity we have in the tube only r and s. However, the situations of r and s in the tube are behaviorally distinguished: r is regularly being cut by Hpa II and reconstituted by the ligase; whereas s is entirely dormant.

I would like to see this thought experiment replicated as an actual wet lab experiment. However, even the thought experiment suggests the following theoretical problem. Find appropriate definitions, in the context of splicing theory, for two new concepts: the *adult language* and the *limit language* of a splicing system. I suggest that these definitions be made in such a way that in the thought experiment above $\{r, s\}$ would be the limit language and $\{s\}$ would be the adult language. For splicing systems generating infinite languages the limit languages might turn out to be quite complex even if the adult languages are not. To define the limit language of a system it may be necessary to augment the splicing system with a structure that allows the expression of a limit on the material available. For this purpose multisets or real valued functions representing concentrations (which may converge to zero) might be appropriate.

4 The Aqueous Computing Concept

After the startling appearance of Leonard Adleman's DNA computation [Ad94], I sketched several tentative plans for DNA computation, which often involved circularizing DNA molecules while holding them attached to a surface through biotin. In the fall of 1997 I had the opportunity to talk through these ideas with Satoshi Kobayashi who immediately suggested letting the molecules go free in solution. I had assumed that this would lead to a jumble of undesired long DNA molecules, but Satoshi said that the Japanese researchers had circularized molecules in solution without ever detecting undesirable concatenations of the molecules. We discussed this issue immediately with Masami Hagiya and Akira Suyama who reinforced Satoshi's statements. Further support for the concept came from Grzegorz Rozenberg and Herman Spaink in Leiden, where we began a wet lab implementation in the summer of 1998. In Binghamton we were honored by Masayuki Yamamura's joining us for the academic year of 1998-1999 during which Susannah Gal taught us the basics of wet lab work so that we three could begin biomolecular computing here. I am very grateful to my colleagues in Tokyo and Leiden for encouragement and expertise in the development of the concept and in its wet lab implementations.

In the remaining three paragraphs of this Section we communicate the essential nature of the aqueous computing concept by treating a tiny instance of one of the standard algorithmic graph problems. In these paragraphs we: (1) give a natural elementary solution; (2) consider how the asymptotic complexity of the procedure might be lowered by parallelism; and (3) tell in the most general terms how this parallelism might be achieved by operating on molecules dissolved in water. In Section 5 a wet solution of a more complicated instance is reported in which the operations are implemented using DNA chemistry.

We wish to find the largest cardinal number that occurs as the cardinal number of an independent subset of the set of vertices of the graph $G = (V, E)$ having vertex set $V = \{a, b, c\}$ and two undirected edges $\{a, b\}$ and $\{b, c\}$. We use bit strings of length three to represent subsets of V. Our method will be to list, as the first step in our algorithm, the bit string 111, which we take to represent V itself. This string 111 will be used as the root node of a binary tree. We take account of the edge $\{a, b\}$ by providing a left child for the root node (i.e., 111) by setting the first bit (representing a) to 0 and a right child by setting the second bit (representing b) to 0. The first two steps result in the first two rows of the binary tree:

$$\begin{array}{cccccccc}
 & & & 111 & & & & \text{(Row 1)} \\
\\
 & 011 & & & & 101 & & \text{(Row 2)} \\
001 & & 010 & & & 101 & 100 & \text{(Row 3)}
\end{array}$$

The third row results from taking account of the edge $\{b, c\}$: For each bit string in row two we provide a left child by setting the second bit (representing b) to 0, and a right child by setting the third bit (representing c) to 0. As the output of this procedure we take the sets encoded by the strings in the final row. In this case the output consists of the four sets $\{c\}$, $\{b\}$, $\{a, c\}$, and $\{a\}$. Since we have treated all (only two, in this case) of the edges of the graph, this procedure assures us that each set in the output is independent. The procedure may not produce all the independent sets (000 has not been produced here). However, it is not difficult to prove that every independent that is not contained in a larger independent set is always produced. Such sets are called *maximal independent sets*. There are two maximal independent sets in the present case: $\{b\}$ and $\{a, c\}$. An independent set that has as its cardinal number the largest cardinal that occurs as the cardinal of an independent set is called a maximum independent set. There is only one *maximum independent set* in the present case: $\{a, c\}$.

If a graph G having n vertices and k edges were to be treated in this way we would generate a tree having: (1) bit strings of length n as nodes, (2) $k + 1$ rows, (3) 2^k leaves, and (4) a total of $2^{(k+1)} - 1$ nodes. The number of set-bit-to-0 operations would be $2^{(k+1)} - 2$. What can be done in parallel? As each new row is generated from its predecessor, all the left children are produced by the same set-bit-to-0 operation. Likewise all the right children are produced by the same set-bit-to-0 operation. By working in water, as we progress from each row to its successor, we are able to generate all left children of the new row in a single step, and likewise for the right children. When this is done the number of computational steps required by such problems grows as a linear function of the number, k, of edges, rather than as an exponential function of k. How do we do this? In Section 5 we report actual wet lab computations done in the aqueous computing manner. Here we sketch the process in the most general terms so as not to wed the concept to specific technologies. The sketch is kept concise by describing the process for our simple graph G having three vertices and two edges:

We choose a water soluble molecule on which we can identify three locations, (i.e., 'stations'), at which we can make an alteration of the molecule in some highly controlled way, (i.e., 'write'), which can be detected later (i.e., 'read'). We decide to interpret the initial condition of the three stations of our molecule as representing 111. We dissolve a large number of these (identical) molecules in a test tube T of water. This tube T is now understood to represent the first row of the binary tree shown above. We may assume that the molecules are (virtually) uniformly distributed throughout T (stir if necessary). Pour T into tubes L and R. In L write 0 at the first station (bit position). (In all cases, each write is done in parallel on all molecules in the tube.) In R write 0 at the second station (bit position). Pour L and R into T. T now contains two varieties of molecules, each of which may be assumed to be uniformly distributed throughout the tube. T is now understood to represent the second row of the binary tree shown above. Pour T into L and R. In L write 0 at the second station. In R write 0 at the third station. Pour L and R into T. T now represents the final row of the binary tree shown above. The reading technology must depend on the details of the writing technology that is used. If the writing of each zero on a molecule changes a physically measurable feature of the molecule, such as its size or weight, then this may allow, as required by the present problem, the molecules having the largest number of remaining ones to be separated out. This is illustrated in detail in the reports of computations in Section 5.

5 Aqueous Solutions Using DNA Plasmids

Maximum Independent Sets in a Graph

The first computation carried out in the aqueous computing manner provided the solution of the instance of the maximum independent set problem (MIS) for the graph $G = (V, E)$ having vertex set $V = \{a, b, c, d, e, f\}$ and four undirected edges: $\{a, b\}, \{b, c\}, \{c, d\}$, and $\{d, e\}$. The problem was to find the cardinal number of a maximum independent set of vertices of G. (A brief inspection confirms that there is only one maximum independent set, $\{a, c, e, f\}$, and it has 4 as its cardinal number.) This particular instance of the MIS was chosen because it is the instance into which the maximal clique problem in [OKLL97] was reformulated and then solved. I am very grateful to Peter Kaplan for a discussion of [OKLL97] and for his encouragement to produce the alternate solution discussed here, which can be viewed as 'doing [OKLL97] upside down'. Comparing these two solutions helps one to see aqueous computing in perspective.

Our aqueous solution to this instance of the MIS was carried out in Leiden in H. Spaink's laboratory and reported in [HRBBLS]. For this and later computations, a special plasmid was constructed to provide a circular memory register molecule P having six 'stations' (one for each vertex of G) at which zeros could be 'written'. The initial condition of P was regarded as a representation of the complete set of six vertices which, as a bit string is written 111111. Writing a zero at a station of P is done by linearizing the plasmid by cutting with the restriction enzyme associated with the station followed by re-ligation into circular form using a ligase. The details of the writing process assure that the circumference of the plasmid (as measured in base pairs) is changed by a known amount. The computation consists of a sequence of four (the number of edges of G) pairs of these cut/paste writing steps that implement the writing steps described in Section 4 (where only two steps were given). The four steps were expected to produce sets of molecules encoding in succession:

$\{111111\} \rightarrow \{011111, 101111\} \rightarrow \{001111, 010111, 10\mathit{1}111, 100111\} \rightarrow \{000111,$
$001011, 010111, 010011, 100111, 101011, \mathit{100111}, 100011\} \rightarrow \{000011, 000101,$
$001011, 001001, 010011, 010101, \mathit{010011}, 010001, 100011, 100101, 101011, 101001,$
$\mathit{100011}, 100001\}$ where repeated bit strings are written in italics.

Thus the final tube was expected to contain the 12 distinct molecular varieties encoding the bit strings that appear in the final set of the sequence above. Note that the numbers of ones remaining in the 12 final bit strings are 2, 3, and 4 and that only one of these strings has four ones, namely: 101011. The actual wet lab reading was done by making a gel separation, reading the lengths in base pairs of the resulting bands and, from the appropriate extreme length, determining how many stations have not been written to zero. This gave the correct answer, namely: 4. (If it is desired to specify the set of vertices that occur in a maximum independent set, two routine ways of doing this are available after cutting the appropriate band from the gel.) Gel photos showing the step by step progress through this aqueous computation to the solution are included in [HRBBLS].

Minimum Dominating Sets in a Graph

Let $G = (V, E)$ be the graph having vertices $V = \{a, b, c, d, e, f\}$ and five undirected edges $\{a, c\}, \{b, c\}, \{c, d\}, \{e, d\}$, and $\{f, d\}$. A subset D of V is a *dominating set* for G if every vertex lies in the neighborhood of a vertex in D. Recall that the *neighborhood* of a vertex v consists of v itself and each vertex that is adjacent to v. Thus the neighborhoods of the vertices a, b, c, d, e, and f of G are $\{a, c\}, \{b, c\}, \{c, a, b, d\}, \{d, c, e, f\}, \{e, d\}$, and $\{f, d\}$, respectively. A dominating set is a minimum dominating set if its cardinal number is the minimum cardinal number that occurs as the cardinal number of a dominating set. A wet lab solution to the following problem has been given. Find the cardinal number of a minimum dominating set (MDS) for G. Notice that a set that contains an element from each of the four neighborhoods $\{a, c\}, \{b, c\}, \{e, d\}$, and $\{f, d\}$ is a dominating set since each of the two remaining neighborhoods contains at least one (in fact two) of these four neighborhoods.

Our solution to this instance of the MDS was also carried out in H. Spaink's laboratory. The same plasmid P used in the MIS instance was used as memory register here. The six stations of P were used to represent the six vertices of G. For the MDS problem it is more natural to regard the original condition of the plasmid P to represent 000000 rather than 111111. Correspondingly, it is more natural to regard our writing as replacing a 0 by a 1. The writing of each 1 represents the choice of a vertex to be included in a dominating set. The writing phase consists again of four pairs of operations each of which is now interpreted as 'write a 1'. Each such step treats one of the four neighborhoods in the short list above and assures that at least one of the vertices in that neighborhood will be included. These four steps were expected to produce sets of molecules encoding in succession:

$\{000000\} \to \{100000, 001000\} \to \{110000, 101000, 011000, 001000\}$
$\to \{110010, 110100, 101010, 101100, 011010, 011100, 001010, 001100\}$
$\to \{110011, 110110, 110101, 110100, 101011, 101110, 101101, 101100,$
$011011, 011110, 011101, 011100, 001011, 001110, 001101, 001100\}$.

Thus the final tube produced in this computation was expected to contain the 16 distinct molecular varieties listed as the final step in the sequence above. Note that the various total numbers of ones occurring in these 16 varieties are: 4, 3, and 2. Note that only one bit string has exactly two ones, namely: 001100. The actual wet lab reading was done by making a gel separation, reading the lengths in base pairs of the resulting bands and, from the appropriate extreme length determining the least number of stations that were written to 1. This provided the correct answer, namely: 2. Gel photos showing the step by step progress of the computation leading to the solution are available although they have not been submitted for publication.

Satisfiability of Sets of Boolean Clauses

In Binghamton S. Gal, M. Yamamura, and I have provided a prototype solution of a three variable, four clause, satisfiability (SAT) problem as an aqueous computation. Let p, q, and r be three Boolean variables. Let p', q', and r' be the negations of these variables. The SAT instance we solved in the wet lab is: Does a truth assignment for the variables p, q, & r exist for which each of the four clauses

$$p \text{ OR } q, \qquad p' \text{ OR } q \text{ OR } r', \qquad q' \text{ OR } r', \qquad p' \text{ OR } r$$

evaluates to true?

We chose to use, as our memory register molecule, a commercially available cloning plasmid, which we will denote as P. We chose six restriction enzyme sites in the multiple cloning site (MCS) of the plasmid each of which served as a station for the computation. We associate the six literals in the order p, p', q, q', r, r' with the six chosen stations of the plasmid. We chose to regard the initial state of P in which the six chosen enzyme sites are functional, as an encoding of 111111 where the first 1 represents the assignment of the truth value 1 (true) to the variable p, the second 1 represents the assignment of 1 (true) to the literal p', the third 1 represents the assignment of 1 (true) to q, etc. Apparently 111111 embodies three logical contradictions. Our computation begins with a sequence of three pairs of writing steps of the same sort as in the MIS and MDS problems treated previously. We write a 0 at a restriction enzyme site (=station) by a process that alters the site permanently so that it can no longer be cut by the enzyme. (Thus each 1 means 'can still be cut here' and each 0 means 'can't be cut here'.) The first three steps eliminate the three contradictions p AND p', q AND q', r AND r'. These three steps were expected to produce sets of molecules encoding in succession the sets of bit strings: $\{111111\} \rightarrow \{011111, 101111\} \rightarrow \{010111, 011011, 100111, 101011\} \rightarrow \{010101, 010110, 011001, 011010, 100101, 100110, 101001, 101010\}$. Thus, it was expected that, after these first three steps, we would have a tube T containing plasmids that represent the eight possible logically consistent truth assignments for the variables and their negations. (Note that this tube can now be used to solve any three variable SAT. Thus plasmids from this tube can be saved, and amplified later in bacteria for use in solving other three variable SATs.)

Next we carry out one step consisting of parallel operations for each of the clauses. These operations, however, are not the previous writing operations. These steps are 'elimination steps'. We eliminate the molecules in T that encode truth assignments that do not satisfy p OR q by pouring T into tubes L and R. In L we linearize the molecules of the form $01XXXX$ (where X can be either 0 or 1) which are the molecules that fail to satisfy p, by applying the enzyme that cuts at the site corresponding to p'. In R we linearize the molecules $XX01XX$ (those that fail to satisfy q) by applying the enzyme that cuts at the site corresponding to q'. From L and R we return *only the plasmids* to T, discarding the linear molecules. Note that the only molecular varieties that appeared in the previous T that do *not* appear in the new T are those of the form $0101XX$. Thus the new T contains molecules of each variety that occurred in the previous T that also satisfy p OR q. Continuing with this T we now eliminate the molecules that encode truth assignments that do not satisfy p' OR q OR r' by pouring T into *three* tubes L, M, and R. In L we linearize the molecules $10XXXX$ (those that fail to satisfy p') by applying the enzyme that cuts at the site corresponding to p. In M we linearize the molecules $XX01XX$ (those that fail to satisfy q) by applying the enzyme that cuts at the site corresponding to q'. In R we linearize the molecules $XXXX10$ (those that fail to satisfy r') by applying the enzyme that cuts at the site corresponding to r. From L, M and R we return *only the plasmids* to T, discarding the linear molecules. The only molecular varieties that appeared in the previous T that do not appear in the new T are those of the form 100110. Thus the new T contains molecules of each variety that occurred in the previous T that also satisfy p' OR q OR r'. Two further elimination steps of this type are required to delete the molecules that fail to satisfy either q' OR r' or p' OR r.

The expected effect of the four elimination steps can be summarized:

$\{010101, 010110, 011001, 011010, 100101, 100110, 101001, 101010\}$
$\rightarrow \{011001, 011010, 100101, 100110, 101001, 101010\} \rightarrow \{011001, 011010,$
$100101, 101001, 101010\} \rightarrow \{011001, 100101, 101001\} \rightarrow \{011001\}.$

Thus we expected that, after the elimination phase is complete, plasmids would remain (of only the single variety representing 011001) and that any remaining plasmid would encode a consistent truth setting for which all four of the given clauses have the value true. The actual wet lab determination was made not only by observing the presence of DNA plasmids in the final tube, but also by a gel separation that produced a single distinct band of the correct length in base pairs to corroborate the conclusion that the four clauses are jointly satisfiable. The confirming gel photos are available but they have not been submitted for publication. A preliminary announcement of this work appeared [HYG99] before the computation had been completed.

(The reader who has access to a molecular biology lab is invited to reproduce our work and develop aqueous computing further. It is an inexpensive process. The only DNA to buy is a standard cloning plasmid. More of the plasmids can be produced using bacteria when needed. Only standard restriction enzymes, a ligase, and a DNA polymerase are used.)

6 Beyond

We close with brief descriptions of one project from each of the following three levels: work in progress (listing all maximal independent subsets of a graph), work planned (writing by methylation), and work being dreamed (writing with light).

Maximal Independent Subsets and the 3×3 Knights Problem

We return to the *maximum* independent set (MIS) problem discussed in Section 5 and continue that computation to create the longer list consisting of all the *maximal* independent subsets of $G = (V, E)$. Recall that the wet lab solution of the MIS concluded with a gel separation that partitioned molecules into three bands on a gel. We expect that the following three step procedure will provide a list of all maximal independent subsets of G: (1) From the gel, cut out the band corresponding to the sets containing 4 vertices. Clone and sequence. As each sequence is obtained list the bit string it encodes (expect only one, namely: 101011). This first step gives the maximum independent set(s). (2) From the gel cut out the band corresponding to the sets containing 3 vertices. Clone and sequence. As each sequence is obtained, list the bit string it encodes, but only if it does not represent a subset (of one) of the maximum set(s) obtained in the previous step (expect: 100101, 010101, 010011). (3) Discard the remainder of the gel, since no set of two vertices could be maximal for this graph G.

Thus the list produced for this graph G is expected to be: 101011, 100101, 010101, 010011. There is an intrinsic subjective element in this procedure that might result in failure to find one or more of the maximal independent sets: Since the same sequence can arise repeatedly, how does one know when *all* the distinct sequences have been determined? This is not a problem when we know the answer in advance, but when the number of distinct sequences is not known one risks missing a sequence no matter how long one continues the clone and sequence process. In a discussion with A. Atanasiu, V. Mitrana, and F. Guzman, a scheme was arrived at that would remove this objection, but only by replacing it with additional error prone biochemical procedures.

Laura Landweber and co-workers at Princeton have treated the problem of producing a list of all the patterns in which knights can be placed on a 3×3 chessboard in such a way that no knight attacks another [FCLL00]. Any number of knights was allowed, from 0 to 9. A careful count will verify that there are exactly 94 distinct ways to place knights on a 3×3 board in this way. The Princeton group used the strategy of producing an initial set of 1024 RNA molecules (512 would be necessary, but 1024 provided for the possibility that an enzyme might not function properly) which are regarded as encoding every possible pattern of placement of knights on the board, attacking or non-attacking. Those molecules that encoded patterns in which an attack was represented were then eliminated. (This procedure follows the concept expressed by Richard Lipton in [Lip95]. We used this concept in the SAT just above, where the first three steps generated all 8 logically consistent truth settings and the last four eliminated all non-solutions.) Ideally, the set of RNA molecules remaining after the elimination process would encode the 94 correct patterns and no other patterns. A sampling procedure detected 30 distinct correct patterns and 1 incorrect pattern [FCLL00].

We have in progress in Binghamton an aqueous computation that treats this same 3×3 knights problem that was treated by the Princeton group. We represent the 3×3 board by:

$$\begin{array}{ccc} a & b & c \\ d & e & f \\ g & h & i \end{array}$$

We view this problem as one of listing all the independent subsets of the vertices of the graph $G = (V, E)$ with 9 vertices: $V = \{a, b, c, d, e, f, g, h, i\}$ and 8 unordered edges: $\{a, f\}, \{f, g\}, \{g, b\}, \{b, i\}, \{i, d\}, \{d, c\}, \{c, h\}, \{h, a\}$. Observe that each edge consists of a pair of squares of the board such that if knights were placed on these two squares they would attack each other. Observe that there is no other pair of squares with this property. Consequently, the acceptable sets of squares at which knights can be placed on the board are precisely the independent subsets of the graph G. It is sufficient to list all the *maximal* independent subsets of G since, from such a list, the complete list can be easily produced by adjoining each proper subset of each maximal independent set to the list. We plan to derive the list of maximal independent sets by generating and reading DNA molecules just as we discussed for the example above. We sketch briefly how we expect this to go:

Since e appears in no edge it must occur in every maximal independent set. On the same plasmid P that we have used previously, we choose eight restriction enzyme sites that lie in the multiple cloning site (MCS) of the plasmid and associate them with the eight vertices a, b, c, d, f, g, h, i where e has not been listed. We consider the initial state of P to represent the pattern in which, at each of the nine squares, a knight sits. We represent this pattern (and the molecular variety encoding it) by 1111e1111, where the e (which will never be altered) remains to ease the reading of the strings. The first 1 says 'knight at a', the second says 'knight at b', etc. We begin (as always in aqueous computing) with a tube T containing a vast number of the plasmids P. We now carry out eight steps (one for each edge of G) of paired write-to-0 operations. Only the first is reviewed: Pour T into L and R. In L produce 0111e1111 and in R produce 1111e0111. Unite L and R into T. Then T will contain the two molecular varieties 0111e1111 and 1111e0111. When all eight steps of paired write-to-0 operations have been completed we expect T to contain 47 distinct molecular varieties (only half of 94, because we haven't considered e). At Binghamton, each of our write-to-0 operations *increases* the circumference of each plasmid written on by a fixed number of base pairs. We will make a gel separation. We expect bands of five different lengths to be produced. One band associated with each of the following number of 'writes': 4, 5, 6, 7, 8. We expect the band at 4 to contain all (and only) the molecules that encode maximum independent sets. Consequently the band at 4 will be cut from the gel, cloned and sequenced, and the resulting maximum independent sets listed. We expect two. All remaining maximum independent sets are expected to appear in the band at 5. Unfortunately, the subsets of the maximum independent sets that contain three vertices will also be in this band. The band at 5 will be cut from the gel, cloned and sequenced, with only those that are maximal being listed. In this band we expect eight maximal independent sets (and eight non-maximal sets). (Following L. Landweber's thoughtful suggestion, we will also cut out from the gel the material at the location at which any molecules that received only 3 writes would occur. DNA present at this location would indicate an error that would correspond to leaving a pair of attacking knight on the board.) The remaining gel may be discarded.

If this computation proceeds as hoped, a list of the 10 correct maximal independent sets will be produced. If these are correct, then they will contain 37 distinct subsets, for a total of 47 sets. Then the list of 94 sets will arise as the list is duplicated with e in one member, but not the other member, of each pair.

Writing by Methylation

A review of our treatment of SAT in Section 5 shows that two distinct processes were used. First came three steps consisting of paired write procedures. These three steps produced eight distinct molecules, all of which had the *same* circumference (length). In all the other computations discussed here, length provided the possibility of reading answers. For SAT, reading was only 'Yes, if plasmids remain' and 'No, if no plasmids remain'. (Followed by cloning and sequencing if plasmids remain and a satisfying truth assignment is required.) The write procedure merely made half of the stations (restriction enzyme sites) of each molecule non-functional (not cutable). Let us call such a process *silencing* the site. The remaining four steps of multiple cuttings then destroyed all plasmids that did not encode truth settings that satisfied all the required clauses.

The implementation for silencing is time consuming and somewhat error prone. We plan to test the use of methylation as a silencing procedure. For each restriction enzyme, there is in nature an enzyme that silences the site of that restriction enzyme. Methylase enzymes silence restriction sites by attaching methyl groups to bases occurring in the site. We plan to repeat the SAT computation described above, but with the silencing done by appropriate methylases. Success with this simple prototype will encourage attacks on larger problems using this 'silence & cut' concept of molecular computing [He00]

Writing with Light

We would like to write with light on molecules in aqueous solution. We wish to construct a memory register molecule having several stations that respond to light of different frequencies. Writing would be done by light sources of different frequencies applied to separate portions of the solution. The previous rhythm of dividing the solution, writing in parallel, and reuniting would proceed as in the previous techniques. Feasibility tests are planned for the underlying operations of our first concept for writing with light in aqueous computing. We believe that the potential exists for freeing aqueous computing from the biochemical context in which it was born.

Acknowledgements

The author acknowledges support during the preparation of this article from DARPA/NSF CCR-9725021. The research work exposited here was supported not only by this same grant, but also by NSF CCR-9509831. Work carried out at Leiden University received support through the Leiden Center for Natural Computing (LCNC & LIACS). I am particularly grateful to Profs. G. Rozenberg and H. Spaink for their support and their encouragement of our joint work. I am likewise grateful to Prof. M. Yamamura who spent the academic year 1998-1999 in joint research with us here in Binghamton. His visit resulted in the development of the Binghamton implementation of aqueous computing.

References

[Ad94]   L. Adleman, Molecular computation of solutions to combinatorial problems, Science 266(1994) 1021-1024.

[CH91]   K. Culik II & T. Harju, Splicing semigroups of dominoes and DNA, Discrete Appl. Math. 31(1991) 261-277.

[DeLR80] A. DeLuca & A. Restivo, A characterization of strictly locally testable languages and its application to subsemigroups of a free semigroup, Inform. & Control 44(1980) 300-319.

[DG89]   K.L. Denninghoff & R. Gatterdam, On the undecidability of splicing systems, Inter. J. Computer Math., 27(1998) 133-145.

[FCLL00] D. Faulhammer, A.R. Cukras, R.J. Lipton, & L. Landweber, Molecular computation: RNA solutions to chess problems, PNAS 97(2000) 1385-1389.

[GJ79]   M.R. Garey & D.S. Johnson, Computers and Intractibility A Guide to the Theory of NP-Completeness, Freeman, New York (1979).

[Ga89]   R. Gatterdam, Splicing systems and regularity, Intern. J. Computer Math., 31(1989) 63-67.

[Ga92]   R. Gatterdam, Algorithms for splicing systems, SIAM J. Computing, 21(1992) 507-520.

[He87]   T. Head, Formal language theory and DNA: an analysis of the generative capacity of specific recombinant behaviors, Bull. Math. Biology, 49(1987) 737-759.

[He99]   T. Head, Circular suggestions for DNA computing, in: Pattern Formation in Biology, Vision and Dynamics, Ed. by, A. Carbone, M. Gromov, & P. Prusinkiewicz, World Scientific, Singapore (2000) 325-335, QH491.C37 1999.
[He00]   T. Head, Writing by methylation proposed for aqueous computing, (to appear).
[He00']   T. Head, Biomolecular realizations of a parallel architecture for solving combinatorial problems, (submitted).
[HPP97]   T. Head, Gh. Paun, & D. Pixton, Language theory and molecular genetics: generative mechanisms suggested by DNA recombination, Chapter 7 of Vol. 2 of: Handbook of Formal Languages, Ed, by G. Rozenberg & A. Salomaa, Springer-Verlag (1997).
[HRBBLS]   T. Head, G. Rozenberg, R. Bladergroen, C.K.D. Breek, & P.H.M. Lommerese, Computing with DNA by operating on plasmids, Bio Systems, (to appear).
[HYG99]   T. Head, M. Yamamura, & S. Gal, Aqueous computing: writing on molecules, in: Proc. Congress on Evolutionary Computation 1999, IEEE Service Center, Piscataway, NJ (1999) 1006-1010.
[HRSW83]   R.W. Hendrix, J.W. Roberts, F.W. Stahl, & R.A. Weisberg, Eds., Lambda-II, Cold Springs Harbor Laboratory, New York (1983).
[HR75]   G.T. Herman & G. Rozenberg, Developmental Systems and Languages, North Holland, New York (1975).
[HU75]   J.E. Hopcroft & J.D. Ullman, Introduction to Automata Theory, Languages, and Computing, Addison-Wesley, Reading, MA (1979).
[LR99]   E. Laun & K.J. Reddy, Wet splicing systems, in: Proc. DIMACS Series in Discrete Math & Theor. Comp. Sci., Vol. 48, Eds., H. Rubin & D. Wood (1999) 73-83.
[Le83]   B. Lewin, Genes, Wiley, New York (1983) [Updated as: Genes II, Genes III, etc.]
[Lin68]   A. Lindenmayer, Mathematical models of cellular interactions in development I, II, J. Theoretical Biology 18(1968) 280-315.
[Lip95]   R.J. Lipton, DNA solution of computational problems, Science 268(1995) 542-545.
[OKLL97]   Q. Ouyang, P.D. Kaplan, P.D.S. Liu, & A. Libchaber, DNA solution of the maximal clique problem, Science 278(1997) 446-449.
[PRS96]   Gh. Paun, G. Rozenberg, & A. Salomaa, Computing by splicing, Theor. Comput. Sci. 168(1996) 321-336.
[PRS98]   Gh. Paun, G. Rozenberg, & A. Salomaa, DNA Computing New Computing Paradigms, Springer-Verlag, Berlin (1998).
[PS96]   Gh. Paun & A. Salomaa, DNA computing based on the splicing operation, Math. Japonica 43(1996) 607-632.
[Pi96]   D. Pixton, Regularity of splicing systems, Discrete Appl. Math. 69(1996) 101-124.
[Pi00]   D. Pixton, Splicing in abstract families of languages, Theoret. Comput. Sci. 234(2000) 135-166.
[PL90]   P. Prusinkiewicz & A. Lindenmayer, The Algorithmic Beauty of Plants, Springer-Verlag, New York (1990).
[RS80]   G. Rozenberg & A. Salomaa, The Mathematical Theory of L-Systems, Academic Press, New York (1980).
[SSD92]   R. Siromoney, K.G. Subramanian & V.R. Dare, Circular DNA and splicing systems, in: LNCS, Ed. by A. Nakamura, A. Saoudi, P.S.P. Wang, & K. Inoue, Springer, New York 654(1992) 260-273.

[YKF97] T. Yokomori, S. Kobayashi, & C. Ferretti, On the power of circular splicing systems, and DNA computability, IEEE Intern. Conf. on Evolutionary Comput. (1997) 219-224.

Some Methods of Computation in White Noise Calculus

Takeyuki Hida

Meijo University, Nagoya, Japan

Abstract. We discuss some computations for random complex systems which are quite different from those for non random systems. There are of course many unusual computations in stochastic calculus; we shall discuss a few selected interesting topics that come from white noise calculus. They are related to the causal calculus where the time development is strictly concerned, to the Feynman path integral where visualized calculus can be seen and to the variational calculus of random fields.

1 Introduction

Computations of random functions need additional considerations compared to the calculation of non-random functions. Behind the computation is the property related to the so-called $\sqrt{n}$ rule (see Schroedinger [12]). In addition, we shall deal with generalized functionals of white noise as well as their addition, multiplication, and differential and integral calculus.

Actually our idea is realized by the following steps:
(1) **Reductionism.** To choose elemental random variables fitting for our analysis.
(2) **Integration, synthesis.** To form functionals of those elemental variables which are suitable representations of the given random complex systems.
(3) **Analysis, emergence.** To establish the analysis of the given functionals to identify the structure of the systems in question.

In what follows the system of elemental random variables is taken to be a *white noise*. Its realization is given by the time derivative $\dot{B}(t)$ of a Brownian motion $B(t)$.

We shall be concerned mainly with the case when the system is Gaussian, hence the operations acting on the system can be linear except the case of the Feynman integral. We will discuss the following topics:

1. Canonical representation of Gaussian processes,
2. Multiple Markov properties of Gaussian processes,
3. A new formulation of the Feynman path integrals,
4. Transformations of stochastic processes, and
5. Random fields and their variations.

These topics are studied in line with white noise analysis in the manner of *innovation approach*.

2 Canonical Representations of Gaussian Systems

First, the topic 1 is discussed. Although the representation theory of Gaussian processes has already been established, we shall revisit and give new observations in connection with stochastic computation.

Given a Gaussian process $X(t)$, we assume that $E(X(t)) = 0$ and $X(t)$ has unit multiplicity, but no remote past. The multiplicity is defined by the resolution of the identity $\{E(t)\}$. Each $E(t)$ is the projection on to the closed linear space spanned by the $X(s), s \leq t$. Then, we have a representation of $X(t)$ in terms of a white noise $\dot{B}(t)$:

$$X(t) = \int^t F(t,u)\dot{B}(u)du.$$

This representation is the *canonical* one, for which

$$E(X(t)/\mathbf{B}_s(X)) = \int^s F(t,u)\dot{B}(u)du$$

holds for any t and s with $t > s$, where $\mathbf{B}_s(X)$ is the σ-field of events determined by $X(u), u \leq s$. In this case F is called the canonical kernel.

Theorem 1. *The canonical representation is unique if it exists. In particular, the canonical kernel is uniquely determined up to sign. In this case, $\dot{B}(t)$ is the innovation.*

There are many representations in terms of white noise, and the canonical case is particularly important. The canonical property may be understood in such a way that

$$\mathbf{B}_t(X) = \mathbf{B}_t(\dot{B})$$

holds for every t. This means that the innovation $\dot{B}(t)$ is formed from the $X(s)$'s with $s \leq t$. This fact is a sort of *causality*.

Stationary Gaussian processes. Assume, in addition, that $(X(t))$ is stationary and mean continuous. Then, the canonical representation always exists and it is of the form

$$X(t) = \int_{-\infty}^t F(t-u)\dot{B}(u)du,$$

where $|\hat{F}(\lambda)|^2 = f(\lambda)$, turns out to be the spectral density function, $\hat{F}$ being the Fourier transform of F.

The multiple Markov properties are quite important in connection with the computation of the best predictor.

Definition 1. *If for any different $t_1, t_2, \ldots, t_N \geq t$, the conditional expectations*

$$E(X(t_j)/\mathbf{B}_t(X)), \quad j = 1, 2, \ldots, N,$$

are linearly independent in $L^2(\Omega)$, while for any $t_1, t_2, \ldots, t_{N+1}$, with $t_j \geq t$,

$$E(X(t_j)/\mathbf{B}_t(X)), \quad j = 1, 2, \ldots, N+1,$$

are linearly dependent, then $X(t)$ is called N-ple Markov.

Note. The 1-ple Markov property is slightly stronger than the simple Markov property.

Theorem 2. *Under the general assumptions, a Gaussian process $X(t)$ is N-ple Markov if and only if the canonical kernel is a Goursat kernel of order N. Namely, we have*

$$F(t, u) = \Sigma_{i=1}^N f_i(t) g_i(u),$$

where $det(f_i(t_j))$ for different t_j's never vanishes, and where g_i's are linearly independent in $L^2([0, s])$, for any s.

For the *computation* of predictors of a multiple Markov Gaussian process we only need a *finite* number of functions of the observed data before the instant t. This fact guarantees computability in the sense shown by the following theorem.

Theorem 3. *Given a canonical representation of an N-ple Markov process $X(t)$ expressed as in Theorem 2. Then, under the condition that the $X(s), s \leq t$, are observed, the best predictor $\hat{X}(t;h)$ of $X(t+h)$ with $h > 0$ is the conditional expectation $E(X(t+h)/\mathbf{B}_t(X))$, which is expressed in the form*

$$\hat{X}(t;h) = \Sigma_{i=1}^N f_i(t+h) U_i(t),$$

where each $U_i(t)$ is an additive process given by

$$U_i(t) = \int_0^t g_i(u) \dot{B}(u) du.$$

Proof. The best predictor is given, as is well known, by the conditional expectation and the canonical property of the representation guarantees that it is given by

$$\int^t F(t+h, u) \dot{B}(u) du.$$

By using the form of the kernel function, we come to the conclusion.

Remark. Here is noted a simple remark that causal operators are always involved in the canonical representation and optimal predictors. Such an observation is useful when we consider the reversibility. See e.g. [11].

Stationary multiple Markov Gaussian processes. We now specify the process $X(t)$ to be stationary and mean continuous in t.

Theorem 4. *The canonical kernel of a stationary N-ple Markov Gaussian process can be expressed in the form*

$$F(t-u) = \Sigma_1^N h_j(t-u), \quad t \geq u,$$

where h_j has the form

$$h_j(u) = const \cdot u^{\alpha_j} e^{-\lambda_j u}, \quad u \geq 0.$$

The spectral density function of the process is a *rational function* of λ^2. (see [2]).

Note on strictly multiple Markov Gaussian processes. The multiple Markov property is one of the concepts used to describe time dependency, so that we do not want to require differentiability. Still, the differential operator, which is a local operator in time, can be used to define the multiple Markov property in a restricted sense. Although this is not a favorable approach, the computation of the predictor becomes simpler.

Definition 2. *Let L_t be an N-th order ordinary differential operator with C^N-class coefficient and let $X(t)$ be $(N-1)$-times differentiable in the mean square sense. If $X(t)$ satisfies the equation*

$$L_t X(t) = \dot{B}(t),$$

then $X(t)$ is called an N-ple Markov process in the restricted sense.

Proposition 1. *An N-ple Markov Gaussian process $X(t)$ in the restricted sense has the canonical representation of the form*

$$X(t) = \int_0^t R(t, u) \dot{B}(u) du,$$

where $R(t, u)$ is Riemann's function associate with L_t.

The Riemann's function is of course a Goursat kernel and it is easy to compute the predictor and filtering with the help of differential operators.

3 Feynman's Path Integrals

To give a plausible interpretation to the Feynman path integral we do not want to use the approximation of the path integral by a piece-wise linear curve nor by the method of analytic continuation of imaginary mass or variance. The basic idea of our approach is to give a visualized expression representing the Feynman's idea. But we have to pay a price, namely we are led to introduce a class of generalized white noise functionals.

Take a white noise as a system of idealized elemental random variables. Hence, from the viewpoint of the reductionism, it is quite reasonable to form elemental functions of the variables $\dot{B}(t)$'s, say polynomials. Unfortunately those polynomials are not ordinary random functions, but being renormalized, they become generalized white noise functionals. Another significant reason to introduce generalized white noise functionals is a probabilistic formulation of the Feynman path integral as is explained below. There are, of course, other examples motivating the introduction of generalized white noise functionals (see [17]). This theory is now called *white noise theory*, which has been developed by many authors. We now have a good interpretation to the Feynman path integral and can speak of its computation.

Let us come to the Feynman's original idea for the path integral. Given a Lagrangian $L(x, \dot{x})$, which is assumed to be expressible as

$$L(x, \dot{x}) = \frac{1}{2}m\dot{x}^2 - V(x),$$

for a particle of mass m moving in a force field of a potential V. Our basic idea is to take a possible trajectory x to be

$$x(s) = y(s) + \frac{\hbar}{m}B(s), 0 \leq s \leq t,$$

which means that a trajectory x is a sum of the sure path y and a Brownian fluctuation term. For the purpose of describing the propagator $G(y_1, Y_2, t)$ we set $y(0) = y_1$ and $y(t) = y_2$, and claim that for a well-behaved potential V we have

$$G(y_1, y_2, t) = E\{N \exp[\frac{im}{\hbar 2} \int_0^t \dot{x}(s)^2 ds$$
$$+ \frac{1}{2} \int_0^t \dot{B}(s)^2 ds] \exp[-\frac{i}{\hbar} \int_0^t V(x(s)) ds] \delta(x(t) - y_2)\},$$

where E means the expectation and N is the renormalizing constant for exponential function; see [15].

We note that in the above expression, there are two factors which involve an awkward term $\dot{B}(t)^2$. However, we can assign a rigorous meaning by introducing the notion of generalized white noise functionals. The (multiplicative) renormalizing constant N is requested for this purpose.

Theorem 5. *The kernel function $G(y_1, y_2, t)$ given above is the quantum mechanical propagator for the dynamical system with Lagrangian L.*

Our method can be applied for a much wider class of potentials (even with some singularity). Such a direction has been developed by L. Streit and others.

Since white noise enjoys the Ergodic Theorem with respect to the time shift transformations, we get

Theorem 6. *The expectation E to get the propagator $G(y_1, y_2, t)$ can be replaced by the time average operation:*

$$\lim T^{-1} \int_0^T .$$

4 Transformations of Stochastic Processes and Subordination

There are various types of transformations of stochastic processes, amplitude modulation, frequency modulation and nonlinear filtering, among others. They have been well established and various methods of computations are known. What we are going to discuss now is quite different, namely, *subordination*. Apart from the rigorous definition, we first explain this notion by specifying the action on white noise and see that it is essentially an infinite dimensional stochastic operation.

Recall that the parameter t of a white noise $\dot{B}(t)$ stands for the time. Let us observe a Brownian motion (the integral of a white noise) at random time instead of continuous t. The random time is now assumed to be the instants when an increasing stable stochastic process jumps. Let the process be a stable process $X_\alpha(t)$ with exponent α, with $0 < \alpha < 1$. Thus the actual process to be observed can be expressed in the following form

$$Y(\alpha, t) = B(X_\alpha(t)) \ .$$

The process $Y(\alpha, t)$ is in fact a stable process of exponent 2α. In terms of the measure on the space of sample functions, we can say that the Wiener measure is transformed to another probability measure, which is singular to the Wiener measure. This transformation is really unconventional and infinite dimensional in the sense that the transformation is done, intuitively speaking, in all directions in the function space uniformly and simultaneously.

We are now in a position to explain the subordination in a typical case which has come from a concrete example; namely, the X-ray photon from Cygnus X-1, which is a black hole candidate. The analysis of the data is discussed in the paper by Negoro et al [8]. See also Si Si [14]. Fitness of the theory to the data will be reported later, but we are optimistic because one of the significant characteristics of the observed data is *self-similarity*.

We now give a plausible reason why a mathematical model of the X-ray photon emission may be given by subordination. The information source can be taken to be a white noise $\dot{B}(t)$. The stable process with exponent $1/2$ is fitting for the random time of observation. The increasing stable process with exponent $1/2$ is viewed as the inverse process of the maximum of Brownian motion. The process describes instants of random observations with exponential holding time. The result of the subordination is another stable process with exponent 1. The proof of this fact is sketched as follows. Consider the characteristic functional of $\dot{Y}(1/2, t)$.

$$E(\exp[i \int \dot{Y}(1/2,t)\xi(t)dt]) = E_Y(E_B(\exp[i \int \dot{B}(Y_{1/2}(t))\dot{Y}_{1/2}(t)\xi(t)dt]))$$

$$= E_Y(E_B(\exp[i \int \dot{B}(s)\xi(Y_{1/2}^{-1}(s))ds]))$$

$$= E_Y(\exp[-\frac{1}{2}\int \xi(Y_{1/2}^{-1}(s))^2 ds])$$

$$= E(\exp[-\frac{1}{2}\int \xi(t)^2 \dot{Y}_{1/2}(t)dt])$$

$$= \exp[-c\int (\xi(t)^2)^{1/2} dt]$$

$$= \exp[-c\int |\xi(t)| dt],$$

which is the characteristic functional of the Cauchy process as is expected. We now recall Lévy's method to construct a Brownian motion by successive interpolation. The subordinated process $Y(\alpha, t)$ can be constructed also by interpolation, but the time interval is chosen at random, namely $\Delta X_\alpha(t)$. The loss of information in this case can be obtained as in [14].

Remark. There are many formulas expressed in terms of an harmonic ratio for the process $Y_{1/2}(t)$ or the maximum of Brownian motion. This suggests to consider projective transformations or conformal transformations acting on the parameter space as part of the harmonic analysis arising from the infinite dimensional rotation group.

5 Variational Calculus

Needless to say, a random field $X(C)$, C being a contour or a surface without boundary, carries more information than $X(t)$, when the parameter varies. We are going to investigate this situation from the view point of computation.

To fix the idea, we consider a random field $X(C)$, which is a linear functional of white noise (hence, a Gaussian random variable) and is indexed by a closed plane curve C. From our stand point that claims reductionism, the innovation for the given $X(C)$ should be obtained. The method has to be computable; in fact we wish to show a method explicitly.

Let (E^*, μ) be a white noise with a two dimensional parameter. Its sample function is denoted by $x(u)$, where $u \in R^2$. We denote a Gaussian random field by $X(C, x)$ or just by $X(C)$, In order to discuss the method in question it is necessary to make some concrete assumptions.

A.1) The parameter C runs through a class **C** which is a collection of smooth ovaloids.

A.2) The $X(C)$ has a canonical representation of the form

$$X(C) = \int_{(C)} F(C, u) x(u) du,$$

where (C) is the domain enclosed by C. For any fixed C the restriction of the kernel $F(C, u)$ to C is non-trivial.

A.3) Deformations of C can be done only by conformal transformations acting on R^2.

Here let us recall the stochastic variational equation (see [7])

$$\delta X(C) = \Phi(X(C'), C' < C, Y(s), s \in C, C, \delta C),$$

where $\{Y(s), s \in C\}$ is the innovation and where $C' < C$ means that C' is in the inside of C.

Our intention is this: Take a variation of the given $X(C)$ and form the innovation for $X(C)$ that appears in the stochastic variational equation. Further, a technical assumption may be set for the actual computation.

A.4) The parameter C is restricted to a subclass $\mathbf{C}_0$ of $\mathbf{C}$ consisting of an image of a conformal map α of the family of concentric circles.

The above assumption implies that $\mathbf{C}$ is an increasing family of smooth convex contours. When a variation $\delta X(C)$ is defined, the infinitesimal deformation of C, symbolically denoted by $C + \delta C$, goes outside. We tacitly understand that once C is fixed, inside of C is the past and the future lies outside. Thus, intuitively speaking, the variation of $X(C)$ is formed towards the future.

Returning to the $X(C)$ given before, we are ready to get the variation assuming that F has the functional derivative F'_n:

$$\delta X(C) = \int_C F(C, s) x(s) \delta n(s) ds + \int_{(C)} \int_C F'_n(C, u)(s) \delta n(s) ds x(u) du,$$

where $\delta n(s)$ is the normal distance between C and $C + \delta C$ measured at s.

Theorem 7. *The variation $\delta X(C)$ and the conformal transformations that keep the C invariant determine the innovation $\{x(s), s \in C\}$.*

Proof. The second term of the above expression for $\delta X(C)$ is the conditional expectation $E(\delta X(C) | \mathbf{B}_C(X))$, where $\mathbf{B}_C(X)$ denotes the σ-field generated by the events determined by the $X(C'), C' < C$. This means that the second term is obtained by the past values of the field $X(C)$. On the other hand, we can manage the first term in such a way that applying the conformal transformations acting on the parameter space R^2 which leaves C invariant (those transformations form a group denoted by G_1), it is possible to get the actual value of $x(s)$, for all $s \in C$.

To show this fact we note that admissible transformations acting on R^2 form a group which is isomorphic (through α) to the transformation group G_0 on R^1 involving, shifts, dilations and reflections. It is known that the unitary representation of the group G_0 on the Hilbert space $L^2(R^1)$ is irreducible. Here we note that the canonical kernel F is not trivial on C, so that the application of the transformations G_1 gives us the values of $x(s)$ on C. Thus, the innovation is obtained.

Two methods of *computability* are involved there; one is the variation which is quite simple since it is isomorphic to the tension of a circle, and the other is the action of the group G_1 (or G_0) which is two dimensional. We may therefore say something optimistic regarding the computability.

Finally we note that similar interesting questions are raised in the case of random fields $X(f)$ depending on a function f instead of a contour C.

Acknowledgments

The author is grateful to the Solvay Institute and the Academic Frontier Project of Meijo University for their supports.

Further Reading

[1] L. Accardi, P. Gibilisco and I.V. Volovich, *The Lévy Laplacian and the Yang-Mills equation*. Volterra Center Publication, Univ. Roma 2, no. 129, 1992.
[2] L. Accardi, T. Hida and Win Win Htay, Boson Fock representation of Stochastic processes. *Mat. Zametkii* 67 (2000), 3–14. (Russian)
[3] R.P. Feynman and A.R. Hibbs, *Quantum Mechanics and Path Integrals*. McGraw-Hill, 1965.
[4] I.M. Gel'fand and N.Ya. Vilenkin, *Generalized Functions* (in Russian), vol.4. 1961. English translation, Academic Press, 1964.
[5] T. Hida, Canonical representations of Gaussian processes and their applications. *Mem. Univ. Kyoto. A*, 33, 1960, 109–155.
[6] T. Hida, Complexity and irreversibility in stochastic analysis. *Les Treilles Conference*, 1999.
[7] T. Hida and Si Si, Innovation for random fields. *Infinite Dimensional Analysis, Quantum Probability and Related Topics* 1, 1998, 499–509.
[8] H. Negoro et al, Spectral variations of the superposed shot as a cause of hard time lags of Cygnus X-1. preprint 2000.
[9] N. Obata, White Noise Calculus and Fock Space. Springer Lecture Notes in Math. no. 1577, 1994.
[10] M. Oda, Fluctuation in astrophysical phenomena. *Mathematical Approach to Fluctuations. Proc. IIAS Workshop* vol. 1, ed. T. Hida, 1994, 115–37.
[11] I. Prigogine, *The End of Certainty. Time, Chaos, and the New Laws of Nature*. The Free Press, 1997.
[12] E. Schroedinger, *What is life?* First published in 1944, also 1992 ed. Cambridge Univ. Press.
[13] Si Si, Topics on random fields. *Quantum Information*. ed. T. Hida and K. Saito, 1999, 179–194.
[14] Si Si, Random irreversible phenomena. Entropy in subordination. *Les Treilles Conference*, 1999.
[15] L. Streit and T. Hida, Generalized Brownian functionals and the Feynman integral, *Stochastic processes and Their Applications* 6, 1983, 55–69.
[16] V. Volterra, *Opere Matematiche*, vol. 5, 1926–1940. In particular, see the paper 1937 proposing the Lotka-Volterra equation.
[17] T. Hida, *Analysis of Brownian Functionals*. Carleton Math. Notes, Carleton University, 1975.

Computing with Membranes: Attacking NP-Complete Problems

Gheorghe Păun

Institute of Mathematics of the Romanian Academy, Bucureşti, Romania

Abstract. The aim of this paper is to introduce to the reader the main ideas of Computing with Membranes, a recent branch of (theoretical) Molecular Computing, with emphasis on some variants which can solve computationally hard problems in polynomial (often, linear) time. In short, in a cell-like system (called a P system), multisets of objects (described by symbols or by strings of symbols) evolve in a membrane structure and compute natural numbers as the result of halting sequences of transitions. The model is parallel, distributed, and nondeterministic.

We present here informally the model, we illustrate by an example the basic definition, then we recall several variants which can solve NP-complete problems in a polynomial time. A complete bibliography of the domain, at the level of the end of August 2000, is also provided.

1 From a Bio-Cell to a Computing-Cell

The incentive to define P systems (initially, in [P17][1], they were called *supper-cell* systems) was the question whether or not the frequent statements (see, e.g., [3], [7]) concerning the fact that "an alive cell is a computer" are just metaphors, or a formal computing device can be abstracted from the cell functioning. As we will immediately see, the latter answer is valid.

Three are the fundamental features of alive cells which will be used in our computing machineries: (1) the **membrane structure**, where (2) **multisets** of chemical compounds evolve according to prescribed (3) **rules**.

A *membrane structure* is a hierarchical arrangement of membranes, all of them placed in a main membrane, called the *skin* membrane. This one delimits the system from its environment. The membranes should be understood as three-dimensional balloons, but a suggestive pictorial representation is by means of planar Euler-Venn diagrams (see Figure 1). Each membrane precisely identifies a *region*, the space between it and all the directly inner membranes, if any exists. A membrane without any membrane inside is said to be *elementary*.

In the regions of a membrane structure we place *multisets* of *objects*. A multiset is a usual set with multiplicities associated with each objects, in the

---

[1] The references of the form [Pn] point to papers in the P system bibliography given at the end of the paper.

form of natural numbers; the meaning is that each object can appear in a number of identical copies in a given region. In turn, for the beginning, the objects are supposed to be *atomic*, in the etymological sense, without "parts", hence we can identify them by symbols from a given alphabet (we always work with finitely many types of objects, that is, with multisets over a finite support-set).

The objects evolve by means of given *rules*, which are associated with the regions (the intuition is that each region has specific chemical reaction conditions, hence the rules from a region cannot necessarily act also elsewhere). These rules specify both object transformation and object transfer from a region to another one. The passing of an object through a membrane is called *communication*.

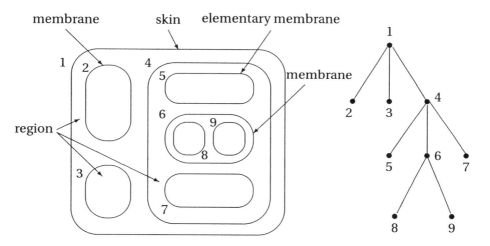

Figure 1: A membrane structure and its associated tree

Here is a typical rule

$$cabb \to caad_{out}d_{in_3},$$

with the following meaning: one copy of the *catalyst* c (note that it is reproduced after the "reaction") together with one copy of object a and two copies of object b react together and produce one copy of c, two of a, and two copies of the object d; one of these latter objects is sent out of the region where the rule is applied, while the second copy is sent to the adjacently inner membrane with the label 3, if such a membrane exists; the objects c, a, a remain in the same membrane (it is supposed that they have associated the communication command *here*, but we do not explicitly write this indication); if there is no membrane with the label 3 directly inside the membrane where the rule is to be applied, then the rule cannot be applied. By a command *out*, an object can be also sent outside the skin membrane, hence it leaves the system and never comes back.

Therefore, the rules perform a multiset processing; in the previous case, the multiset represented by *cabb* is subtracted from the multiset of objects in a given region, objects *caa* are added to the multiset in that region, while copies of *d* are added to the multisets in the upper and lower regions.

To a rule as above we can also add the *action-symbol* δ; when a rule which contains δ is used, then the membrane is *dissolved*, all its objects remain free in the immediately upper membrane and its rules are lost. The skin membrane is never dissolved. A sort of dual action can also be considered, τ, which makes a membrane thicker, thus inhibiting the communication through it; a thicker membrane can be open again to communications after using a rule which contains the symbol δ. By using the actions δ, τ one can control the communication through membranes in a very effective way, but we do not enter here into details (the interested reader is referred to [P19]).

The rules are used *in a nondeterministic maximally parallel manner*: the objects to evolve and the rules to be applied to them are chosen in a nondeterministic manner, but no object which can evolve at a given step should remain. Sometimes, a priority relation among rules is considered, hence the rules to be used and the objects to be processed are selected in such a way that only rules which have maximal priority among the applicable rules are used.

The membrane structure together with the multisets of objects and the sets of evolution rules present in its regions constitute a *P system*. The membrane structure and the objects define a *configuration* of a given P system. By using the rules as suggested above, we can define *transitions* among configurations. A sequence of transitions is called a *computation*. We accept as successful computations only the *halting* ones, those which reach a configuration where no further rule can be applied.

With a successful computation we can associate a *result*, for instance, by counting the objects present in the halting configuration in a specified elementary membrane, i_o. More precisely, we can use a P system for solving three types of tasks: as a *generative* device (start from an initial configuration and collect all vectors of natural numbers describing the multiplicities of objects present in membrane i_o at the end of all successful computations), as a *computing* device (start with some input placed in an initial configuration and read the output at the end of a successful computation, by considering the contents of membrane i_o), and as a *decidability* device (introduce a problem in an initial configuration and look for the answer in a specified number of steps, in the contents of membrane i_o).

Of course, instead of using an elementary membrane for reading the output, we can consider the objects which leave the skin membrane and we can "compose" the result by using them.

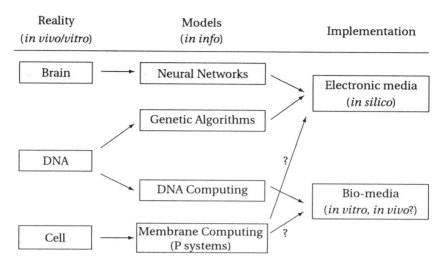

Figure 2: The four bio-domains of Natural Computing

Some of these notions will be partially given in a more formal manner in the subsequent sections. We anticipate an important observation: *various computing devices as sketched above are computationally complete, in the Turing sense; for instance, in the generative case, they can generate all recursively enumerable sets of vectors of natural numbers.*

This means computing *competence*. What about computing *performances*, in the sense of the (time) complexity of solving problems? In Sections 5, 6, 7 we will see that certain classes of P systems, with an enhanced parallelism, are able to solve NP-complete problems in linear time. Of course, this means trading space for time and using an exponential space, in the form of membranes and objects present in the configurations of the system during a computation (this is usual in DNA Computing, for instance, where we can count on the massive parallelism made possible by the fact that billions of molecules can find room in a tiny test tube).

We will return to this important topic, the main subject of this paper, after formally defining the notion of a P system.

We close this section with the remark that up to now no implementation of P systems was done, but there is a fundamental question whether or not this should be done in a biochemical media or on an electronic support (as it is the case with other successful areas of Natural Computing, such as Neural Networks and Genetic Algorithms, which are inspired from the alive nature but are implemented on usual computers). Figure 2 illustrates this dilemma.

Attempts to simulate P systems on an electronic computer were already done, see [Madrid], [P12], [P32], but the results are not yet answering the question of the practical usefulness of this approach.

Of course, we have also to be prepared for the case that no implementation of P systems will be done (that is, no implementation of a real practical interest), that the attempt will remain forever *in info*, as a subject for mathematical investigations. Anyway, it is too early to ask such questions, serious interdisciplinary investigations should precede any sensible answer.

2 A Formal Definition of a P System

A membrane structure can be mathematically represented by a tree, in the natural way, or by a string of matching parentheses. The tree of the structure in Figure 1 is given in the same figure, while the parenthetic representation of that structure is the following:

$$[_1 [_2]_2 [_3]_3 [_4 [_5]_5 [_6 [_8]_8 [_9]_9]_6 [_7]_7]_4]_1.$$

The tree representation makes possible considering various parameters, such as the *depth* of the membrane structure, and also suggests considering membrane structures of particular types (described by linear trees, star trees, etc.).

A multiset over an alphabet $V = \{a_1, \ldots, a_n\}$ is a mapping μ from V to $\mathbf{N}$, the set of natural numbers, and it can be represented by any string $w \in V^*$ such that $\Psi_V(w) = (\mu(a_1), \ldots, \mu(a_n))$[2]. Operations with multisets are defined in the natural manner.

With these simple prerequisites, we can define a P system (of *degree* m, $m \geq 1$) as a construct

$$\Pi = (V, T, C, \mu, w_1, \ldots, w_m, (R_1, \rho_1), \ldots, (R_m, \rho_m), i_o),$$

where:

(i) V is an alphabet; its elements are called *objects*;
(ii) $T \subseteq V$ (the *output* alphabet);
(iii) $C \subseteq V, C \cap T = \emptyset$ (*catalysts*);
(iv) μ is a membrane structure consisting of m membranes, with the membranes and the regions labeled in a one-to-one manner with elements of a given set H; in this section we use the labels $1, 2, \ldots, m$;
(v) $w_i, 1 \leq i \leq m$, are strings representing multisets over V associated with the regions $1, 2, \ldots, m$ of μ;
(vi) $R_i, 1 \leq i \leq m$, are finite sets of *evolution rules* over V associated with the regions $1, 2, \ldots, m$ of μ; ρ_i is a partial order relation over R_i, $1 \leq i \leq m$, specifying a *priority* relation among rules of R_i.
An evolution rule is a pair (u, v), which we will usually write in the form $u \to v$, where u is a string over V and $v = v'$ or $v = v'\delta$, where v' is a string over $\{a_{here}, a_{out}, a_{in_j} \mid a \in V, 1 \leq j \leq m\}$, and δ is a special symbol not in V. The length of u is called *the radius* of the rule $u \to v$.

---

[2] The elements of formal language theory which we use here can be found, e.g., in [9]. In particular, Ψ_V is the Parikh mapping associated with V.

(vii) i_o is a number between 1 and m.

When presenting the evolution rules, the indication "here" is in general omitted.

If Π contains rules of radius greater than one, then we say that Π is a system *with cooperation*. Otherwise, it is a *non-cooperative* system. A particular class of cooperative systems is that of *catalytic* systems: the only rules of a radius greater than one are of the form $ca \to cv$, where $c \in C, a \in V - C$, and v contains no catalyst; moreover, no other evolution rules contain catalysts (there is no rule of the form $c \to v$ or $a \to v_1 c v_2$, for $c \in C$).

The $(m+1)$-tuple $(\mu, w_1, \ldots, w_m)$ constitutes the *initial configuration* of Π. In general, any sequence $(\mu', w'_{i_1}, \ldots, w'_{i_k})$, with μ' a membrane structure obtained by removing from μ all membranes different from $i_1, \ldots, i_k$ (of course, the skin membrane is not removed), with w'_j strings over V, $1 \leq j \leq k$, and $\{i_1, \ldots, i_k\} \subseteq \{1, 2, \ldots, m\}$, is called a *configuration* of Π.

It should be noted the important detail that the membranes preserve the initial labeling in all subsequent configurations; in this way, the correspondence between membranes, multisets of objects, and sets of evolution rules is well specified by the subscripts of these elements.

For two configurations $C_1 = (\mu', w'_{i_1}, \ldots, w'_{i_k})$, $C_2 = (\mu'', w''_{j_1}, \ldots, w''_{j_l})$, of Π we write $C_1 \Longrightarrow C_2$, and we say that we have a *transition* from C_1 to C_2, if we can pass from C_1 to C_2 by using the evolution rules appearing in $R_{i_1}, \ldots, R_{i_k}$ in the following manner (rather than a completely cumbersome formal definition we prefer an informal one, explained by examples).

Consider a rule $u \to v$ in a set R_{i_t}. We look to the region of μ' associated with the membrane i_t. If the objects mentioned by u, with the multiplicities at least as large as specified by u, appear in w'_{i_t}, then these objects can evolve according to the rule $u \to v$. The rule can be used only if no rule of a higher priority exists in R_{i_t} and can be applied at the same time with $u \to v$. More precisely, we start to examine the rules in the decreasing order of their priority and assign objects to them. A rule can be used only when there are copies of the objects whose evolution it describes and which were not "consumed" by rules of a higher priority and, moreover, there is no rule of a higher priority, irrespective which objects it involves, which is applicable at the same step. Therefore, all objects to which a rule *can* be applied *must* be the subject of a rule application. All objects in u are "consumed" by using the rule $u \to v$.

The result of using the rule is determined by v. If an object appears in v in the form a_{here}, then it will remain in the same region i_t. If an object appears in v in the form a_{out}, then a will exit the membrane i_t and will become an element of the region which surrounds membrane i_t. In this way, it is possible that an object leaves the system: if it goes outside the skin of the system, then it never comes back. If an object appears in the form a_{in_q}, then a will be added to the multiset from membrane q, providing that a is adjacent to the membrane q. If a_{in_q} appears in v and membrane q is not one of the membranes delimiting "from below" the region i_t, then the application of the rule is not allowed.

If the symbol δ appears in v, then membrane i_t is removed (we say *dissolved*) and at the same time the set of rules R_{i_t} (and its associated priority relation) is removed. The multiset from membrane i_t is added (in the sense of multisets union) to the multiset associated with the region which was directly external to membrane i_t. We do not allow the dissolving of the skin membrane, because this means that the whole "cell" is lost, we do no longer have a correct configuration of the system.

All these operations are performed in parallel, for all possible applicable rules $u \to v$, for all occurrences of multisets u in the regions associated with the rules, for all regions at the same time. No contradiction appears because of multiple membrane dissolving, or because simultaneous appearance of symbols of the form a_{out} and δ. If at the same step we have a_{in_i} outside a membrane i and δ inside this membrane, then, because of the simultaneity of performing these operations, again no contradiction appears: we assume that a is introduced in membrane i at the same time when it is dissolved, thus a will remain in the region surrounding membrane i; that is, from the point of view of a, the effect of a_{in_i} in the region outside membrane i and δ in membrane i is a_{here}.

A sequence of transitions between configurations of a given P system Π is called a *computation* with respect to Π. A computation is *successful* if and only if it halts, that is, there is no rule applicable to the objects present in the last configuration, and, if the membrane i_o is present as an elementary one in the last configuration of the computation. (Note that the output membrane was not necessarily an elementary one in the initial configuration.) The result of a successful computation is $\Psi_T(w)$, where w describes the multiset of objects from T present in the output membrane in a halting configuration. The set of such vectors $\Psi_T(w)$ is denoted by $Ps(\Pi)$ (from "Parikh set") and we say that it is *generated* by Π.

The family of sets of vectors of natural numbers $Ps(\Pi)$ generated by P systems with priority, catalysts, and the membrane dissolving action, and of degree at most $m, m \geq 1$, using target indications of the form *here, out, in_j* (but not using the action τ, which will be defined in Section 4), is denoted by $NP_m(Pri, Cat, tar, \delta, n\tau)$; when one of the features $\alpha \in \{Pri, Cat, \delta\}$ is not present, we replace it with $n\alpha$. When the number of membranes is not bounded, we replace the subscript m with $*$.

3 An Example

Before recalling some results about the computing power of P systems of the types introduced in the previous section, we shall consider one example, of a P system with a generative task. Many P systems used to solve decidability problems will be considered in the subsequent sections.

Example 1. Consider the P system of degree 4

$$\Pi_1 = (V, T, C, \mu, w_1, w_2, w_3, w_4, (R_1, \rho_1), (R_2, \rho_2), (R_3, \rho_3), (R_4, \rho_4), 4),$$

$V = \{a, b, b', c, e, f\}, T = \{e\}, C = \emptyset,$
$\mu = [_1[_2[_3]_3[_4]_4]_2]_1,$
$w_1 = \lambda, R_1 = \emptyset, \rho_1 = \emptyset,$
$w_2 = \lambda, R_2 = \{b' \to b, b \to be_{in_4}, r_1 : ff \to f, r_2 : f \to \delta\}, \rho_2 = \{r_1 > r_2\},$
$w_3 = af, R_3 = \{a \to ab', a \to b'\delta, f \to ff\}, \rho_3 = \emptyset,$
$w_4 = \lambda, R_4 = \emptyset, \rho_4 = \emptyset.$

The initial configuration of the system is presented in Figure 3. No object is present in membrane 2, hence no rule can be applied here. The only possibility is to start in membrane 3, using the objects a, f, present in one copy each. By using the rules $a \to ab', f \to ff$, in parallel for all occurrences of a and f currently available, after n steps, $n \geq 0$, we get n occurrences of b' and 2^n occurrences of f. In any moment, instead of $a \to ab'$ we can use $a \to b'\delta$ (note that we always have only one copy of a). In that moment we have $n + 1$ occurrences of b' and 2^{n+1} occurrences of f and we dissolve membrane 3.

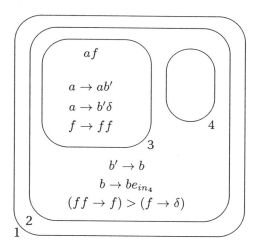

Figure 3: A P system generating $n^2, n \geq 1$

The rules of the former membrane 3 are lost, the rules of membrane 2 can now be applied to the objects left free in its region. Due to the priority relation, we have to use the rule $ff \to f$ as much as possible. In one step, we pass from b'^{n+1} to b^{n+1}, while the number of f occurrences is divided by two. In the next step, $n+1$ occurrences of e are introduced in membrane 4: each occurrence of the symbol b introduces one occurrence of e. At the same time, the number of f occurrences is divided again by two. We can continue. At each step, further $n + 1$ occurrences of e are introduced in the output membrane. This can be done $n + 1$ steps: n times when the rule $ff \to f$ is used (thus diminishing the number of f occurrences to one), and one when using the rule $f \to \delta$ (it may now be used). In this moment, membrane 2 is dissolved, which entails the fact that its rules are removed. No further step is possible, the computation stops.

In total, we have $(n+1) \cdot (n+1)$ copies of e in membrane 4, for some $n \geq 0$, that is,
$$Ps(\Pi) = \{(n^2) \mid n \geq 1\}.$$

4 The Computational Completeness

At the end of Section 1 we have mentioned that P systems as introduced above are equivalent with Turing Machines. More precisely, we have the following theorem, whose (rather complex) proof can be found in [P17] (the families of recursively enumerable sets of vectors of natural numbers is denoted by $PsRE$):

Theorem 1. $NP_2(Pri, Cat, tar, n\delta, n\tau) = PsRE$.

In the aim of obtaining more "realistic" systems, we can consider weaker communication commands. The weakest one is to add no label to *in*: if an object a_{in} is introduced in some region of a system, then a will go to any of the adjacent lower membranes, nondeterministically chosen; if no inner membrane exists, then a rule which introduces a_{in} cannot be used.

An intermediate possibility is to associate both with objects and membranes *electrical charges*, indicated by $+, -, 0$ (positive, negative, neutral). The charges of membranes are given in the initial configuration and are not changed during computations, the charge of objects are given by the evolution rules, in the form $a \to b^+d^-$. A charged object will immediately go into one of the directly lower membrane of the opposite polarization, nondeterministically chosen, the neutral objects remain in the same region or will exit it.

In both previous cases, we still use the indications *here* and *out* for objects with a neutral charge.

Because communication by means of *in* commands and by electrical charges is expected to be weaker than communication by commands of the form in_j, we compensate the loss in power by making use of the actions δ, τ in order to control membrane thickness (hence permeability). This is done as follows. Initially, all membranes are considered of thickness 1. If a rule in a membrane of thickness 1 introduces the symbol τ, then the membrane becomes of thickness 2. A membrane of thickness 2 does not become thicker by using further rules which introduce the symbol τ, but no object can enter or exit it. If a rule which introduces the symbol δ is used in a membrane of thickness 1, then the membrane is dissolved; if the membrane had thickness 2, then it returns to thickness 1. If at the same step one uses rules which introduce both δ and τ in the same membrane, then the membrane does not change its thickness. These actions of the symbols δ, τ are illustrated in Figure 4.

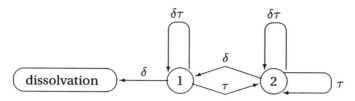

Figure 4: The effect of actions δ, τ

The use of commands *here, out, in* is indicated in the notations of families of generated sets of vectors by writing i/o, and the use of electrical charges is indicated by writing $\pm$ instead of *tar*, respectively.

The proof of the following result can be found in [P19].

Theorem 2. $NP_*(nPri, Cat, \pm, \delta, \tau) = PsRE.$

Unfortunately, the proof gives no bound on the number of membranes, so the problem whether or not the degree of systems of the type in Theorem 2 gives an infinite hierarchy remains *open*.

It is also *open* the problem of the size of the family $NP_*(nPri, Cat, i/o, \delta, \tau)$. However, if instead of using catalysts as considered up to now we use a sort of "catalysts with a short term memory", then again a characterization of $PsRE$ can be obtained. Such catalysts (we call them *bi-stable*) have two states each, c and $\bar{c}$, and they can enter rules of the forms $ca \to \bar{c}v, \bar{c}a \to cv$ (always changing from c to $\bar{c}$ and back). We write $2Cat$ instead of Cat when denoting the generated families of vector sets. The proof of the next result can be found in [P23]; note that this time the hierarchy on the number of membranes collapses at the first level, which shows the excessive power of bi-stable catalysts.

Theorem 3. $NP_1(nPri, 2Cat, i/o, n\delta, n\tau) = PsRE.$

5 P Systems with Active Membranes

In alive cells, many reactions take place in a direct connection with (the proteins embedded in) membranes; moreover, the cells can be divided. These observations suggest to consider systems with the membranes explicitly involved in rules, also provided with the possibility of dividing membranes. This is a very powerful feature, because the working space can be exponentially increased.

We recall from [Pdivid] the definition of *P systems with active membranes* in their general form. Such a system is a construct

$$\Pi = (V, H, \mu, w_1, \ldots, w_m, R),$$

where:

(i) $m \geq 1$;
(ii) V is an alphabet;
(iii) H is a finite set of *labels* for membranes;
(iv) μ is a *membrane structure*, consisting of m membranes, labeled (not necessarily in a one-to-one manner) with elements of H; all membranes in μ are supposed to be neutral;
(v) $w_1, \ldots, w_m$ are strings over V, describing the *multisets of objects* placed in the m regions of μ;
(vi) R is a finite set of *developmental rules*, of the following forms:
   (a) $[_h a \to v]_h^\alpha$,
       for $h \in H, \alpha \in \{+, -, 0\}, a \in V, v \in V^*$
       (object evolution rules, associated with membranes and depending on the label and the charge of the membranes, but not directly involving the membranes, in the sense that the membranes are neither taking part to the application of these rules nor are they modified by them);
   (b) $a[_h \,]_h^{\alpha_1} \to [_h b]_h^{\alpha_2}$,
       for $h \in H, \alpha_1, \alpha_2 \in \{+, -, 0\}, a, b \in V$
       (communication rules; an object is introduced in the membrane, maybe modified during this process; also the polarization of the membrane can be modified, but not its label);
   (c) $[_h a]_h^{\alpha_1} \to [_h \,]_h^{\alpha_2} b$,
       for $h \in H, \alpha_1, \alpha_2 \in \{+, -, 0\}, a, b \in V$
       (communication rules; an object is sent out of the membrane, maybe modified during this process; also the polarization of the membrane can be modified, but not its label);
   (d) $[_h a]_h^\alpha \to b$,
       for $h \in H, \alpha \in \{+, -, 0\}, a, b \in V$
       (dissolving rules; in reaction with an object, a membrane can be dissolved, while the object specified in the rule can be modified);
   (e) $[_h a]_h^{\alpha_1} \to [_h b]_h^{\alpha_2} [_h c]_h^{\alpha_3}$,
       for $h \in H, \alpha_1, \alpha_2, \alpha_3 \in \{+, -, 0\}, a, b, c \in V$
       (division rules for elementary membranes; in reaction with an object, the membrane is divided into two membranes with the same label, maybe of different polarizations; the object specified in the rule is replaced in the two new membranes by possibly new objects; all objects different from a are duplicated in the two new membranes);
   (f) $[_{h_0} [_{h_1} \,]_{h_1}^{\alpha_1} \ldots [_{h_k} \,]_{h_k}^{\alpha_1} [_{h_{k+1}} \,]_{h_{k+1}}^{\alpha_2} \ldots [_{h_n} \,]_{h_n}^{\alpha_2}]_{h_0}^{\alpha_0}$
       $\to [_{h_0} [_{h_1} \,]_{h_1}^{\alpha_3} \ldots [_{h_k} \,]_{h_k}^{\alpha_3}]_{h_0}^{\alpha_5} [_{h_0} [_{h_{k+1}} \,]_{h_{k+1}}^{\alpha_4} \ldots [_{h_n} \,]_{h_n}^{\alpha_4}]_{h_0}^{\alpha_6}$,
       for $k \geq 1, n > k, h_i \in H, 0 \leq i \leq n$, and $\alpha_0, \ldots, \alpha_6 \in \{+, -, 0\}$ with $\{\alpha_1, \alpha_2\} = \{+, -\}$; if this membrane with the label h_0 contains other membranes than those with the labels $h_1, \ldots, h_n$ specified above, then they must have neutral charges in order to make this rule applicable;
       (division of non-elementary membranes; this is possible only if a membrane contains two immediately lower membranes of opposite polarizations, $+$ and $-$; the membranes of opposite polarizations

are separated in the two new membranes, but their polarization can change; always, all membranes of opposite polarizations are separated by applying this rule; all membranes of neutral polarization are duplicated and then are part of the contents of both copies of the membrane h_0).

Note that in all rules of types (a) – (e) only one object is specified (that is, the objects do not directly interact) and that, with the exception of rules of type (a), always single objects are transformed into single objects (the two objects produced by a division rule of type (e) are placed in two different regions). Also, it is important to note that rules of type (e) refer to elementary membranes.

A system as above is said to be *non-cooperative*. If we also allow rules of type (a) of the general form $[_h u \to v]_h^\alpha$, for $h \in H, \alpha \in \{+, -, 0\}, u \in V^+, v \in V^*$, then we say that we have a *cooperative* system. Note that such rules, with more than one object in their "left hand side" are only allowed for the type (a), not for the other types (although the cooperation can be easily extended to all types of rules, here we do not need such a powerful feature). In all sections below, excepting Section 6, we use only non-cooperative rules.

These rules are applied according to the following *principles*:

1. All the rules are applied in parallel: in one step, the rules of type (a) are applied to all objects to which they can be applied, all other rules are applied to all membranes to which they can be applied; an object can be used by only one rule, non-deterministically chosen (there is no priority relation among rules), but any object which can evolve by a rule of any form, must do it.
2. If a membrane is dissolved, then all the objects in its region are left free in the surrounding region.
3. All objects and membranes not specified in a rule and which do not evolve are passed unchanged to the next step. For instance, if a membrane with the label h is divided by a rule of type (e) which involves an object a, then all other objects in membrane h which do not evolve are introduced in each of the two resulting membranes h. Similarly, when dividing a membrane h by means of a rule of type (f), the neutral membranes are reproduced in each of the two new membranes with the label h, unchanged if no rule is applied to them (in particular, the contents of these neutral membranes is reproduced unchanged in these copies, providing that no rule is applied to their objects).
4. If at the same time a membrane h is divided by a rule of type (e) and there are objects in this membrane which evolve by means of rules of type (a), then in the new copies of the membrane we introduce the result of the evolution; that is, we may suppose that first the evolution rules of type (a) are used, changing the objects, and then the division is produced, so that in the two new membranes with label h we introduce copies of the changed objects. Of course, this process takes only one step. The same

assertions apply to the division by means of a rule of type (f): always we assume that the rules are applied "from bottom-up", in one step, but first the rules of the innermost region and then level by level until the region of the skin membrane.

5. The rules associated with a membrane h are used for all copies of this membrane, irrespective whether or not the membrane is an initial one or it is obtained by division. At one step, a membrane h can be the subject of only one rule of types (b) – (f).
6. The skin membrane can never divide. As any other membrane, the skin membrane can be "electrically charged".

The membrane structure of the system at a given time, together with all multisets of objects associated with the regions of this membrane structure is the *configuration* of the system at that time. The $(m+1)$-tuple $(\mu, w_1, \ldots, w_m)$ is the *initial configuration*. We can pass from a configuration to another one by using the rules from R according to the principles given above. We say that we have a (direct) *transition* among configurations.

Note that during a computation the number of membranes can increase and decrease but the labels of these membranes are always among the labels of membranes present in the initial configuration (by division we only produce membranes with the same label as the label of the divided membrane).

During a computation, objects can leave the skin membrane (by means of rules of type (c)). The result of a halting computation consists of the vector describing the multiplicity of all objects which are sent out of the system during the computation; a non-halting computation provides no output. The set of all vectors of natural numbers produced in this way (we say *computed*) by a system Π is denoted by $N(\Pi)$. The family of all such sets $N(\Pi)$ of vectors of natural numbers is denoted by NPA.

In [Pdivid] one proves that $PsRE = NPA$ and that the SAT problem (the satisfiability of propositional formulas in the conjunctive normal form) can be solved in linear time by systems as above (the time, consisting of parallel steps in our system, for solving a formula with n variables and m clauses is $2n + 2m + 1$). In the proofs of both these results from [Pdivid] rules of type (f) are used, but it was recently proved that this is not necessary.

First, in order to obtain the equality $PsRE = NPA$ it suffices to use rules of types (a) – (d) (hence no membrane division is necessary).

Secondly, a linear solution of SAT was obtained in [Pz3] by using only rules of types (a) – (e).

A generalization of the model above was considered in [Pkr]: when dividing a membrane, k new membranes can be produced, not necessarily two as in the basic variant. By using systems of this type, in [Pkr] it is proved that both the Hamiltonian Path Problem (HPP) and the Node Coloring Problem (NCP) can be solved in linear time, while in [Pkr2] it is shown that the Data Encryption Standard (DES) can be broken in linear time.

In some sense, this generalization is not necessary: it can proved that HPP can be solved in polynomial time by systems with 2-division of the form in

[P..], that is, using rules (a) – (f), or by systems which use rules (a) – (d), plus rules of type (e) which can also divide non-elementary membranes.

It is worth mentioning that in [Pz3] it is proved that HPP can be solved in linear time also by a P system which uses only rules of types (a) – (e), providing that the k-division of membranes is possible.

6 Using Cooperative Rules

Working with non-cooperative rules is natural from a mathematical point of view, because it is natural to look for *minimalistic* models, but from a biochemical point of view this is not only non-necessary, but also non-realistic: most chemical reactions involve two or more chemical compounds (and also produce two or more compounds). "Reactions" of the form $a \to bc$ corresponds to breaking a molecule a into two smaller molecules, b and c, but many reactions are of the form $ab \to cd$, where a and b interact in producing c and d.

The cooperative rules are very powerful in what it concerns the computing capacity of P systems, in the sense that by using such rules it is very easy to obtain computational universality results ([Pcell1], [Pcelletol], [PCCA]). We are not interested here in the computing capacity, but in the computing efficiency. Not entirely surprisingly, we find that using such rules we can solve many NP-complete problems in a rather uniform manner, that is, by P systems which are very similar to each other.

The general structure (and functioning) of these systems is as follows:

1. Always we start with two membranes, always the central one is divided into an exponential number of copies.
2. In a central "parallel engine" one generates, making use of the membrane division, a "data pool" of an exponential size; due to the parallelism, this takes, however, only a linear time. In parallel with this process, a "timer" is simultaneously tick-ing, in general, for synchronization reasons.
3. After finishing the generation of the "data pool", one checks whether or not any solution to our problem exists; this is the step where we need cooperative rules.
4. A message is sent out of the system at a precise moment telling whether or not the problem has a solution. In all cases, the last two steps are done in a constant number of time units, again making use of the parallelism.

Ten problems from [4] were solved in [P25] in this rather uniform way, five from logic and five from graph theory. For the reader convenience, we recall from [P25] the case of the SAT problem.

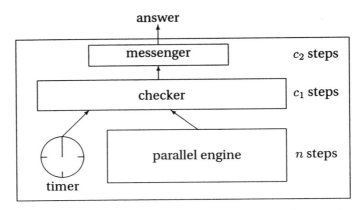

Figure 5: The shape of P systems solving NP-complete problems

INSTANCE: A collection C of clauses $C_i = y_{i,1} \vee \ldots \vee y_{i,p_i}$, $1 \leq i \leq m$, for some $m \geq 1, p_i \geq 1$, and $y_{i,j} \in \{x_k, \neg x_k \mid 1 \leq k \leq n\}, n \geq 1$, for each $1 \leq i \leq m, 1 \leq j \leq p_i$.

QUESTION: Is there a satisfying truth-assignment for C?

P SYSTEM:

$V = \{a_i, t_i, f_i \mid 1 \leq i \leq n\} \cup \{c_i \mid 1 \leq i \leq n+1\} \cup \{k_i \mid 1 \leq i \leq m\} \cup \{yes\}$,

$\mu = [_1 [_0 \ ]_0]_1$,

$w_0 = c_1 a_1 \ldots a_n$,

$R = \{[_0 a_i]_0 \to [_0 t_i]_0 [_0 f_i]_0 \mid 1 \leq i \leq n\}$ (*parallel engine*),

$\cup \ \{[_0 c_i \to c_{i+1}]_0 \mid 1 \leq i \leq n\}$ (*timer*)

$\cup \ \{[_0 c_{n+1} t_i \to k_1 t_i]_0 \mid x_i \text{ appears in } C_1, 1 \leq i \leq n\}$

$\cup \ \{[_0 c_{n+1} f_i \to k_1 f_i]_0 \mid \neg x_i \text{ appears in } C_1, 1 \leq i \leq n\}$

$\cup \ \{[_0 k_j t_i \to k_{j+1} t_i]_0 \mid x_i \text{ appears in } C_{j+1}, 1 \leq i \leq n, 1 \leq j \leq m-1\}$

$\cup \ \{[_0 k_j t_i \to k_{j+1} f_i]_0 \mid x_i \text{ appears in } C_{j+1}, 1 \leq i \leq n, 1 \leq j \leq m-1\}$

(*truth-checking rules*)

$\cup \ \{[_0 k_m]_0 \to [_0 \ ]_0 yes, \ [_1 yes]_1 \to [_1 \ ]_1 yes\}$ (*messenger rules*).

In n steps, we generate all 2^n truths-assignments of the n variables (a rule $[_0 a_i]_0 \to [_0 t_i]_0 [_0 f_i]_0$ divides the membrane with label 0; the object a_i is replaced in the obtained membranes by t_i, f_i, respectively, indicating the *true* and *false* values of variable x_i); at the same time, the counter c_i arrives at c_{n+1} and starts checking the truth of clauses; if a clause C_j is valid, then the object k_j is introduced. Checking all clauses takes m steps. (The checking rules are "localized", they are applied in the 2^n copies of the membrane with label 0; this is indicated by writing the rules in the form $[_0 \alpha \to \beta]_0$.) If in any copy of the membrane 0 we can obtain the object k_m, then the "message" *yes* is sent to membrane 1 and from here out of the system. This means that the set C of clauses is satisfiable if and only if at step $n+m+2$ we get the object *yes* outside the system.

Note that for each clause we have as many checking rules as many literals we have in the clause and that the cooperative rules always involve two objects in their left hand sides.

7 P Systems with Worm-Objects

In P systems with symbol-objects we work with multisets and the result of a computation is a natural number or a vector of natural numbers. It is also possible to consider P systems with string-objects, where we work with sets of strings (hence languages) and the result of a computation is a string. We can combine the two ideas: we can work with multisets of strings and consider as the result of a computation the number of strings present in the halting configuration in a given membrane. To this aim, we need operations with strings which can increase and decrease the number of occurrences of strings.

The following four operations were considered in [P1] (they are slight variants of the operations used in [10]):

1. *Replication.* If $a \in V$ and $u_1, u_2 \in V^+$, then $r : a \to u_1 \| u_2$ is called a *replication rule.* For strings $w_1, w_2, w_3 \in V^+$ we write $w_1 \Longrightarrow_r (w_2, w_3)$ (and we say that w_1 is replicated with respect to rule r) if $w_1 = x_1 a x_2$, $w_2 = x_1 u_1 x_2$, $w_3 = x_1 u_2 x_2$, for some $x_1, x_2 \in V^*$.
2. *Splitting.* If $a \in V$ and $u_1, u_2 \in V^+$, then $r : a \to u_1 | u_2$ is called a *splitting rule.* For strings $w_1, w_2, w_3 \in V^+$ we write $w_1 \Longrightarrow_r (w_2, w_3)$ (and we say that w_1 is splitted with respect to rule r) if $w_1 = x_1 a x_2$, $w_2 = x_1 u_1$, $w_3 = u_2 x_2$, for some $x_1, x_2 \in V^*$.
3. *Mutation.* A *mutation rule* is a context-free rewriting rule, $r : a \to u$, over V. For strings $w_1, w_2 \in V^+$ we write $w_1 \Longrightarrow_r w_2$ if $w_1 = x_1 a x_2$, $w_2 = x_1 u x_2$, for some $x_1, x_2 \in V^*$.
4. *Recombination.* Consider a string $z \in V^+$ (as a *crossing-over block*) and four strings $w_1, w_2, w_3, w_4 \in V^+$. We write $(w_1, w_2) \Longrightarrow_z (w_3, w_4)$ if $w_1 = x_1 z x_2$, $w_2 = y_1 z y_2$, and $w_3 = x_1 z y_2$, $w_4 = y_1 z x_2$, for some $x_1, x_2, y_1, y_2 \in V^*$.

Let us note that replication and splitting increase the number of strings, while mutation and recombination not; by sending strings out of the system, we can also decrease the number of objects.

A *P system* (of degree $m, m \geq 1$) *with worm-objects* is a construct

$$\Pi = (V, \mu, A_1, \ldots, A_m, (R_1, S_1, M_1, C_1), \ldots, (R_m, S_m, M_m, C_m), i_o),$$

where:

- V is an alphabet;

- μ is a membrane structure of degree m;
- $A_1, \ldots, A_m$ are multisets of finite support over V^*, associated with the regions of μ;
- for each $1 \leq i \leq m$, R_i, S_i, M_i, C_i are finite sets of replication rules, splitting rules, mutation rules, and crossing-over blocks, respectively, given in the following forms:
    a. replication rules: $(a \rightarrow u_1 \| u_2; tar_1, tar_2)$, for $tar_1, tar_2 \in \{here, out\} \cup \{in_j \mid 1 \leq j \leq m\}$;
    b. spliting rules: $(a \rightarrow u_1 | u_2; tar_1, tar_2)$, for $tar_1, tar_2 \in \{here, out\} \cup \{in_j \mid 1 \leq j \leq m\}$;
    c. mutation rules: $(a \rightarrow u; tar)$, for $tar \in \{here, out\} \cup \{in_j \mid 1 \leq j \leq m\}$;
    d. crossing-over blocks: $(z; tar_1, tar_2)$, for $tar_1, tar_2 \in \{here, out\} \cup \{in_j \mid 1 \leq j \leq m\}$;
- $i_o \in \{1, 2, \ldots, m\}$ specifies the *output membrane* of the system; it should be an elementary membrane of μ.

The $(n+1)$-tuple $(\mu, A_1, \ldots, A_m)$ constitutes the *initial configuration* of the system. By applying the operations defined by the components $(R_i, S_i, M_i, C_i), 1 \leq i \leq m$, we can define transitions from a configuration to another one. This is done as usual in P systems area, according to the following additional rules: A string which enters an operation is "consumed" by that operation, its multiplicity is decreased by one. The multiplicity of strings produced by an operation is accordingly increased. A string is processed by only one operation. For instance, we cannot apply two mutation rules, or a mutation rule and a replication one, to the same string.

The result of a halting computation consists of the number of strings in region i_o at the end of the computation. A non-halting computation provides no output. For a system Π, we denote by $N(\Pi)$ the set of numbers computed in this way. By $NWP_m, m \geq 1$, we denote the sets of numbers computed by all P systems with worm-objects with at most m membranes. The family of all recursively enumerable sets of natural numbers is denoted by nRE.

The proof of the following result can be found in [P13] (in [P1] one characterizes nRE without obtaining a bound on the number of membranes).

Theorem 4. $nRE = NWP_6$.

It is an *open problem* whether or not the bound 6 in this theorem can be improved; we expect a positive answer.

By using P systems with worm-objects we can again solve NP-complete problems in polynomial time. This was shown in [P1] for HPP; we recall some details.

Consider a directed graph $\gamma = (U, E)$ without cycles $(i, i), i \in U$, and two distinct nodes from U, i_{in} and i_{out}. Assume that U contains n nodes, identified with the numbers $1, 2, \ldots, n$ (hence the Hamiltonian paths will consist of $n-1$ arcs) and that the maximum outdegree of the graph (the number of arcs having the origin in a given node) is equal to k. By a simple renumbering, assume that $i_{in} = 1$ and that $i_{out} = n$.

We construct the P system Π_γ (of degree n), associated with γ, with the following components:

$V = \{\langle i,r \rangle, [i,r], \langle i,r; j_1, \ldots, j_s \rangle \mid 1 \leq i \leq n, 0 \leq r \leq k,$
$\quad \{j_1, \ldots, j_s\} \subseteq \{j \in U \mid (i,j) \in E\}\},$
$\mu = [_{n-1}[_n]_n[_{n-2} \cdots [_2[_0]_0]_2 \cdots]_{n-2}]_{n-1},$
$A_0 = \{(\langle 1,0 \rangle, 1)\};$ all other initial multisets are empty,
$R_0 = \{(\langle i,r \rangle \to [i,r] \langle j_1, r+1 \rangle || \langle i,r; j_2, \ldots, j_s \rangle; here, here),$
$\quad (\langle i,r; j_h, \ldots, j_s \rangle \to [i,r] \langle j_h, r+1 \rangle || \langle i,r; j_{h+1}, \ldots, j_s \rangle; here, here),$
$\quad (\langle i,r; j_{s-1}, j_s \rangle \to [i,r] \langle j_{s-1}, r+1 \rangle || [i,r] \langle j_s, r+1 \rangle; here, here) \mid$
$\quad$ for all $1 \leq i \leq n-1, 0 \leq r \leq n-2,$ where
$\quad j_1, \ldots, j_s$ are all the nodes such that $(i, j_l) \in E, 1 \leq l \leq s\},$
$M_0 = \{(\langle n, n-1 \rangle \to [n, n-1]; out)\}, S_0 = C_0 = \emptyset,$
$R_i = S_i = C_i = \emptyset,$ for all $2 \leq i \leq n-1,$
$M_i = \{([i,j] \to [i,j]; out) \mid 1 \leq j \leq n-1\},$ for all $2 \leq i \leq n-2,$
$M_{n-1} = \{([n, n-1] \to [n, n-1]; in_n)\},$
$R_n = S_n = M_n = C_n = \emptyset.$

The membrane structure of this system is given in Figure 6. We start from the unique object $\langle 1, 0 \rangle$, present in the inner membrane, that with the label 0, by repeatedly using replication rules. These rules always prolong a string which represents a path in the graph starting from node 1. No path can be continued if either it reaches node n or we have already made $n - 1$ steps (hence the path already passes through n nodes and any further step will surely repeat a node). In this way, we can generate all paths in the graph γ starting in node 1 and of length (as the number of arcs) at most $n - 1$.

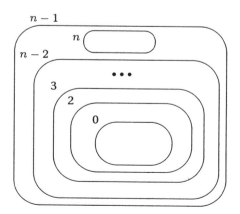

Figure 6: The membrane structure for HPP

Only strings which end with $\langle n, n-1 \rangle$ can be sent outside membrane 0.

In each membrane $i, 2 \leq i \leq n-2$, a string can be only processed if it contains a symbol of the form $[i, t]$, for some $1 \leq t \leq n-1$. (This means that node i was visited at time t.) If this is the case, then the string is sent unmodified to the next membrane. If a string reaches the skin membrane, then it can be sent to the output membrane if and only if it contains the symbol $[n, n-1]$ (this means that the corresponding path ends in node n).

It is easy to see that we have an answer whether or not the Hamiltonian Path Problem for γ has a solution after at most $n^2 - n - 1$ steps performed by the P system Π_γ.

The two phases of our computation, generating all candidate paths (in membrane 0) and then checking whether or not at least a path is Hamiltonian (in membranes $2, 3, \ldots, n-1$), are similar to the two phases of Adleman's algorithm [1], with the difference that we need a quadratic time for producing the candidate solutions (Adleman performs this in a constant parallel biochemical time); however, we grow only candidate solutions of a prescribed length.

Now, we can proceed as Lipton [6], extending the previous procedure to the SAT problem. Because the truth-assignments to n variables $x_1, \ldots, x_n$ correspond to the paths from a given node to another given node in a graph with outdegree 2 (see [6]), the solution is obtained in linear time: n steps for generating all truths-assignments and m steps to check the truth value of each of the m clauses.

A variant of P systems with string-objects, working in a quite different manner, was considered in [Pmm], under the name of P systems with valuations. (The string-objects are processed by local mutations, insertion, and deletion of single symbols, and have associated valuations, integer values computed by a morphism into the set of integers; the sign of the valuation of a string determines the communication of that string from a membrane to another one. Further details can be found in the mentioned paper.) It was shown in [Pmm] that the 3-Colorability Problem and and the Bounded Post Correspondence Problem (both known to be NP-complete) can be solved in a linear time by such systems.

Note that in both cases discussed in this section one use no membrane division, the exponential workspace is obtained by handling the number of string-objects in a membrane structure which can at most dissolve membranes (this is the case of systems in [Pmm]).

8 Concluding Remarks

The variants and the results considered in this paper prove, on the one hand, the versatility of the P system framework, which is a general environment for devising computing models and algorithms dealing with multisets in a distributed, parallel, nondeterministic manner, and, on the other hand, the usefulness of these systems, at least from a theoretical point of view. In a quite natural manner, one can devise polynomial (often, linear) algorithms for many NP-complete problems, in certain cases (for instance, when cooperative rules are used) of a rather uniform shape.

There are many results in P systems area, other than referring to characterizations of the power of Turing machines and to solving hard problems in a feasible (parallel) time. The bibliography which follows is meant to help the reader in having an image about the current state of the domain. On the other hand, many open problems and research topics wait for further investigations. Technical open problems can be found in most of the papers available up to now, while [P21] is entirely devoted to listing some research directions as they have appeared of interest in January 2000.

Of course, the main research topic concerns the implementation of P systems, in bio-media or on an electronic support, but, we have also said it before, this is a long range interdisciplinary task.

Bibliography of P systems (August 2000)

[P1] A. Atanasiu, Arithmetic with membranes, *Pre-proc. of Workshop on Multiset Processing*, Curtea de Argeş, Romania, 2000.

[P2] A. V. Baranda, J. Castellanos, F. Arroyo, R. Gonzalo, Data structures for implementing P systems in silico, *Pre-proc. of Workshop on Multiset Processing*, Curtea de Argeş, Romania, 2000.

[P3] C. Calude, Gh. Păun, *Computing with Cells and Atoms*, Taylor and Francis, London, 2000 (Chapter 3: "Computing with Membranes").

[P4] J. Castellanos, Gh. Păun, A. Rodriguez-Paton, Computing with membranes: P systems with worm-objects, *IEEE Conf. SPIRE 2000*, La Coruña, Spain, and *Auckland University, CDMTCS Report* No 123, 2000 (www.cs.auckland.ac.nz/CDMTCS).

[P5] J. Dassow, Gh. Păun, On the power of membrane computing, *J. of Universal Computer Sci.*, 5, 2 (1999), 33–49 (www.iicm.edu/jucs).

[P6] J. Dassow, Gh. Păun, Concentration controlled P systems, submitted, 1999.

[P7] R. Freund, Generalized P systems, *Fundamentals of Computation Theory, FCT'99*, Iaşi, 1999 (G. Ciobanu, Gh. Păun, eds.), *LNCS* 1684, Springer, 1999, 281–292.

[P8] R. Freund, Generalized P systems with splicing and cutting/recombination, *Workshop on Formal Languages*, FCT99, Iaşi, 1999.

[P9] R. Freund, F. Freund, Molecular computing with generalized homogeneous P systems, *Proc. Conf. DNA6* (A. Condon, G. Rozenberg, eds.), Leiden, 2000, 113–125.

[P10] P. Frisco, Membrane computing based on splicing: improvements, *Pre-proc. of Workshop on Multiset Processing*, Curtea de Argeş, Romania, 2000.

[P11] M. Ito, C. Martin-Vide, Gh. Păun, A characterization of Parikh sets of ET0L languages in terms of P systems, submitted, 2000.

[P12] S. N. Krishna, Computing with simple P systems, *Pre-proc. Workshop on Multiset Processing*, Curtea de Argeş, Romania, 2000.

[P13] S. N. Krishna, R. Rama, A variant of P systems with active membranes: Solving NP-complete problems, *Romanian J. of Information Science and Technology*, 2, 4 (1999), 357–367.

[P14] S. N. Krishna, R. Rama, On the power of P systems with sequential and parallel rewriting, *Intern. J. Computer Math.*, 77, 1-2 (2000), 1–14.

[P15] S. N. Krishna, R. Rama, Computing with P systems, submitted, 2000.

[P16] S. N. Krishna, R. Rama, On simple P systems with external output, submitted, 2000.
[P17] M. Maliţa, Membrane computing in Prolog, *Pre-proc. of Workshop on Multiset Processing*, Curtea de Argeş, Romania, 2000.
[P18] C. Martin-Vide, V. Mitrana, P systems with valuations, submitted, 2000.
[P19] C. Martin-Vide, Gh. Păun, Computing with membranes. One more collapsing hierarchy, *Bulletin of the EATCS*, to appear.
[P20] C. Martin-Vide, Gh. Păun, String objects in P systems, *Proc. of Algebraic Systems, Formal Languages and Computations Workshop*, Kyoto, 2000, RIMS Kokyuroku Series, Kyoto Univ., 2000.
[P21] C. Martin-Vide, Gh. Păun, G. Rozenberg, Plasmid-based P systems, submitted, 2000.
[P22] A. Păun, On P systems with membrane division, submitted, 2000.
[P23] A. Păun, M. Păun, On the membrane computing based on splicing, submitted, 2000.
[P24] Gh. Păun, Computing with membranes, *Journal of Computer and System Sciences*, 61 (2000), in press, and *Turku Center for Computer Science-TUCS Report* No 208, 1998 (www.tucs.fi).
[P25] Gh. Păun, Computing with membranes. An introduction, *Bulletin of the EATCS*, 67 (Febr. 1999), 139–152.
[P26] Gh. Păun, Computing with membranes – A variant: P systems with polarized membranes, *Intern. J. of Foundations of Computer Science*, 11, 1 (2000), 167–182.
[P27] Gh. Păun, P systems with active membranes: Attacking NP complete problems, *J. Automata, Languages, and Combinatorics*, 5 (2000), in press, and *Auckland University, CDMTCS Report* No 102, 1999 (www.cs.auckland.ac.nz/CDMTCS).
[P28] Computing with membranes (P systems): Twenty six research topics, *Auckland University, CDMTCS Report* No 119, 2000 (www.cs.auckland.ac.nz/CDMTCS).
[P29] Gh. Păun, G. Rozenberg, A. Salomaa, Membrane computing with external output, *Fundamenta Informaticae*, 41, 3 (2000), 259–266.
[P30] Gh. Păun, Y. Sakakibara, T. Yokomori, P systems on graphs of restricted forms, submitted, 1999.
[P31] Gh. Păun, Y. Suzuki, H. Tanaka, P Systems with energy accounting, submitted, 2000.
[P32] Gh. Păun, Y. Suzuki, H. Tanaka, T. Yokomori, On the power of membrane division in P systems, *Proc. Words, Semigroups, Combinatorics Conf.*, Kyoto, 2000.
[P33] Gh. Păun, G. Thierrin, Multiset processing by means of systems of finite state transducers, *Pre-Proc. of Workshop on Implementing Automata* WIA99, Potsdam, August 1999, Preprint 5/1999 of Univ. Potsdam (O. Boldt, H. Jürgensen, L. Robbins, eds.) XV 1-17.
[P34] Gh. Păun, T. Yokomori, Membrane computing based on splicing, *Preliminary Proc. of Fifth Intern. Meeting on DNA Based Computers* (E. Winfree, D. Gifford, eds.), MIT, June 1999, 213–227.
[P35] Gh. Păun, T. Yokomori, Simulating H systems by P systems, *Journal of Universal Computer Science*, 6, 2 (2000), 178–193 (www.iicm.edu/jucs).
[P36] Gh. Păun, S. Yu, On synchronization in P systems, *Fundamenta Informaticae*, 38, 4 (1999), 397–410.
[P37] I. Petre, A normal form for P systems, *Bulletin of the EATCS*, 67 (Febr. 1999), 165–172.
[P38] I. Petre, L. Petre, Mobile ambients and P systems, *J. Universal Computer Sci.*, 5, 9 (1999), 588–598.

[P39] Y. Suzuki, H. Tanaka, On a LISP implementation of a class of P systems, *Romanian J. of Information Science and Technology*, 3, 2 (2000).
[P40] Y. Suzuki, H. Tanaka, Artificial life and P systems, *Pre-proc. Workshop on Multiset Processing*, Curtea de Argeş, Romania, 2000.
[P41] Cl. Zandron, Cl. Ferretti, G. Mauri, Two normal forms for rewriting P systems, submitted, 2000.
[P42] Cl. Zandron, Cl. Ferretti, G. Mauri, Priorities and variable thickness of membranes in rewriting P systems, submitted, 2000.
[P43] Cl. Zandron, Cl. Ferretti, G. Mauri, Solving NP-complete problems using P systems with active membranes, submitted, 2000.
[P44] Cl. Zandron, Cl. Ferretti, G. Mauri, Using membrane features in P systems, *Pre-proc. of Workshop on Multiset Processing*, Curtea de Argeş, Romania, 2000.

References

1. L. M. Adleman, Molecular computation of solutions to combinatorial problems, *Science*, 226 (November 1994), 1021–1024.
2. G. Berry, G. Boudol, The chemical abstract machine, *Theoretical Computer Sci.*, 96 (1992), 217–248.
3. D. Bray, Protein molecules as computational elements in living cells, *Nature*, 376 (1995), 307–312
4. M. R. Garey, D. J. Johnson, *Computers and Intractability. A Guide to the Theory of NP-Completeness*, W. H. Freeman and Comp., San Francisco, 1979.
5. T. Head, Formal language theory and DNA: An analysis of the generative capacity of specific recombinant behaviors, *Bulletin of Mathematical Biology*, 49 (1987), 737–759.
6. R. J. Lipton, Using DNA to solve NP-complete problems, *Science*, 268 (April 1995), 542–545.
7. W. R. Loewenstein, *The Touchstone of Life. Molecular Information, Cell Communication, and the Foundations of Life*, Oxford Univ. Press, New York, Oxford, 1999.
8. Gh. Păun, G. Rozenberg, A. Salomaa, *DNA Computing. New Computing Paradigms*, Springer-Verlag, Berlin, 1998.
9. G. Rozenberg, A. Salomaa, eds., *Handbook of Formal Languages*, 3 volumes, Springer-Verlag, Berlin, 1997.
10. M. Sipper, Studying Artificial Life using a simple, general cellular model, *Artificial Life Journal*, 2, 1 (1995), 1–35.

DNA Processing in Ciliates – The Wonders of DNA Computing *in vivo*

Grzegorz Rozenberg

[1] Leiden Center for Natural Computing, Leiden University, The Netherlands
[2] Department of Computer Science, University of Colorado, Boulder, U.S.A.

Natural Computing is computing inspired by (or gleaned from) nature. During the last decade the area of Natural Computing became one of the most active areas of computer science, and without doubt it will become central for computing in the coming years. Characteristic for Natural Computing is the metaphorical use of concepts, principles and mechanisms underlying natural systems. Thus, e.g., evolutionary algorithms use the concepts of mutation, recombination, and natural selection from biology, while neural networks are inspired by the highly interconnected neural structures in the brain and nervous system. In both areas, the algorithms are implemented on traditional silicon-based computers. On the other hand, DNA Computing (or more generally, Molecular Computing) is much bolder: it is using novel paradigms (coming from molecular biology), *but* it also aims at the implementation of algorithms based on these paradigms in biological hardware (bioware).

Although DNA Computing (see, e.g., [1] and [8]) is centered around Computer Science, it is a genuinely interdisciplinary research area, where computer scientists cooperate with biologists, chemists, crystallographers, ... to develop novel, "molecular" methods of computing. It is concerned with computational properties of (bio)molecules. Its attractiveness for computing stems from the fact that it takes the science of computing into the world of nanotechnology, and it provides massive parallelism of astounding scale.

One can distinguish two strands of research on DNA Computing: DNA Computing *in vitro* and DNA Computing *in vivo*. Most of the current research in DNA Computing falls in the former category, where the computations are performed in a test tube, outside of living cells. A typical computation scheme in DNA Computing *in vivo* consists of injecting into a living cell some "input", and then by observing the behavior of the cell (e.g., the change of some metabolic processes) one "reads out" the output of the computation. Another branch of DNA Computing *in vivo* studies computational processes *actually taking place* in the living cells. Such a study provides one with an insight into the computational properties of biomolecules in their favorite environment - a living cell. One hopes that this insight may lead to the harnessing of the computational processes in living cells for the sake of computing (for humans).

In particular, DNA processing in ciliates has attracted the attention of the DNA Computing community (see, e.g., [6], [7], [2], [3], [4], [9]). Ciliates are a very ancient group of organisms (their origin is estimated at about 2×10^9

years ago). They comprise tens of thousands of genetically different organisms. Through the evolution, ciliates have developed nuclear dualism: they have two kinds of nuclei - a micronucleus and a macronucleus. Micronucleus is a sort of the storage nucleus where the genome is stored until it is needed in the process of sexual reproduction. Macronucleus is the "expression nucleus", and it provides the RNA transcripts needed to operate the cell. When starved, ciliates proceed to sexual reproduction, and during this process the micronuclear genome is transformed into its macronuclear form. This transformation process turns out to be fascinating, also from the computational point of view.

The form of a gene in the macronuclear genome is different (and it can be *drastically* different) from its form in the micronuclear genome (see, e.g., [4] and [5]). Roughly speaking, one can see the macronuclear gene as consisting of a finite number of consecutive overlapping sequences $M_1, \ldots, M_n$ (called MDSs). Then, the micronuclear gene is a permutation $\rho = M_{i_1}, \ldots, M_{i_n}$ of these MDSs, with some of them possibly inverted, where any two consecutive MDSs in ρ are divided by "padding sequences" called IESs. The process of assembly of the macronuclear gene from its micronuclear form consists then of cutting out the padding IESs, and splicing the MDSs in their orthodox order $M_1, \ldots, M_n$ suitable for transcription.

We investigate the (macronuclear) gene assembly from the operational point of view. To this aim, we consider three molecular operations (see [3] and [7]): *ld-excision, hi-excision/reinsertion,* and *dlad-excision/reinsertion* which form the basis for the computational framework for gene assembly. This set of operations turns out to be assembly-universal in the sense that, using these operations any micronuclear form of a gene can be transformed into its macronuclear form (see [2]).

We propose two formal systems for the investigation of gene assembly: one based on strings (string reduction system) and one based on graphs (graph reduction system). We demonstrate that although the graph reduction system is more abstract than the string reduction system, the assembly strategies can be translated from one system to the other.

The study of gene assembly in ciliates has certainly deepened our understanding of computing with biomolecules. It has also confirmed our deep conviction that complex biological phenomena are based on computational processes. Learning about the nature of these computations may substantially change the way we compute, and the way we reason about computation.

References

1. Condon, A., and Rozenberg, G. (eds.), *Proceedings of the Sixth International Meeting on DNA Based Computers*, Lecture Notes in Computer Science, Springer-Verlag (2000).
2. Ehrenfeucht, A., Petre, I., Prescott, D.M., and Rozenberg, G., Universal and simple operations for gene assembly in ciliates, to appear in *Words, Sequences, Languages: Where computer science, biology and linguistics come across*, Mitrana, V., Martin-Vide, C. (eds.) (2000).

3. Ehrenfeucht, A., Petre, I., Prescott, D.M., and Rozenberg, G., String and graph reduction systems for gene assembly in ciliates, manuscript (2000).
4. Ehrenfeucht, A., Prescott, D.M., and Rozenberg, G., Computational aspects of gene (un)scrambling in ciliates. In *Evolution as Computation*, Landweber, L., Winfree, E. (eds), 45-86, Springer-Verlag, Berlin, Heidelberg (2000).
5. Ehrenfeucht, A., Harju, T., Petre, I., Prescott, D.M., and Rozenberg, G., Formal systems for gene assembly in ciliates, manuscript (2000).
6. Landweber, L.F., and Kari, L., The evolution of cellular computing: nature's solution to a computational problem. *Proceedings of the 4th DIMACS meeting on DNA based computers*, Philadelphia 3-15 (1998).
7. Landweber, L.F., and Kari, L., Universal molecular computation in ciliates. In *Evolution as Computation*, Landweber, L., Winfree, E. (eds), Springer-Verlag, Berlin, Heidelberg, to appear (2000).
8. Paun, G., Rozenberg, G., and Salomaa, A., *DNA Computing*, Springer-Verlag, Berlin, Heidelberg (1998).
9. Prescott, D.M., Ehrenfeucht, A., and Rozenberg, G., Molecular operations for DNA processing in Hypotrichous ciliates, manuscript (1999).

Macroscopic Molecular Computation with Gene Networks

Hava T. Siegelmann and Asa Ben-Hur

Faculty of Industrial Engineering Technion, Haifa, Israel

In recent years scientists have been looking for new paradigms for constructing computational devices. These include quantum computation, DNA computation, neural networks, neuromorphic engineering and other analog VLSI devices. Since the 60's genetic regulatory systems are thought of as "circuits" or "networks" of interacting components. The genetic material is the "program" that guides protein production in a cell. Protein levels determine the evolution of the network at subsequent times, and thus serve as its "memory". This analogy between computing and the process of gene expression was pointed out in various papers. Bray suggests that protein based circuits are the device by which unicellular organisms react to their environment, instead of a nervous system. However, until recently this was only a useful metaphor for describing gene networks. Recent papers describe the successful fabrication of synthetic networks, i.e. *programming* of a gene network. Furthermore, it was shown both theoretically and experimentally that chemical reactions can be used to implement Boolean logic and neural networks.

Protein concentrations are continuous variables that evolve continuously in time. Moreover, biological systems do not have timing devices, so a description in terms of a *map* that simultaneously updates the system variables is inadequate. It is thus necessary to model gene networks by differential equations, and consider them as analog computers. The particular model of genetic networks we consider here (originally proposed as a model of biological oscillations) assumes switch-like behavior, so that protein concentrations are described by piecewise linear equations.

Computation with biological hardware is also the issue in the field of DNA computation. As a particular example, the guided homologous recombination that takes place during gene rearrangement in ciliates was interpreted as a process of computation. This process, and DNA computation in general, are symbolic, and describe computation at the molecular level, whereas gene networks are analog representations of the macroscopic evolution of protein levels in a cell.

We present a new paradigm based on genetic regulatory networks. Using a model of gene expression by piecewise linear differential equations we show that the evolution of protein concentrations is a cell can be considered a process of computation. This is demonstrated by showing that this model can simulate memory bounded Turing machines.

The relation between digital models of computation and analog models is explored in a recent book (H.T. Siegelmann. *Neural Networks and Analog*

Computation: Beyond the Turing Limit, Springer Verlag, 1999), mainly from the perspective of neural networks. It is shown there that analog models are potentially stronger than digital ones, assuming an ideal noiseless environment. In this talk, on the other hand, we consider the possibility of noise and propose a design principle which makes the model equations robust. We found that the recently proposed gene network follows this principle. It seems that the "theoretical design of complex and practical gene networks is a realistic and achievable goal."

In Vitro Transcriptional Circuits

Erik Winfree

Depts. of CS and CNS, Caltech, Pasadena, U.S.A.

The structural similarity of neural networks and genetic regulatory networks to digital circuits, and hence to each other, was noted from the very beginning of their study [7, 5]. Generic properties of both types of network have been analyzed using the same class of abstract models [3]. The same rate equations proposed for recurrent neural networks [4] have been used, with a few embellishments, to model genetic regulatory circuits controlling development of Drosophila embryos [6]. However, whereas research in neural networks has a long history of both analytic approaches, aimed at understanding the brain, and synthetic approaches, aimed at creating intelligent machines, research in genetic regulatory networks has been entirely analytic prior to very recent work implementing small synthetic genetic regulatory networks in E. Coli [8, 1, 2]. Although genetic regulatory systems are not fast, they are small; E. coli is one cubic micron yet it contains thousands of regulated genes.

Our goal has been to find the simplest possible biochemical system that mimics the architecture of genetic regulation, and to develop an in-vitro implementation where arbitrary circuits can be created. We consider only two enzymes in addition to DNA and RNA molecules: T7 RNA polymerase (RNAP) and a ribonuclease. RNAP transcribes synthetic DNA templates to produce RNA transcripts, which in turn function as "binding proteins" to regulate transcription of other DNA templates. Regulation of transcription is controlled by binding of the RNA to the DNA template, changing its secondary structure to expose or hide the promoter sequence; only templates with exposed promoter sequences will be transcribed and produce RNA.

Computational networks consist of several mutually interacting species in the same reaction. Let R_i be the ith RNA species, D_{ij} be the DNA template that produces RNA species i in response to RNA species j, and $\sigma(x)$ be the transcription rate as a function of the switching species. Then, as an approximation valid within specified circumstances,

$$\frac{d}{dt}[R_j] = -[R_j] + \sum_i [D_{ij}]\sigma([R_j]).$$

The first term accounts for ribonuclease degradation of the RNA species, while the second term sums all transcription from templates that produce RNA species j. The behavior of this network is programmed by choice of the concentrations $[D_{ij}]$, which are formally analogous to the choice of synaptic weights in a fully connected neural network.

The simplest non-trivial network is a ring oscillator: it consists of three DNA templates, each producing a repressor against the next template in the cycle. Here we report simulations of the full chemical kinetics that suggest that multiple cycles of the oscillator should be observed in a standard transcription reaction, prior to depletion of the rNTP energy source. Similarly, small associative networks and feedforward networks are shown by simulation to be feasible.

References

1. Michael B. Elowitz and Stanislas Leibler, *A synthetic oscillatory network of transcriptional regulators*, Nature **403** (2000), 335–338.
2. Timothy S. Gardner, Charles R. Cantor, and James J. Collins, *Construction of a genetic toggle switch in escherichia coli*, Nature **403** (2000), 339–342.
3. Leon Glass and Stuart A. Kauffman, *The logical analysis of continuous, non-linear biochemical control networks*, Journal of Theoretical Biology **39** (1973), 103–129.
4. John J. Hopfield, *Neurons with graded response have collective computational properties like those of two-state neurons*, Proc. Nat. Acad. Sci. USA **81** (1984), no. 10, 3088–3092.
5. W. S. McCulloch and W. Pitts, *A logical calculus of the ideas immanent in nervous activity*, Bull. Math. Biophys. **5** (1943), 115.
6. E. Mjolsness, D. H. Sharp, and J. Reinitz, *A connectionist model of development*, Journal of Theoretical Biology **152** (1991), 429–453.
7. J. Monod and F. Jacob, *General conclusions: teleonomic mechanisms in cellular metabolism, growth and differentiation*, Cold Spring Harb. Symp. Quant. Biol. **26** (1961), 389–401.
8. Ron Weiss, George E. Homsy, and Thomas F. Knight, Jr., *Toward* in vivo *digital circuits*, Proceedings of the DIMACS Workshop on Evolution as Computation, held at Princeton, January 11-12, 1999 (Laura F. Landweber and Erik Winfree, eds.), Springer-Verlag, to appear.

Parallelizing with Limited Number of Ancillae

Hideaki Abe and Shao Chin Sung

School of Information Science, Japan Advanced Institute of Science and Technology, Ishikawa, Japan

Abstract. In this paper, parallelization methods on a condition are proposed for three types of quantum circuits, where the condition is that the number of available ancillae is limited and parallelization means that a given quantum circuit is reconstructed as one with smaller depth. As a by-product, for the three types of n-input quantum circuits, upper bounds on the number of ancillae for parallelizing to logarithmic depth are reduced to $1/\log n$ of the previous upper bounds.

1 Introduction

Quantum circuits are proposed as a parallel model of quantum computation by Deutsch [3], in which computing devices, called *quantum gates*, are connected acyclicly. When quantum circuits are implemented it is not realistic to implement all quantum gates which can be considered. In order to realize an arbitrary quantum computation on quantum circuits, the concept of universality of quantum gates is discussed in the literature [1–6]. One of the best known results shown by Barenco *et al* [2]. They showed that the set consisting of all one-input one-output quantum gates and a two-input two-output quantum gate, called *controlled-not gate*, are universal. Here we are concerned with parallelization for three types of quantum circuits which are quantum circuits consisting of controlled-not gates, quantum circuits consisting of controlled-not gates and a type of one-input one-output quantum gates, called *phase-shift gates*, and quantum circuits consisting of controlled-not gates and a type of one-input one-output quantum gates, called *Walsh-Hadamard* gates.

As a previous work, Moore and Nilsson [7] showed that the three types of quantum circuits can be parallelizing to logarithmic depth by using a certain number of ancillae. We parallelize these three types of quantum circuits where the number of available ancillae is less than used by Moore and Nilsson [7]. As a by-product, for the three types of n-input quantum circuits, upper bounds on the number of ancillae for parallelizing to logarithmic depth are reduced to $1/\log n$ of the upper bounds shown by Moore and Nilsson [7].

2 Preliminaries

Let us introduce some terminology on quantum circuits (see [3, 8] for details). Let

$$|0\rangle = \begin{pmatrix} 1 \\ 0 \end{pmatrix} \text{ and } |1\rangle = \begin{pmatrix} 0 \\ 1 \end{pmatrix}.$$

For $x = (x_1, x_2, \ldots, x_n) \in \{0,1\}^n$ with $n \geq 1$, let

$$|x\rangle = |x_1, x_2, \ldots, x_n\rangle = |x_1\rangle \otimes |x_2\rangle \otimes \cdots \otimes |x_n\rangle,$$

where $\otimes$ is the tensor product operation. Notice that $|x\rangle$ is the unit vector of length 2^n for each $x \in \{0,1\}^n$. A matrix U is called *unitary* if $UU^\dagger = U^\dagger U = I$, where $U^\dagger$ is the transposed conjugate of U, and I is a unit matrix.

In quantum computers, a bit is represented by a two-state physical system, and is called a *qubit*. For $n \geq 1$, a state $|\psi\rangle$ of an n-qubit is represented as a superposition of $|x\rangle$s with $x \in \{0,1\}^n$, i.e.,

$$|\psi\rangle = \sum_{x \in \{0,1\}^n} a_x |x\rangle,$$

where all a_xs are complex numbers satisfying $\sum_{x \in \{0,1\}^n} |a_x|^2 = 1$. Each a_x is called the *amplitude* of $|x\rangle$ in $|\psi\rangle$, and b_x, which satisfies $a_x = |a_x|e^{ib_x}$, is called the *phase* of $|x\rangle$ in state $|\psi\rangle$.

A *quantum circuit* is a directed network connecting quantum gates acyclicly. A quantum gate has the same number of inputs and outputs. A k-input k-output quantum gate G is specified by a $2^k \times 2^k$ unitary matrix $U_G = [u_{xy}]$ for $x, y \in \{0,1\}^k$ and realizes a mapping of states of its inputs to states of its outputs as follows:

$$|x\rangle \mapsto \sum_{y \in \{0,1\}^k} u_{xy} |y\rangle.$$

Here, we consider three types of quantum gates, which are *phase-shift gates*, *Walsh-Hadamard gates*, and *controlled-not gates* and are respectively specified by unitary matrices PS$_\theta$ for $\theta : \{0,1\} \to [0, 2\pi)$, WH, and CN, which are defined as follows:

$$\text{PS}_\theta = \begin{pmatrix} e^{i\theta(0)} & 0 \\ 0 & e^{i\theta(1)} \end{pmatrix}, \quad \text{WH} = \frac{1}{\sqrt{2}} \begin{pmatrix} 1 & 1 \\ 1 & -1 \end{pmatrix}, \quad \text{CN} = \begin{pmatrix} 1 & 0 & 0 & 0 \\ 0 & 1 & 0 & 0 \\ 0 & 0 & 0 & 1 \\ 0 & 0 & 1 & 0 \end{pmatrix}.$$

A PS gate PS$_\theta$ and a WH gate realize the mappings $|x\rangle \mapsto e^{i\theta(x)} |x\rangle$ and $|x\rangle \mapsto |0\rangle + (-1)^x |1\rangle$ for $x, y \in \{0,1\}$, respectively. Applying these gates on the j-th qubit of the n-qubit, the following mappings are respectively realized:

$$|x_1, x_2, \ldots, x_n\rangle \mapsto e^{i\theta(x_j)} |x_1, x_2, \ldots, x_n\rangle$$

and

$$|x_1, x_2, \ldots, x_n\rangle \mapsto \frac{1}{\sqrt{2}} \sum_{y \in \{0,1\}} (-1)^{yx_j} |x_1, \ldots, x_{j-1}, y, x_{j+1}, \ldots, x_n\rangle.$$

A CN gate realizes the mapping $|x_1, x_2\rangle \mapsto |x_1, x_1 \oplus x_2\rangle$. The first input of a CN gate is called the *control-bit*, and the second input of a CN gate is called the *target-bit.* Applying it on the n-qubit for which the j-th qubit is its first input and the k-th qubit as its second input, the following mapping is realized:

$$|x_1, x_2, \ldots, x_n\rangle \mapsto |x_1, \ldots, x_{k-1}, x_k \oplus x_j, x_{k+1}, \ldots, x_n\rangle.$$

The *depth* of a quantum circuit is the length (i.e., number of quantum gates) of the longest directed path in it. *Ancilla* is a qubit which is in state $|0\rangle$ at the beginning and the end of computation. We say that for some $\gamma \geq 0$ and $x, y \in \{0,1\}^n$, a quantum circuit with γ ancillae *realizes* a mapping

$$|x\rangle \mapsto \sum_{y \in \{0,1\}^n} \alpha_y |y\rangle,$$

if it realizes the mapping

$$|x, 0^\gamma\rangle \mapsto \sum_{y \in \{0,1\}^n} \alpha_y |y, 0^\gamma\rangle.$$

3 Main Results

We obtain the following results. Let γ be the number of available ancillae.

- Quantum circuits consisting of CN gates can be parallelized to

$$O\left(\frac{n^2}{\gamma} + \log \frac{\gamma}{n}\right)$$

depth.
- Quantum circuits consisting of CN gates and PS gates can be parallelized to

$$O\left(\frac{n^2 + nm}{\gamma} + \log \frac{\gamma}{n} + \log \frac{\gamma}{m}\right)$$

depth, where m is the number of PS gates in the quantum circuit to be parallelized.
- Quantum circuits consisting of CN gates and WH gates can be parallelized to

$$O\left(\frac{n^2}{\gamma} + \log \frac{\gamma}{n}\right)$$

depth.

From these results, we have the following upper bounds of number of available ancillae for obtaining a quantum circuit with logarithmic depth. These upper bounds are $1/\log n$ of the bounds shown by Moore and Nilsson [7].

- Quantum circuits consisting of CN gates can be parallelized to $O(\log n)$ depth if $n^2/\log n$ ancillae are available.
- Quantum circuits consisting of CN gates and m PS gates can be parallelized to $O(\log n + \log m)$ depth if $(n^2 + nm)/\log n$ ancillae are available.
- Quantum circuits consisting of CN gates and WH gates can be parallelized to $O(\log n)$ depth if $n^2/\log n$ ancillae are available.

4 Parallelizing Quantum Circuits Consisting of CN Gates

Recall that a CN gate on a n-qubit with the j-th qubit as its control-bit and the k-th qubit as its target-bit realizes the mapping

$$|x_1, x_2, \ldots, x_n\rangle \mapsto |x_1, \ldots, x_{k-1}, x_k \oplus x_j, x_{k+1}, \ldots, x_n\rangle.$$

It follows that a quantum circuit consisting of CN gates realizes a mapping

$$|x_1, x_2, \ldots, x_n\rangle \mapsto \left| \bigoplus_{j=1}^n x_j a_{j1}, \bigoplus_{j=1}^n x_j a_{j2}, \ldots, \bigoplus_{j=1}^n x_j a_{jn} \right\rangle$$

for some $n \times n$ 0-1 matrix $A = [a_{jk}]$, i.e., the mapping realized by the quantum circuit can be specified by A. By applying one more CN gate, the matrix which specifies the resulting mapping can be obtained by adding one row to another row of A over mod 2. Notice that the rank of such a matrix is the same as the rank of A. Also notice that A is a unit matrix if the quantum circuit consists of no gate. Therefore, a mapping can be realized by a quantum circuit consisting of CN gates if and only if the mapping can be specified by an $n \times n$ 0-1 matrix with rank n. It follows that, by applying the Gauss-Jordan elimination method, every quantum circuit consisting of CN gates can be parallelized to linear depth without using any ancillae.

Before explaining how we can parallelize quantum circuits consisting of CN gates by using ancillae, let us consider a restricted class of mappings which can be realized by quantum circuits consisting of CN gates. Let $\text{CN}(n, m)$ be a set of mappings over states of an $(n + m)$-qubit such that each mapping of $\text{CN}(n, m)$ can be specified by an $n \times m$ 0-1 matrix $A = [a_{jk}]$ such that

$$|x_1, x_2, \ldots, x_n, y_1, y_2, \ldots, y_m\rangle \mapsto |x_1, x_2, \ldots, x_n, z_1, z_2, \ldots, z_m\rangle,$$

where $z_k = y_k \oplus (\bigoplus_{j=1}^n x_j a_{jk})$ for each $1 \le k \le m$. Every mapping of $\text{CN}(n, m)$ can be realized by a quantum circuit consisting of at most nm CN gates. That is, a mapping which is specified by an $n \times m$ 0-1 matrix $A = [a_{jk}]$ can be realized by a quantum circuit consisting of CN gates such that for $1 \le j \le n$ and $1 \le k \le m$, if $a_{jk} = 1$, then there is a CN gate on an $(n + m)$-qubit with the j-th qubit as its control-bit and the $(n + k)$-th qubit as its target-bit. Such a quantum circuit can be constructed in linear depth.

Lemma 1. *Every mapping of* $\text{CN}(n, m)$ *can be realized by a quantum circuit consisting of* CN *gates with* $O(n + m)$ *depth and without ancillae.*

Proof. The quantum circuit realizing a mapping of $\text{CN}(n, m)$ which is specified by an $n \times m$ 0-1 matrix $A = [a_{jk}]$ is constructed in $l = \max\{n, m\}$ stages. Let $L(j, k) = ((j + k) \bmod l) + 1$. If $a_{jk} = 1$, then a CN gate on an $(n + m)$-qubit with the j-th qubit as its control-bit and the $(n + k)$-th qubit as its target-bit is allocated in the $L(j, k)$-th stage. It is obvious that the quantum circuit realizes the desired mapping. Since $L(j, k) \ne L(j', k)$ for $1 \le j, j' \le n$ with $j \ne j'$ and

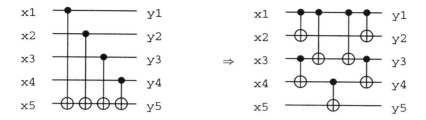

Fig. 1. A realization for a mapping of CN$(n,1)$.

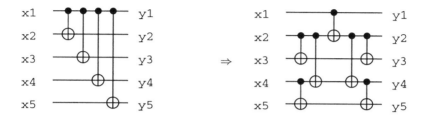

Fig. 2. A realization for a mapping of CN$(1,m)$.

$L(j,k) \neq L(j,k')$ for $1 \leq k,k' \leq m$ with $k \neq k'$, each stage can be realized by a quantum circuit consisting of CN gates with at most one depth. Therefore, from $1 \leq L(j,k) \leq l$, the depth of the quantum circuit is at most $l = O(n+m)$. □

For some special case, a mapping of CN(n,m) can be realized more efficiently without using ancillae. For example, every mapping of CN$(n,1)$ can be realized by a quantum circuit consisting of CN gates with $O(\log(n+1))$ depth (see Fig. 1), and every mapping of CN$(1,m)$ can be realized by a quantum circuit consisting of CN gates with $O(\log(m+1))$ depth (see Fig. 2). When the number of available ancillae is αn, every mappings of CN(n,m) can be realized in the following lemma.

Lemma 2. *Every mapping of* CN(n,m) *can be realized by a quantum circuit consisting of* CN *gates with*

$$O\left(n + \frac{m}{\alpha+1} + \log(\alpha+1)\right)$$

depth and with αn *ancillae for* $\alpha \geq 0$.

Proof. The quantum circuit realizing a mapping of CN(n,m) which is specified by an $n \times m$ 0-1 matrix $A = [a_{jk}]$ is constructed in three stages. The first stage realizes a mapping

$$|x_1, x_2, \ldots, x_n, 0^{\alpha n}, y_1, y_2, \ldots, y_m\rangle \mapsto |(x_1, x_2, \ldots, x_n)^{\alpha+1}, y_1, y_2, \ldots, y_m\rangle.$$

This mapping can be realized by applying a mapping $|x_j, 0^\alpha\rangle \mapsto |(x_j)^{\alpha+1}\rangle$ of $\text{CN}(1, \alpha)$ for all $1 \leq j \leq n$ in parallel. As mentioned above, all such mappings can be realized by a quantum circuit consisting of CN gates with $O(\log(\alpha + 1))$ depth.

The second stage realizes a mapping

$$|(x_1, x_2, \ldots, x_n)^{\alpha+1}, y_1, y_2, \ldots, y_m\rangle \mapsto |(x_1, x_2, \ldots, x_n)^{\alpha+1}, z_1, z_2, \ldots, z_m\rangle,$$

where $z_k = y_k \oplus (\bigoplus_{j=1}^{n} x_j a_{jk})$ for each $1 \leq k \leq m$. Let $l = \lceil m/(\alpha + 1) \rceil$. The mapping can be realized by applying $\alpha + 1$ mappings in parallel such that for $1 \leq k \leq \alpha$, the k-th mapping is

$$|x_1, x_2, \ldots, x_n, y_{(k-1)l+1}, y_{(k-1)l+2}, \ldots, y_{kl}\rangle$$
$$\mapsto |x_1, x_2, \ldots, x_n, z_{(k-1)l+1}, z_{(k-1)l+2}, \ldots, z_{kl}\rangle,$$

and the $(\alpha + 1)$-th mapping is

$$|x_1, x_2, \ldots, x_n, y_{\alpha l+1}, y_{\alpha l+2}, \ldots, y_m\rangle \mapsto |x_1, x_2, \ldots, x_n, z_{\alpha l+1}, z_{\alpha l+2}, \ldots, z_m\rangle.$$

From Lemma 1, each of these $\alpha + 1$ mappings can be realized by a quantum circuit consisting of CN gates with $O(n + l) = O(n + m/(\alpha + 1))$ depth.

Finally, the third stage realizes the inverse mapping of which is realized in the first stage, i.e.,

$$|(x_1, x_2, \ldots, x_n)^{\alpha+1}, z_1, z_2, \ldots, z_m\rangle \mapsto |x_1, x_2, \ldots, x_n, 0^{\alpha n}, z_1, z_2, \ldots, z_m\rangle.$$

It is clear that the desired mapping is realized by these three stages, and the total depth of these three stages is $O(n + m/(\alpha + 1) + \log(\alpha + 1))$. □

When the number of available ancillae is βm, every mappings of $\text{CN}(n, m)$ can also be realized as following lemma.

Lemma 3. *Every mapping of $\text{CN}(n, m)$ can be realized by a quantum circuit consisting of CN gates with*

$$O\left(\frac{n}{\beta + 1} + m + \log(\beta + 1)\right)$$

depth and with βm ancillae for $\beta \geq 0$.

Proof. The quantum circuit realizing a mapping of $\text{CN}(n, m)$ which is specified by an $n \times m$ 0-1 matrix $A = [a_{jk}]$ is constructed in three stages. Let $l = \lceil n/(\beta + 1) \rceil$. The first stage realizes a mapping

$$|x_1, \ldots, x_n, 0^{\beta m}, y_1, \ldots, y_m\rangle$$
$$\mapsto |x_1, \ldots, x_n, p_1, \ldots, p_{\beta m}, y_1 \oplus p_{\beta m+1}, \ldots, y_m \oplus p_{(\beta+1)m}\rangle,$$

where $p_{(k'-1)m+k} = \bigoplus_{j=(k'-1)l+1}^{k'l} x_j a_{jk}$ for $1 \leq k' \leq \beta$ and $1 \leq k \leq m$, and $p_{\beta m+k} = \bigoplus_{j=\beta l+1}^{n} x_j a_{jk}$ for $1 \leq k \leq m$. It can be done by applying $\beta + 1$ mappings in parallel such that for $1 \leq k' \leq \beta$, the k'-th mapping is

$$|x_{(k'-1)l+1}, \ldots, x_{k'l}, 0^m\rangle \mapsto |x_{(k'-1)l+1}, \ldots, x_{k'l}, p_{(k'-1)m+1}, \ldots, p_{k'm}\rangle,$$

and the $(\beta + 1)$-th mapping is

$$|x_{\beta l+1}, \ldots, x_n, y_1, \ldots, y_m\rangle \mapsto |x_{\beta l+1}, \ldots, x_n, y_1 \oplus p_{\beta m+1}, \ldots, y_m \oplus p_{(\beta+1)m}\rangle.$$

From Lemma 1, each of these mappings can be realized by a quantum circuit consisting of CN gates with $O(l + m) = O(n/(\beta + 1) + m)$ depth.

Notice that for $1 \leq k \leq m$,

$$p_k \oplus p_{m+k} \oplus \cdots \oplus p_{(\beta-1)m+k} \oplus (y_k \oplus p_{\beta m+k}) = y_k \oplus \bigoplus_{j=1}^{n} x_j a_{jk}.$$

The second stage realizes m mappings such that for $1 \leq k \leq m$, the k-th mapping is

$$|p_k, p_{m+k}, \ldots, p_{(\beta-1)m+k}, y_k \oplus p_{\beta m+k}\rangle$$
$$\mapsto \left|p_k, p_{m+k}, \ldots, p_{(\beta-1)m+k}, y_k \oplus \bigoplus_{j=1}^{n} x_j a_{jk}\right\rangle.$$

These mappings are mappings of $\mathrm{CN}(\beta, 1)$, and thus each of which can be realized by a quantum circuit with $O(\log(\beta + 1))$ depth.

Finally, the third stage constructs the mappings

$$\left|x_1, \ldots, x_n, p_1, \ldots, p_{\beta m}, y_1 \oplus \bigoplus_{j=1}^{n} x_j a_{j1}, \ldots, y_m \oplus \bigoplus_{j=1}^{n} x_j a_{jm}\right\rangle$$
$$\mapsto \left|x_1, \ldots, x_n, 0^{\beta m}, y_1 \oplus \bigoplus_{j=1}^{n} x_j a_{j1}, \ldots, y_m \oplus \bigoplus_{j=1}^{n} x_j a_{jm}\right\rangle,$$

and is realized by applying β mappings in parallel such that for $1 \leq k \leq \beta$, the k-th mapping is

$$|x_{(k'-1)l+1}, \ldots, x_{k'l}, p_{(k'-1)m+1}, \ldots, p_{k'm}\rangle \mapsto |x_{(k'-1)l+1}, \ldots, x_{k'l}, 0^m\rangle.$$

Therefore these three stages realize the desired mapping, and the total depth is $O(n/(\beta + 1) + m + \log(\beta + 1))$. □

By applying both Lemma 2 and Lemma 3 for realizing mappings of $\mathrm{CN}(n, m)$ we obtain the following lemma.

Lemma 4. *Every mapping of* $\mathrm{CN}(n,m)$ *can be realized by a quantum circuit consisting of* CN *gates with*

$$O\left(\frac{nm}{\gamma} + \log\frac{\gamma}{n} + \log\frac{\gamma}{m}\right)$$

depth and with γ ancillae for $\gamma \geq \min\{n,m\}$.

Proof. From Lemma 2, if $\gamma/2 = \alpha n$ ancillae are available, every mapping of $\mathrm{CN}(n,m)$ can be realized by a quantum circuit consisting of CN gates with $O(n + m/(\alpha+1) + \log(\alpha+1))$ depth. Notice that in this setting $\alpha+1$ mappings are applied in parallel, each of which is a mapping of $\mathrm{CN}(n,l)$ for some $l \leq \lceil m/(\alpha+1) \rceil$. For realizing the $\alpha+1$ mappings, the remaining $\gamma/2$ ancillae can be used such that $\gamma/(2(\alpha+1)) = \beta(\lceil m/(\alpha+1) \rceil$ ancillae can be used for each of the $\alpha+1$ mappings. Therefore, the total depth is reduced to

$$O\left(\frac{n}{\beta+1} + \frac{m}{\alpha+1} + \log(\alpha+1) + \log(\beta+1)\right) = O\left(\frac{nm}{\gamma} + \log\frac{\gamma}{n} + \log\frac{\gamma}{m}\right).$$

□

Form this lemma, the following theorem is obtained.

Theorem 1. *Quantum circuits consisting of* CN *gates can be parallelized to*

$$O\left(\frac{n^2}{\gamma} + \log\frac{\gamma}{n}\right)$$

depth by using γ ancillae for $\gamma \geq n$.

Proof. Suppose a quantum circuit which realizes a mapping

$$|x_1, x_2, \ldots, x_n\rangle \mapsto \left|\bigoplus_{j=1}^{n} x_j a_{j1}, \bigoplus_{j=1}^{n} x_j a_{j2}, \ldots, \bigoplus_{j=1}^{n} x_j a_{jn}\right\rangle$$

for some $n \times n$ 0-1 matrix $A = [a_{jk}]$ is given. This mapping can be realized in two stages. The first stage realizes a mapping of $\mathrm{CN}(n,n)$ which is specified by A, i.e.,

$$|x_1, \ldots, x_n, 0^n\rangle \mapsto \left|x_1, \ldots, x_n, \bigoplus_{j=1}^{n} x_j a_{j1}, \ldots, \bigoplus_{j=1}^{n} x_j a_{jn}\right\rangle.$$

Finally, the second stage realizes a mapping of $\mathrm{CN}(n,n)$ which is specified by the inverse matrix A^{-1} of A, i.e.,

$$\left|x_1, \ldots, x_n, \bigoplus_{j=1}^{n} x_j a_{j1}, \ldots, \bigoplus_{j=1}^{n} x_j a_{jn}\right\rangle$$

$$\mapsto \left| x_1 \oplus x_1, \ldots, x_n \oplus x_n, \bigoplus_{j=1}^{n} x_j a_{j1}, \ldots, \bigoplus_{j=1}^{n} x_j a_{jn} \right\rangle.$$

As mentioned above, the rank of A is n. Thus, the inverse matrix A^{-1} of A exists.

Therefore, by realizing these two stages one by one, the desired mapping follows:

$$|x_1, x_2, \ldots, x_n, 0^n\rangle \mapsto \left| 0^n, \bigoplus_{j=1}^{n} x_j a_{j1}, \bigoplus_{j=1}^{n} x_j a_{j2}, \ldots, \bigoplus_{j=1}^{n} x_j a_{jn} \right\rangle.$$

From Lemma 4, when $\gamma - n$ ancillae are available, every mapping of CN(n,n) can be realized by a quantum circuit consisting of CN gates with $O(n^2/\gamma + \log(\gamma/n))$ depth. It follows that the total depth is also $O(n^2/\gamma + \log(\gamma/n))$. □

5 Parallelizing Quantum Circuits Consisting of CN Gates and PS Gates and of CN Gates and WH Gates

Recall that a PS gate PS$_\theta$ on a j-th qubit of the n-qubit realizes the mapping

$$|x_1, x_2, \ldots, x_n\rangle \mapsto e^{i\theta(x_j)} |x_1, x_2, \ldots, x_n\rangle.$$

Thus, a quantum circuit consisting of CN gates and m PS gates PS$_{\theta_1}$, PS$_{\theta_2}$, ..., PS$_{\theta_m}$ realizes the mapping

$$|x_1, x_2, \ldots, x_n\rangle \mapsto e^{i \sum_{k=1}^{m} \theta_k(y_k)} |z_1, z_2, \ldots, z_n\rangle, \quad (1)$$

such that there exist an $n \times m$ 0-1 matrix $A = [a_{jk}]$ satisfying $y_k = \bigoplus_{j=1}^{n} x_j a_{jk}$ for $1 \leq k \leq m$ and an $n \times n$ 0-1 matrix $B = [b_{jk}]$ satisfying $z_l = \bigoplus_{j=1}^{n} x_j b_{jk}$ for $1 \leq l \leq n$.

Theorem 2. *Quantum circuits consisting CN gates and PS gates can be parallelized to*

$$O\left(\frac{n^2 + nm}{\gamma} + \log \frac{\gamma}{n} + \log \frac{\gamma}{m}\right)$$

depth by using γ ancillae for $\gamma \geq m$, where m is the number of PS gates in the quantum circuit.

Proof. Suppose a quantum circuit consisting of CN gates and m PS gates PS$_{\theta_1}$, PS$_{\theta_2}$, ..., PS$_{\theta_m}$ is given and the quantum circuit realizes the mapping shown in (1). This mapping can be realized in four stages. The first stage realizes a mapping

$$|x_1, x_2, \ldots, x_n, 0^m\rangle \mapsto |x_1, x_2, \ldots, x_n, y_1, y_2, \ldots, y_m\rangle.$$

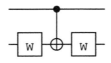

Fig. 3. A realization for a π-shift.

This is a mapping of CN(n, m), and thus from Lemma 4 it can be realized by a quantum circuit consisting of CN gates with $O(nm/\gamma + \log(\gamma/n) + \log(\gamma/m))$ depth by using the remaining $\gamma - m$ ancillae.

The second stage realizes the mapping

$$|y_1, y_2, \ldots, y_m\rangle \mapsto e^{i\sum_{k=1}^{m} \theta_k(y_k)} |y_1, y_2, \ldots, y_m\rangle$$

by applying the PS gates PS$_{\theta_1}$, PS$_{\theta_2}$, ..., PS$_{\theta_m}$ in parallel. Thus, it can be done in one depth. The third stage constructs the inverse mapping of the mappping constructed in the first stage, i.e.,

$$|x_1, x_2, \ldots, x_n, y_1, y_2, \ldots, y_m\rangle \mapsto |x_1, x_2, \ldots, x_n, 0^m\rangle.$$

Again depth of this stage is $O(nm/\gamma + \log(\gamma/n) + \log(\gamma/m))$.

Finally, the fourth stage realizes the mapping

$$|x_1, x_2, \ldots, x_n\rangle \mapsto |z_1, z_2, \ldots, z_n\rangle.$$

From Theorem 1, this stage can be done by a quantum circuit consisting of CN gates with $O(n^2/\gamma + \log(\gamma/n))$ depth. Therefore, the total depth of these four stages is $O((n^2 + nm)/\gamma + \log(\gamma/n) + \log(\gamma/m))$. □

Before explaining how we can parallelize quantum circuits consisting of CN gates and WH gates, let us define a two-input two-output PS gate, called a π-shift. This gate is specified by the following unitary matrix:

$$U_\pi = \begin{pmatrix} 1 & 0 & 0 & 0 \\ 0 & 1 & 0 & 0 \\ 0 & 0 & 1 & 0 \\ 0 & 0 & 0 & -1 \end{pmatrix}.$$

A π-shift can be realized by using WH gates and a CN gate (see Fig. 3), since $U_\pi = (I \otimes \text{WH}) \cdot \text{CN} \cdot (I \otimes \text{WH})$.

When the number of available ancillae is n, for quantum circuits consisting of π-shifts, we obtain the following lemma.

Lemma 5. *By using n ancillae a quantum circuit consisting of π-shifts can be reconstructed as a quantum circuit with a constant number of subcircuits each of which is either a quantum circuit consisting of CN gates or a quantum circuit consisting of WH gates.*

Proof. Suppose a quantum circuits consisting of π-shifts is given. The quantum circuit can be reconstructed as a quantum circuit with five subcircuits by using n ancillae. The first subcircuit is a quantum circuit consisting of CN gates, which realizes the mapping

$$|x_1, x_2, \ldots, x_n, 0^n\rangle \mapsto |x_1, x_2, \ldots, x_n, x_1, x_2, \ldots, x_n\rangle.$$

Without loss of generality, for each π-shift, let the π-shift be applied on a n-qubit for which the j_l-th qubit as its first input and the k_l-th qubit as its second input, where $1 \leq l \leq m$ with m is the number of π-shifts in the quantum circuit. The π-shift can be realized by applying a π-shift on the $2n$-qubit for which the j_l-th qubit is its first input and the $(k_l + n)$-th qubit is its second input, since a π-shift is a PS gate. As mentioned above, the π-shift can be realized by applying a WH gates on the $(k_l + n)$-th qubit of a $2n$-qubit, a CN gate on a $2n$-qubit for which the j_l-th qubit as its control-bit and the $(k_l + n)$-th qubit as its target-bit, and a WH gate on the $(k_l + n)$-th qubit of a $2n$-qubit. Therefore, from WH · WH $= I$, the second and fourth subcircuits are quantum circuits consisting of WH gates and the third subcircuit is a quantum circuit consisting of CN gates.

Finally, the fifth subcircuit is a quantum circuit consisting of CN gates, which realizes the inverse mapping of which realized by the first subcircuit, i.e.,

$$|x_1, x_2, \ldots, x_n, x_1, x_2, \ldots, x_n\rangle \mapsto |x_1, x_2, \ldots, x_n, 0^n\rangle.$$

$\square$

In previous work for quantum circuits consisting of CN gates and WH gates, Moore and Nilsson [7] showed that every quantum circuit consisting of CN gates and WH gates can be reconstructed by using an ancilla with state $|1\rangle$ as a quantum circuit with a constant number of subcircuits each of which is either a quantum circuit consisting of CN gates, a quantum circuit consisting of WH gates, or a quantum circuit consisting of π-shifts. Thus, from Lemma 5, every quantum circuit consisting of CN gates and WH gates can be reconstructed by using an ancilla with state $|1\rangle$ as a quantum circuit with constant number of subcircuits each of which is either a quantum circuit consisting of CN gates or a quantum circuit consisting of WH gates. Notice that every quantum circuit consisting of WH gates can be realized by such a quantum circuit with at most one depth, since WH · WH $= I$. Therefore, from Theorem 1, we immediately obtain the following theorem.

Theorem 3. *Quantum circuits consisting of* CN *gates and* WH *gates can be parallelized to depth*

$$O\left(\frac{n^2}{\gamma} + \log \frac{\gamma}{n}\right)$$

by using γ ancillae for $\gamma \geq n$.

6 Concluding Remarks

We have proposed parallelization methods for three types of quantum circuits when the number of available ancillae is limited. However, we still do not know any non-trivial lower bound on depth for realizing a mapping on the three types of quantum circuits. For more general quantum circuits, a parallelization method is still not known. From Barenco *et al* [2], in order to realize universal quantum computation, it is sufficient to consider quantum circuits consisting of CN gates, PS gates, and one-input one-output quantum gates, which is specified by

$$\begin{pmatrix} \cos\rho & \sin\rho \\ \sin\rho & -\cos\rho \end{pmatrix},$$

for $0 \leq \rho < 2\pi$, while the WH gate can be specified by this matrix with $\rho = \pi/4$.

References

1. A. Barenco. A universal two-bit gate for quantum computation. preprint (1994).
2. A. Barenco, C.H. Bennett, R. Cleve, D.P. DiVincenzo, N. Margolus, P. Shor, T. Sleator, J. Smolin, and H. Weinfurter. Elementary gates for quantum computation. *Phys. Rev* **A** (52), 3457–3467, (1995).
3. D. Deutsch. Quantum computational networks. *Proc. Roy. Soc. London Ser. A* **425** 73–90 (1989).
4. D. Deutsch, A. Barenco, and A. Ekert. Universality in quantum computation. submitted to *Proc. Roy. Soc. London Ser.* (1995).
5. D.P. DiVincenzo. Two-bit gates are universal. *Phys. Rev. A* **50** 1015 (1995).
6. S. Lloyed. Almost any quantum logic gate is universal. preprint (1994).
7. C. Moore and M. Nilsson. Parallel quantum computation and quantum codes. *SIAM J. Comput.* (to appear) available at lanl e-print quant-ph/9808027.
8. A. Yao. Quantum circuit complexity. in Proc. 34th Annual IEEE Symp. on Foundations of Computer Science 352–361 (1993).

Upper and Lower Bounds on Continuous-Time Computation

Manuel Lameiras Campagnolo[1] and Cristopher Moore[2,3,4]

[1] D.M./I.S.A., Universidade Técnica de Lisboa, Tapada da Ajuda, Lisboa, Portugal
[2] Computer Science Department, University of New Mexico, Albuquerque, U.S.A.
[3] Physics and Astronomy Department, University of New Mexico, Albuquerque, U.S.A.
[4] Santa Fe Institute, Santa Fe, U.S.A.

Abstract. We consider various extensions and modifications of Shannon's General Purpose Analog Computer, which is a model of computation by differential equations in continuous time. We show that several classical computation classes have natural analog counterparts, including the primitive recursive functions, the elementary functions, the levels of the Grzegorczyk hierarchy, and the arithmetical and analytical hierarchies.

1 Introduction

The theory of analog computation, where the internal states of a computer are continuous rather than discrete, has enjoyed a recent resurgence of interest. This stems partly from a wider program of exploring alternative approaches to computation, such as quantum and DNA computation; partly as an idealization of numerical algorithms where real numbers can be thought of as quantities in themselves, rather than as strings of digits; and partly from a desire to use the tools of computation theory to better classify the variety of continuous dynamical systems we see in the world (or at least in its classical idealization).

However, in most recent work on analog computation (e.g. [BSS89], [Mee93], [Sie98], [Moo98]) time is still discrete. Just as in standard computation theory, the machines are updated with each tick of a clock. If we are to make the states of a computer continuous, it makes sense to consider making its progress in time continuous too. While a few efforts have been made in the direction of studying computation by continuous-time dynamical systems [Moo90], [Moo96], [Orp97b], [Orp97a], [SF98], [Bou99], [Bou99b], [CMC99], [CMC00], [BSF00], no particular set of definitions has become widely accepted, and the various models do not seem to be equivalent to each other. Thus analog computation has not yet experienced the unification that digital computation did through Turing's work in 1936.

In this paper, we take as our starting point Shannon's General Purpose Analog Computer (GPAC), a natural model of continuous-time computation defined in terms of differential equations. By extending it with various operators and oracles, we show that a number of classical computation classes have natural analog counterparts, including the primitive recursive and elementary functions, the levels of the Grzegorczyk hierarchy, and (if some physically unreasonable operators are allowed) the arithmetical and analytical hierarchies. We review recent results on these extensions, place them in a unified framework, and suggest directions for future research.

The paper is organized as follows. In Section 2 we review the standard computation classes over the natural numbers. In Section 3 we review Shannon's GPAC, and in Section 4 we show that a simple extension of it can compute all primitive recursive functions. In Section 5 we restrict the GPAC to linear differential equations, and show that this allows us to compute exactly the elementary functions, or the levels of the Grzegorczyk hierarchy if we allow a certain number of nonlinear differential equations as well. In Section 6 we show that allowing zero-finding on the reals yields much higher classes in the arithmetical and analytical hierarchies, and in Section 7 we conclude.

2 Recursive Function Classes Over $\mathbb{N}$

In classical recursive function theory, where the inputs and outputs of functions are in the natural numbers $\mathbb{N}$, computation classes are often defined as the smallest set containing a basis of initial functions and closed under certain operations, which take one or more functions in the class and create new ones. Thus the set consists of all those functions that can be generated from the initial ones by applying these operations a finite number of times. Typical operations include (here x represents a vector of variables, which may be absent):

1. *Composition:* Given f and g, define $(f \circ g)(x) = f(g(x))$.
2. *Primitive recursion:* Given f and g of the appropriate arity, define h such that $h(x, 0) = f(x)$ and $h(x, y+1) = g(x, y, h(x, y))$.
3. *Iteration:* Given f, define h such that $h(x, y) = f^{[y]}(x)$, where $f^{[0]}(x) = x$ and $f^{[y+1]}(x) = f(f^{[y]}(x))$.
4. *Limited recursion:* Given f, g and b, define h as in primitive recursion but only on the condition that $h(x, y) \leq b(x, y)$. Thus h is only allowed to grow as fast as another function already in the class.
5. *Bounded sum:* Given $f(x, y)$, define $h(x, y) = \sum_{z<y} f(x, z)$.
6. *Bounded product:* Given $f(x, y)$, define $h(x, y) = \prod_{z<y} f(x, z)$.
7. *Minimization* or *Zero-finding:* Given $f(x, y)$, define $h(x) = \mu_y f(x, y)$ as the smallest y such that $f(x, y) = 0$ provided that $f(x, z)$ is defined for all $z \leq y$. If no such y exists, h is undefined.
8. *Bounded minimization:* Given $f(x, y)$, define $h(x, y_{\max})$ as the smallest $y < y_{\max}$ such that $f(x, y) = 0$ and as $y_{\max}$ if no such y exists.

Note that minimization is the only one of these that can create a partial function; all the others yield total functions when applied to total functions.

In bounded minimization, we only check for zeroes less than $y_{\max}$, and return $y_{\max}$ if we fail to find any.

Along with these operations, we will start with basis functions such as

1. The zero function, $\mathcal{O}(x) = 0$
2. The successor function, $\mathcal{S}(x) = x + 1$
3. The projections, $U_i^n(x_1, \ldots, x_n) = x_i$
4. Addition
5. Multiplication
6. Cut-off subtraction, $x \dotdiv y = x - y$ if $x \geq y$ and 0 if $x < y$

Then by starting with various basis sets and demanding closure under various properties, we can define the following classical complexity classes:

1. The *elementary* functions $\mathcal{E}$ are those that can be generated from zero, successor, projections, addition, and cut-off subtraction, using composition, bounded sum, and bounded product.
2. The *primitive recursive* functions $\mathcal{PR}$ are those that can be generated from zero, successor, and projections using composition and primitive recursion. We get the same class if we replace primitive recursion with iteration.
3. The *partial recursive* functions are those that can be generated from zero, successor, and projections using composition, primitive recursion and minimization.
4. The *recursive* functions are the partial recursive functions that are total.

A number of our results regard the class $\mathcal{E}$ of elementary functions, which was introduced by Kálmar [Kál43]. For example, multiplication and exponentiation over $\mathbb{N}$ are both in $\mathcal{E}$, since they can be written as bounded sums and products respectively: $xy = \sum_{z<y} x$ and $x^y = \prod_{z<y} x$. Since $\mathcal{E}$ is closed under composition, for each m the m-times iterated exponential $\exp^{[m]}(x)$ is in $\mathcal{E}$, where $\exp^{[0]}(x) = x$ and $\exp^{[m+1]}(x) = 2^{\exp^{[m]}(x)}$. In fact, these are the fastest-growing functions in $\mathcal{E}$, in the sense that no elementary function can grow faster than $\exp^{[m]}$ for some fixed m. The following bound will be useful to us below [Cut80]:

Proposition 1 *If $f \in \mathcal{E}$, there is a number m such that, for all x, $f(x) \leq \exp^{[m]}(\|x\|)$ where $\|x\| = \max_i x_i$.*

The elementary functions also correspond to a natural time-complexity class:

Proposition 2 *The elementary functions are exactly the functions computable by a Turing machine in elementary time, or equivalently in time bounded by $\exp^{[m]}(|x|)$ for some fixed m.*

The class $\mathcal{E}$ is therefore very large, and many would argue that it contains all *practically* computable functions. It includes, for instance, the connectives of propositional calculus, functions for coding and decoding sequences of natural numbers such as the prime numbers and factorizations, and most of the useful number-theoretic and metamathematical functions. It is also closed under limited recursion and bounded minimization [Cut80,Ros84].

However, $\mathcal{E}$ does not contain all recursive functions, or even all primitive recursive ones. For instance, Proposition 1 shows that it does not contain the iterated exponential $\exp^{[m]}(x)$ where the number of iterations m is a variable, since any function in $\mathcal{E}$ has an upper bound where m is fixed. To include such functions, we need to include the higher levels of the *Grzegorczyk hierarchy* [Grz53,Ros84]. This hierarchy was used as an early stratification of the primitive recursive functions according to their computational complexity:

Definition 3 (The Grzegorczyk hierarchy) *Let $\mathcal{E}^0$ denote the smallest class containing zero, the successor function, and the projections, and which is closed under composition and limited recursion. Let $\mathcal{E}^{n+1}$ be defined similarly, except with the function E_n added to the list of initial functions, where E_n is defined as follows:*

$$E_0(x, y) = x + y$$
$$E_1(x) = x^2 + 2$$
$$E_{n+1}(x) = E_n^{[x]}(2)$$

where by $f^{[x]}$ we mean f iterated x times.

The functions E_n are, essentially, repeated iterations of the successor function, and each one grows qualitatively more quickly than the previous one. $E_1(x)$ grows quadratically, and composing it with itself produces functions that grow as fast as any polynomial. $E_2(x)$ grows roughly as 2^{2^x}, and composing it yields functions as large as $\exp^{[m]}$ for any fixed m. $E_3(x)$ grows roughly as $\exp^{[2x]}(2)$, and so on. (These somewhat awkward definitions of E_0 and E_1 are the historical ones.)

We will use the fact that for $n \geq 3$, we can replace limited recursion in the definition of $\mathcal{E}^n$ with bounded sum and bounded product [Ros84]:

Proposition 4 *For $n \geq 3$, $\mathcal{E}^n$ is the smallest class containing zero, successor, the projections, cut-off subtraction, and E_{n-1}, which is closed under composition, bounded sum, and bounded product.*

One consequence of this is that the elementary functions are simply the third level of the Grzegorczyk hierarchy [Ros84], i.e. $\mathcal{E} = \mathcal{E}^3$. Moreover, the union of all the levels of the Grzegorczyk hierarchy is simply the class $\mathcal{PR}$ of primitive recursive functions:

Proposition 5 $\mathcal{PR} = \cup_n \mathcal{E}^n$.

It is known that the class of primitive recursive functions can be defined using iteration instead of primitive recursion [Odi89, p.72]. This means that iteration cannot be used freely in the Grzegorczyk hierarchy. Rather, as the definitions suggest, iteration moves a function one level up. As a matter of fact, iteration of E_{n-1} for a *fixed* number of times gives a bound on any function in $\mathcal{E}^n$, but unbounded iteration of E_{n-1} defines E_n and generates precisely $\mathcal{E}^{n+1}$. In this sense, the Grzegorczyk hierarchy stratifies the primitive recursive functions according to how many levels of iteration are needed to define them, or equivalently how many nested FOR-loops are required to compute them in a simplified programming language.

3 Differential Equations, Differentially Algebraic Functions and Shannon's GPAC

An ordinary differential equation of order n is an equation of the form

$$F(x, y(x), y'(x), \ldots y^{(n)}(x)) = 0.$$

If F is a polynomial this equation is called *differentially algebraic* (d.a.) and its solutions are called differentially algebraic functions. The set of d.a. functions includes the polynomials, e^x, and trigonometric functions, as well as sums, products, compositions and solutions of differential equations formed from these such as $f' = \sin f$. Examples of functions which are not d.a. include Euler's Γ function and Riemann's ζ function [Rub89b]

The General Purpose Analog Computer (GPAC) is a simple model of a computer evolving in continuous time. It was originally defined as a mathematical model of an analog device, the Differential Analyser, the fundamental principles of which were described first by Lord Kelvin in 1876 [Kel76] and later by Vannevar Bush [Bow96]. The outputs are generated from the inputs by means of a dependence defined by a finite directed graph (not necessarily acyclic) where each node is either an adder, a unit that outputs the sum of its inputs, or an integrator, a unit with two inputs u and v that outputs the Riemann-Stieltjes integral $\int u \, dv$. These components are used to form circuits like the one in Figure 1, which calculates the function $\sin t$.

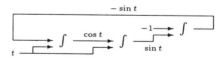

Fig. 1. A simple GPAC circuit that calculates $\sin t$. Its initial conditions are $\sin(0) = 0$ and $\cos(0) = 1$. The output w of the integrator unit $\int$ obeys $dw = u \, dv$ where u and v are its upper and lower inputs respectively.

Shannon [Sha41] showed that the class of functions generable in this abstract model is the set of solutions of systems of the following system of quasi-linear differential equations,

$$A(\boldsymbol{x}, \boldsymbol{y}) \, \boldsymbol{y}' = B(\boldsymbol{x}, \boldsymbol{y}), \tag{1}$$

satisfying some initial condition $y(x_0) = y_0$. Here A and B are $n \times n$ and $m \times m$ matrices linear in 1 and the variables $x_1, ..., x_m, y_1, ..., y_n$, and y' is the $n \times m$ matrix of the derivatives of y with respect to x. Later, Pour-El [PE74] made this definition more precise by requiring the solution to be unique for all initial values belonging to a closed set with non-empty interior called the *domain of generation* of the initial condition. We call the set of such solutions of (1) the class of GPAC-computable functions.

The following fundamental result [Sha41,PE74,LR87] establishes that the GPAC-computable functions essentially coincide with the differentially algebraic ones:

Proposition 6 (Shannon, Pour-El, Lipshitz, Rubel) *Let I and J be closed intervals of $\mathbb{R}$. If y is GPAC-computable on I then there is a closed subinterval $I' \subset I$ and a polynomial $P(x, y, y', ..., y^{(n)})$ such that $P = 0$ on I'. If $y(x)$ is the unique solution of $P(x, y, y', ..., y^{(n)}) = 0$ satisfying a certain initial solution on J then there is a closed subinterval $J' \subset J$ on which $y(x)$ is GPAC-computable.*

We will use $\mathcal{G}$ to denote the class of GPAC-computable functions, or equivalently the class of d.a. functions.

Now we show that $\mathcal{G}$ lacks an important closure property: it is not closed under iteration. The proof relies on a result of differential algebra on the iterated exponential function $\exp^{[n]}(x)$ defined by $\exp^{[0]}(x) = x$ and $\exp^{[n]}(x) = e^{\exp^{[n-1]}(x)}$. The following lemma follows from a more general theorem of Babakhanian [Bab73]:

Lemma 7 *For $n \geq 0$, $\exp^{[n]}(x)$ satisfies no non-trivial algebraic differential equation of order less than n.*

Proposition 6, Lemma 7 and our previous remarks are combined in [CMC99] to prove that:

Proposition 8 *The class $\mathcal{G}$ is not closed under iteration. Specifically, there is no GPAC-computable function $F(x, n)$ of two variables that matches the iterated exponential $\exp^{[n]}(x)$ for integer values of n.*

In the next section, we will show that while the GPAC is not closed under iteration, a natural extension of it is, and that this extension therefore includes all the primitive recursive functions.

4 Extending the GPAC

In analogy with oracles in classical computation theory, we can ask what functions become GPAC-computable if we add one or more additional basis functions φ. In terms of Shannon's circuit model, what things become GPAC-computable when we have "black boxes" that compute φ, which we can plug in to our circuit along with integrators and adders? We will refer to the resulting class as $\mathcal{G} + \varphi$.

One such extension explored in [CMC99] is the family of functions $\theta_k(x) = x^k \theta(x)$, where $\theta(x)$ is the Heaviside step function

$$\theta(x) = \begin{cases} 1 \text{ if } x \geq 0 \\ 0 \text{ if } x < 0 \end{cases}$$

For each k, we can think of $\theta_k(x)$ as a $(k-1)$-times differentiable way of testing whether $x \geq 0$. We claim that this is a physically realistic way to allow our computer to sense inequalities without introducing discontinuities.

In addition, we can show that allowing those functions is equivalent to relaxing slightly the definition of GPAC by solving first-order differential equations with two boundary values instead of just an initial condition. Thus $\mathcal{G} + \theta_k$ is a natural extension of the GPAC. Specifically,

Definition 9 *The function $y = f(x)$ belongs to the class θ if it is the unique solution on $I = [x_1, x_2] \subset \mathbb{R}$ of the differential equation $(a_0 + a_1 x + a_2 y) y' = b_0 + b_1 x + b_2 y$ with boundary values $y(x_1) = y_1$ and $y(x_2) = y_2$.*

For instance, the differential equation $xy' = 2y$ with boundary values $y(1) = 1$ and $y(-1) = 0$ has a unique solution on $I = [-1, 1]$, namely $y = x^2 \theta(x)$. Then, we can prove [CMC99] that adding any function in θ to the set of basis functions of $\mathcal{G}$ is equivalent to adding θ_k for some k:

Proposition 10 *For any $\varphi \in \theta - \mathcal{G}$ there is a k such that $\mathcal{G} + \varphi = \mathcal{G} + \theta_k$.*

The main property of $\mathcal{G} + \theta_k$ we show here is the following:

Proposition 11 *$\mathcal{G} + \theta_k$ is closed under iteration for any $k > 1$. That is, if f of arity n belongs to $\mathcal{G} + \theta_k$ then there exists a function F of arity $n+1$ also in $\mathcal{G} + \theta_k$, such that $F(\boldsymbol{x}, t) = f^{[t]}(\boldsymbol{x})$ for $t \in \mathbb{N}$.*

The proof is constructive. To iterate a function we use a pair of "clock" functions to control the evolution of two "simulation" variables, similar to the approach in [Bra95,Moo96]. Both simulation variables have the same value x at $t = 0$. The first variable is iterated during half of a unit period while the second remains constant (its derivative is kept at zero by the corresponding clock function). Then, the first variable remains steady during the following half unit period and the second variable is brought up to match it. Therefore, at time $t = 1$ both variables have the same value $f(x)$. This process is repeated until the desired number of iterations is obtained.

If we denote the simulation variables by y_1 and y_2, and the clock functions by $\theta_k(\sin 2\pi t)$ and $\theta_k(-\sin 2\pi t)$, then the function that iterates f is the unique solution of:

$$\begin{aligned} |\cos \pi t|^{k+1} y_1' &= -\pi(y_1 - f(y_2)) \theta_k(\sin 2\pi t) \theta_k(t) \\ |\sin \pi t|^{k+1} y_2' &= -\pi(y_2 - y_1) \theta_k(-\sin 2\pi t) \theta_k(t) \end{aligned} \quad (2)$$

where $|x|^k$ can defined in $\mathcal{G} + \theta_k$ as $|x|^k = \theta_k(x) + \theta_k(-x)$.

For general k, the proof that $y_1(t) = f^{[t]}(x)$ relies on the local behavior of Equation (2) in the neighborhood of $x = t$ and $x = t+1$ for $t \in \mathbb{N}$. For instance, as $t \to 1$ from below, (2) becomes

$$\epsilon y_1' = -2^{k+1}(y_1 - f(y_2))$$

to first order in $\epsilon = 1 - t$. The solution of this is

$$y_1(\epsilon) = C\epsilon^{2^{k+1}} + f(y_2)$$

for constant C, and y_1 rapidly approaches $f(y_2)$ no matter where it starts on the real line. Similarly, y_2 rapidly approaches y_1 as $t \to 2$, and so on, so for any integer $t > 1$, $y_1(t) = y_2(t) = f^{[t]}(x)$. This shows that $F(x,t) = y_1(t)$ can be defined in $\mathcal{G} + \theta_k$, so $\mathcal{G} + \theta_k$ is closed under iteration. Details can be found in [CMC99].

As an example, in Figure 2 we iterate the exponential function, which as we pointed out in Proposition 8 cannot be done in $\mathcal{G}$. Note that this is a numerical integration of Eq. (2) using a standard package (Mathematica), so this system of differential equations actually works in practice.

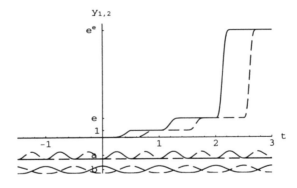

Fig. 2. A numerical integration on $[-1.5, 3]$ of the system of equations (2) for iterating the exponential function. Here $k = 2$. The values of y_1 and y_2 at $t = 0, 1, 2, 3$ are $0, 1, e$, and e^e respectively. On the graph below we show (a) the clock functions $\theta_2(\sin(2\pi t))$, $\theta_2(\sin(-2\pi t))$ and (b) the functions $|\cos \pi t|^3$, $|\sin \pi t|^3$. Note that the term $\theta_k(t)$ on the right of Equation (2) assures that $y_1(t) = y_2(t) = 0$ for all $t < 0$ and, therefore, that the solution is unique on $\mathbb{R}$.

Let us set the convention that $\mathcal{G} + \theta_k$ contains a function on $\mathbb{N}$ if it contains some extension of it to $\mathbb{R}$. Since $\mathcal{G} + \theta_k$ contains zero, successor, and projections, and is closed under composition and iteration, it follows that:

Proposition 12 $\mathcal{G} + \theta_k$ *contains all primitive recursive functions.*

In fact, it is known that flows in three dimensions, or iterated functions in two, can simulate arbitrary Turing machines. In two dimensions, these functions can be infinitely differentiable [Moo90], piecewise-linear [Moo90,KCG94], or closed-form analytic and composed of a finite number of trigonometric terms [KM99]. (In [KM99] a simulation in one dimension is achieved at the cost of an exponential slowdown.) However, Proposition 12 is in some sense more elegant than these constructions, since it uses the operators of recursion theory directly instead of relying on a particular simulation or encoding of a Turing machine.

Furthermore, since for any Turing machine $\mathcal{M}$, the function $F(x,t)$ that gives the output of $\mathcal{M}$ on input x after t steps is primitive recursive, and since $\mathcal{G} + \theta_k$ is closed under composition, we can say that $\mathcal{G} + \theta$ is *closed under time complexity* in the following sense:

Proposition 13 *If a Turing machine $\mathcal{M}$ computes the function $h(x)$ in time bounded by $T(x)$, with T in $\mathcal{G} + \theta_k$, then h belongs to $\mathcal{G} + \theta_k$.*

Since any function computable in primitive recursive time is primitive recursive, Proposition 13 alone does not show that $\mathcal{G} + \theta_k$ contains any non-primitive recursive functions on the integers. However, if $\mathcal{G} + \theta_k$ contains a function such as the Ackermann function which grows more quickly than any primitive recursive function, this proposition shows that $\mathcal{G} + \theta_k$ contains many other non-primitive recursive functions as well.

It is believed, but not known [Hay96], that all differentially algebraic functions on the complex plane are bounded by some elementary function, i.e. $\exp^{[n]}(x)$ for some n, whenever they are defined for all $x > 0$. For real solutions of d.a. equations the conjecture is known to be false due to a theorem of Vijayaraghavan [Vij32,BBV37,Ban75]. However, the examples of d.a. functions that grow arbitrarily quickly are solutions of equations whose parameters are defined by limit processes, and this gives rise to non-primitive recursive constants. If we restrict ourselves to a model where the GPAC only has access to rational constants in its initial conditions and parameters, we believe the following is true:

Conjecture 14 *Functions $f(x)$ in $\mathcal{G} + \theta_k$ have primitive recursive upper bounds whenever they are defined for all $x > 0$, if the parameters and initial values of their defining differential equations are rational.*

We might try proving this conjecture by using numerical integration to approximate GPAC-computable functions with recursive ones. However, strictly speaking this approximation only works when a bound on the derivatives is known *a priori* [VSD86] or on arbitrarily small domains [Rub89]. If this conjecture is false, then Proposition 13 shows that $\mathcal{G} + \theta_k$ contains a wide variety of non-primitive recursive functions.

We close this section by noting that since all functions in $\mathcal{G} + \theta_k$ are $(k-1)$-times continuously differentiable, $\mathcal{G} + \theta_k$ is a near-minimal departure from analyticity. In fact, if we wish to sense inequalities in an infinitely-differentiable way, we can add a C^∞ function such as $\theta_\infty(x) = e^{-1/x}\theta(x)$ to $\mathcal{G}$ and get the same results. The most general version of Proposition 11 is the following:

Proposition 15 *If $\varphi(x)$ has the property that it coincides with an analytic function $f(x)$ over an open interval (a,b), but that $\int_b^c (\varphi(x) - f(x))\, dx \neq 0$ for some $c > b$, then $\mathcal{G} + \varphi$ is closed under iteration and contains all the primitive recursive functions.*

We prove this by replacing $\theta_k(x)$ with $\varphi(x+b) - f(x+b)$, and we leave the details to the reader. Thus any departure from analyticity over an open interval creates a system powerful enough to contain all of $\mathcal{PR}$.

5 Linear Differential Equations, Elementary Functions and the Grzegorczyk hierarchy

In this section, we show that restricting the kind of differential equations we allow the GPAC to solve yields various subclasses of the primitive recursive functions: namely, the elementary functions $\mathcal{E}$ and the levels $\mathcal{E}^n$ of the Grzegorczyk hierarchy.

Let us first look at the special case of linear differential equations. If a first-order ordinary differential equation can be written as

$$\boldsymbol{y}'(x) = A(x)\, \boldsymbol{y}(x) + \boldsymbol{b}(x), \tag{3}$$

where $A(x)$ is a $n \times n$ matrix whose entries are functions of x, and $\boldsymbol{b}(x)$ is a vector of functions of x, then it is called a first-order linear differential equation. If $\boldsymbol{b}(x) = 0$ we say that the system is homogeneous. We can reduce a non-homogeneous system to a homogeneous one by introducing an auxiliary variable.

The fundamental existence theorem for differential equations guarantees the existence and uniqueness of a solution in a certain neighborhood of an initial condition for the system $\boldsymbol{y}' = f(\boldsymbol{y})$ when f is Lipschitz. For linear differential equations, we can strengthen this to global existence whenever $A(x)$ is continuous, and establish a bound on $\boldsymbol{y}$ that depends on $\|A(x)\|$:

Proposition 16 ([Arn96]) *If $A(x)$ is defined and continuous on an interval $I = [a,b]$ where $a \leq 0 \leq b$, then the solution of a homogeneous linear differential equation with initial condition $\boldsymbol{y}(0) = \boldsymbol{y}_0$ is defined and unique on I. Furthermore, if $A(x)$ is increasing then this solution satisfies*

$$\|\boldsymbol{y}(x)\| \leq \|\boldsymbol{y}_0\|\, e^{\|A(x)\|\, x}. \tag{4}$$

Given functions f and g, we can form the function h such that $h(\boldsymbol{x}, 0) = f(\boldsymbol{x})$ and $\partial_y h(\boldsymbol{x}, y) = g(\boldsymbol{x}, y)\, h(\boldsymbol{x}, y)$. We call this operation *linear integration*, and write $h = f + \int gh\, dy$ as shorthand. Then we can define an analog class $\mathcal{L}$ which is closed under composition and linear integration. As before, we can define classes $\mathcal{L} + \varphi$ by allowing additional basis functions φ as well. Specifically, we will consider the class $\mathcal{L} + \theta_k$:

Definition 17 *A function* $h : \mathbb{R}^m \to \mathbb{R}^n$ *belongs to* $\mathcal{L} + \theta_k$ *if its components can be inductively defined from the constants* $0, 1, -1,$ *and* π, *the projections, and* θ_k, *using composition and linear integration.*

The reader will note that we are including π as a fundamental constant. We will need this for Lemma 21. We have not found a way to derive π from linear differential equations alone; perhaps the reader can find a way to do this, or a proof that we cannot. (Since π can easily be generated in $\mathcal{G}$, we have $\mathcal{L} + \theta_k \subseteq \mathcal{G} + \theta_k$.)

We wish to show that for any $k > 2$, $\mathcal{L} + \theta_k$ is an analog characterization of the elementary functions. First, note that by Proposition 16 all functions in $\mathcal{L} + \theta_k$ are total. In addition, their growth is bounded by a finitely iterated exponential, $\exp^{[m]}$ for some m. The following is proved in [CMC00], using the fact that if f and g are bounded by a finite tower of exponentials then their composition and linear integration $h = f + \int gh\,dy$ are as well:

Proposition 18 *Let h be a function in $\mathcal{L} + \theta_k$ of arity m. Then there is a constant d and constants A, B, C, D such that, for all $x \in \mathbb{R}^m$,*

$$\|h(x)\| \le A \exp^{[d]}(B\|x\|)$$
$$\|\partial_{x_i} h(x)\| \le C \exp^{[d]}(D\|x\|) \text{ for all } i = 1, \ldots, m$$

where $\|x\| = \max_i |x_i|$.

Note the analogy with Proposition 1 for elementary functions. In fact, we will now show that the relationship between $\mathcal{E}$ and $\mathcal{L} + \theta_k$ is very tight: all functions in $\mathcal{L} + \theta_k$ can be approximated by elementary functions, and all elementary functions have extensions to the reals in $\mathcal{L} + \theta_k$.

We say that a function over the reals is computable if it fulfills Grzegorczyk and Lacombe's, or equivalently, Pour-El and Richards' definition of computable continuous real function [Grz55,Grz57,Lac55,PR89]. Furthermore, we say that it is elementary computable if the corresponding functional is elementary, according to the definition proposed by Grzegorczyk or Zhou [Grz55,Zho97]. Conversely, as in the previous section we say that $\mathcal{L} + \theta_k$ contains a function on $\mathbb{N}$ if it contains some extension of it to the reals.

First, it is possible to approximate effectively any function in $\mathcal{L} + \theta_k$ in elementary time. Proposition 2 implies then that the discrete approximation is an elementary function as well. The constructive inductive proof is given in [CMC00] and is based on numerical techniques to integrate any function definable in $\mathcal{L} + \theta_k$. The elementary bound on the time complexity of numerical integration follows from Proposition 18. Thus:

Proposition 19 *If f belongs to $\mathcal{L} + \theta_k$ for any $k > 2$, then f is elementarily computable.*

Moreover, we can approximate any $\mathcal{L} + \theta_k$ function that sends integers to integers to error less than $1/2$ and obtain its value exactly in elementary time:

Proposition 20 *If a function $f \in \mathcal{L}+\theta_k$ is an extension of a function $\tilde{f} : \mathbb{N} \to \mathbb{N}$, then $\tilde{f}$ is elementary.*

We can also show the converse of this, i.e. that $\mathcal{L} + \theta_k$ contains all elementary functions, or rather, extensions of them to the reals.

First, we show that $\mathcal{L} + \theta_k$ contains (extensions to the reals of) the basis functions of $\mathcal{E}$. Successor and addition are easy to generate in $\mathcal{L}$. So are $\sin x$, $\cos x$ and e^x, since each of these are solutions of simple linear differential equations, and arbitrarily rational constants as shown in [CMC00]. With θ_k we can define cut-off subtraction $x \dotdiv y$ as follows. We first define a function $s(z)$ such that $s(z) = 0$ when $z \leq 0$ and $s(z) = 1$ when $z \geq 1$, for all $z \in \mathbb{Z}$. This can be done in $\mathcal{L} + \theta_k$ by setting $s(0) = 0$ and $\partial_z s(z) = c_k \theta_k(z(1-z))$, where $c_k = 1/\int_0^1 z^k(1-z)^k\,dz$ is a rational constant depending on k. Then $x \dotdiv y = (x-y)s(x-y)$ is an extension to the reals of cut-off subtraction.

Now, we just have to show that $\mathcal{L} + \theta_k$ has the same closure properties as $\mathcal{E}$, namely the ability to form bounded sums and products.

Lemma 21 *Let f be a function on $\mathbb{N}$ and let g be the function on $\mathbb{N}$ defined from f by bounded sum or bounded product. If f has an extension to the reals in $\mathcal{L} + \theta_k$ then g does also.*

First of all, for any $f \in \mathcal{L} + \theta_k$ there is a function $F \in \mathcal{L} + \theta_k$ that matches f on the integers, and whose values are constant on the interval $[j, j+1/2]$ for integer j [CMC00]. Bounded sum of f is then easily defined in $\mathcal{L} + \theta_k$ by linear integration: simply write $g(0) = 0$ and $g'(t) = c_k F(t) \theta_k(\sin 2\pi t)$, where c_k is a constant definable in $\mathcal{L} + \theta_k$. Then $g(t) = \sum_{z<n} f(z)$ whenever $t \in [n-1/2, n]$.

Defining the bounded product $g_n = \prod_{j<n} f_j$ of f in $\mathcal{L} + \theta_k$ is more difficult. We can approximate the iteration $g_{j+1} = g_j f_j$ using synchronized clock functions as in proof of Proposition 11. However, since the model we propose here only allows linear integration, the simulated functions cannot coincide exactly with the bounded product. Nevertheless, we can define a sufficiently close approximation because f and g have bounded growth by Proposition 18. Then since f and g have integer values, the accumulated error on $[0, n]$ resulting from this approximation can be removed with a suitable continuous step function ϕ definable in $\mathcal{L} + \theta_k$ by $\phi(0) = 0$ and $\phi'(t) = c_k \theta_k(-\cos 2\pi t)$, where c_k is a constant depending only on k. The function ϕ is such that $\phi(t) = j$ if $t \in [j-1/4, j+1/4]$ for all integer j and so, ϕ returns the integer closest to t as long as the error is $1/4$ or less.

If we define a two-component function $y(\tau, t)$ where $y_1(\tau, 0) = y_2(\tau, 0) = 1$,

$$\partial_t y_1 = (y_2 F(t) - y_1)\, c_k \theta_k(\sin 2\pi t)\, \beta(\tau)$$
$$\partial_t y_2 = (y_1 - y_2)\, c_k \theta_k(-\sin 2\pi t)\, \beta(\tau) \qquad (5)$$

and $\beta(\tau)$ is an increasing function of τ, then $g_n = \phi(y_1(n, n))$. We can show that if β grows fast enough (roughly as fast as the bound on f given in Proposition 18), then by setting $\tau = n$ we can make the approximation error

$|y_1(n,n) - g_n|$ as small as we like, and then remove it with ϕ. Note that the system 5 is linear in y_1 and y_2. Details are given in [CMC00].

We illustrate this construction in Figure 3. We approximate the bounded product of the identity function, i.e. the factorial $(n-1)! = \prod_{j<n} j$. As before, we numerically integrated Equation (5) using a standard package.

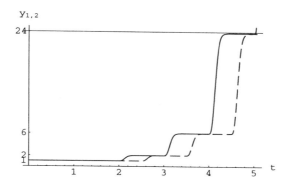

Fig. 3. A numerical integration of Equation (5), where f is a $\mathcal{L} + \theta_k$ function such that $f(0) = 1$ and $f(x) = x$ for $x \geq 1$. Here, $k = 2$. We obtain an approximation of an extension to the reals of the factorial function. In this example, where we chose a small $\tau < 4$, the approximation is just sufficient to remove the error with ϕ and obtain exactly $\prod_{n<5} n = 4! = \phi(y_1(5))$.

We do not know whether $\mathcal{L} + \theta_k$ is closed under bounded product for functions with real, rather than integer, values. We conjecture that it is not, but we have no proof of this. In any case, we have proved that:

Proposition 22 *If f is an elementary function, then $\mathcal{L} + \theta_k$ contains an extension of f to the reals.*

Taken together, Propositions 19, 20 and 22 show that the analog class $\mathcal{L} + \theta_k$ corresponds to the elementary functions in a natural way. It is interesting that linear integration alone gives extensions to the reals of all elementary functions, since these are all the functions that can be computed by any practically conceivable digital device. In terms of dynamical systems, $\mathcal{L} + \theta_k$ corresponds to cascades of finite depth, each level of which depends linearly on its own variables and the output of the level before it. We find it surprising that such systems, as opposed to highly non-linear ones, have so much computational power.

Next, we will extend the above results to the higher levels of the Grzegorczyk hierarchy, $\mathcal{E}^n$ for $n \geq 3$, by allowing the GPAC to solve a certain number of nonlinear differential equations.

Definition 23 (The hierarchy $\mathcal{G}_n + \theta_k$) *Let $\mathcal{G}_3 + \theta_k = \mathcal{L} + \theta_k$ be the smallest class containing the constants $0, 1, -1,$ and π, the projections, and θ_k, which is closed* under composition and linear integration. For $n \geq 3$, $\mathcal{G}_{n+1} + \theta_k$ is defined as the class which contains $\mathcal{G}_n + \theta_k$ and solutions of Equation (2) applied to functions f in $\mathcal{G}_n + \theta_k$, and which is closed under composition and linear integration.

Note that $\mathcal{L} + \theta_k$ contains $\exp^{[2]}(x)$, which grows roughly as fast as E_2 as noted in Section 2. Since $\mathcal{G}_{n+1} + \theta_k$ contains iterations of functions in $\mathcal{G}_n + \theta_k$, it contains at least one function that grows roughly as E_n. From Proposition 4 and using techniques similar to the proofs of Propositions 19, 20 and 22 we can then show (see [CMC00] for details) that:

Proposition 24 *The following correspondences exist between $\mathcal{G}_n + \theta_k$ and the levels of the Grzegorczyk hierarchy, $\mathcal{E}^n$ for all $n \geq 3$:*

1. *Any function in $\mathcal{G}_n + \theta_k$ is computable in $\mathcal{E}^n$.*
2. *If $f \in \mathcal{G}_n + \theta_k$ is an extension to the reals of some $\tilde{f}$ on $\mathbb{N}$, then $\tilde{f} \in \mathcal{E}^n$.*
3. *Conversely, if $f \in \mathcal{E}^n$ then some extension of it to the reals is in $\mathcal{G}_n + \theta_k$.*

A few remarks are in order. First, since we only have to solve Equation (2) $n - 3$ times to obtain a function that grows as fast as E_{n-1}, this analog model contains exactly the nth level of the Grzegorczyk hierarchy if it is allowed to solve $n - 3$ non-linear differential equations of the form (2).

Secondly, notice that Proposition 24 implies that $\cup_n (\mathcal{G}_n + \theta_k)$ includes all primitive recursive functions since, as mentioned in Proposition 5, $\cup_n \mathcal{E}^n = \mathcal{PR}$.

Finally, instead of allowing the GPAC to solve Equation (2), we can keep everything linear and define $\mathcal{G}_n + \theta_k$ by adding a new basis function which is an extension to the reals of E_{n-1}. While this produces a smaller set of functions on $\mathbb{R}$, it produces extensions to $\mathbb{R}$ of the same set of functions on $\mathbb{N}$ as the class defined here [CMC00].

6 Zero-Finding on the Reals

In [Moo96] another definition of analog computation is proposed, the $\mathbb{R}$-*recursive* functions. These are the functions that can be generated from the constants 0 and 1 from composition, integration of differential equations of the form $h(\boldsymbol{x}, 0) = f(\boldsymbol{x})$ and $\partial_y h(\boldsymbol{x}, y) = g(h(\boldsymbol{x}, y), \boldsymbol{x}, y)$ on whatever interval the result $h = f + \int g \, dy$ is unique and well-defined, and the following minimization or zero-finding operator:

Definition 25 (Zero-finding) *If f is $\mathbb{R}$-recursive, then $h(\boldsymbol{x}) = \mu_y f(\boldsymbol{x}, y) = \inf\{y \in \mathbb{R} \mid f(\boldsymbol{x}, y) = 0\}$ is $\mathbb{R}$-recursive whenever it is well-defined, where the infimum is defined to find the zero of $f(\boldsymbol{x}, \cdot)$ closest to the origin, that is, to minimize $|y|$. If both $+y$ and $-y$ satisfy this condition we return the negative one by convention.*

To what extent is this the correct extension of recursion theory to the reals? Integration of differential equations seems to be the closest continuous analog to primitive recursion: we define $h(y + dy)$, rather than $h(y + 1)$, in terms of $h(y)$. This definition of zero-finding also seems fairly intuitive; however, it is hard to imagine how a physical process could locate a zero of a function unless it is differentiable, or at least continuous.

Although it is not explicitly recognized as such in [Moo96], the definition of $\mathbb{R}$-recursive relies on another operator, namely the assumption that $x \cdot 0 = 0$ even when x is undefined. While this can be justified by defining

$$x \cdot y = \int_0^y x \, dy$$

it actually deserves to be thought of as an operator in its own right, since it can convert partial functions into total ones. We can then combine μ with a "compression trick" to search over the integers to see if a function over $\mathbb{N}$ has any zeroes. This allows us to solve the Halting Problem; however, in physical terms, it corresponds to having our device run faster and faster, until it accomplishes an infinite amount of computation in finite time. This would require an infinite amount of energy, infinite forces, or both. Thus the *physical Church-Turing thesis*, that physically feasible devices can only compute recursive functions, remains intact.

In fact, by iterating this construction, we can compute functions in any level of the *arithmetical hierarchy* Σ_ω^0, of which the recursive and partial recursive functions are just the 0th and 1st levels respectively, Σ_0^0 and Σ_1^0. Sets in the jth level of this hierarchy can be defined with j alternating quantifiers over the set of integers, $\exists$ and $\forall$, applied to recursive predicates [Odi89]. We note that Bournez uses a similar recursion to show that systems with piecewise-constant derivatives can compute various levels of the hyperarithmetical hierarchy [Bou99,Bou99b].

If we quantify over functions instead of integers we define another hierarchy of even larger classes called the *analytical hierarchy* Σ_ω^1 [Odi89]. These functions are $\mathbb{R}$-recursive as well, since we can encode sequences of integers as continued fractions and then search over the reals for a zero [Moo96].

Since this μ-operator is unphysical, in [Moo96] we stratify the class of $\mathbb{R}$-recursive functions according to how many nested uses of the μ-operator are needed to define a given function. Define M_j as the set of functions definable from the constants $0, 1, -1$ with composition, integration, and j or fewer nested uses of μ. (We allow -1 as fundamental since otherwise we would have to define it as $\mu_y[y + 1]$. This way, $\mathbb{Z}$ and $\mathbb{Q}$ are contained in M_0.) We call this the μ-hierarchy.

For functions over $\mathbb{R}$ we believe that the μ-hierarchy is distinct. For instance, the characteristic function $\chi_\mathbb{Q}$ of the rationals is in M_2 but not in M_1. The recursive and partial recursive functions over $\mathbb{N}$ have extensions to the reals in M_2 and M_3 respectively. For higher j, M_j contains various levels of the analytical hierarchy as shown in Table 1, but we have no upper bounds for these classes.

However, in the classical setting, Kleene showed that any partial recursive function can be written in the form $h(x) = U(\mu_y T(x,y))$, where U and T are primitive recursive functions. Moreover, U and T can be elementary, or taken from an even smaller class [Odi89,Cut80,Ros84]. Thus the set of partial recursive functions can be defined even if we only allow one use of the zero-finding operator, and the μ-hierarchy collapses to its first level. Since the class $\mathcal{L} + \theta_k$ discussed in the previous section includes the elementary functions, this also means that combining a single use of μ with linear integration gives, at a minimum, the partial recursive functions.

If μ is not used at all we get M_0, the "primitive $\mathbb{R}$-recursive functions." These include the differentially algebraic functions $\mathcal{G}$ discussed above, as well as constants such as e and π; however, since the definition of integration in [Moo96] is somewhat more liberal than that of the GPAC where we require the solution to be unique for a domain of generation with a non-empty interior, M_0 also includes functions with discontinuous derivatives like $|x| = \sqrt{x^2}$ and the sawtooth function $\sin^{-1}(\sin x)$.

7 Conclusion

We have explored a variety of models of analog computation inspired by Shannon's GPAC. By allowing various basis functions and operators, we obtain analog counterparts of various function classes familiar from classical recursion theory. We summarize these in Table 1.

There are many open questions waiting to be addressed. These include:

1. Can we obtain upper bounds on classes like $\mathcal{G} + \theta_k$ and M_j, that so far we only have lower bounds for?
2. Is there more physical version of the μ operator which nonetheless extends $\mathcal{G}$ in a non-trivial way?
3. Do lower-level classes like P and NP have natural analog counterparts?

We look forward to addressing these with the enthusiastic reader.

Acknowledgements. We thank José Félix Costa for his collaboration on several results described in this paper, and Christopher Pollett and Ilias Kastanas for helpful discussions. This work was partially supported by grants from the Fundação para a Ciência e Tecnologia (PRAXIS XXI/BD/18304/98) and the Luso-American Development Foundation (754/98). MLC also thanks the Santa Fe Institute for hosting a visit that made this work possible.

References

[Arn96] V. I. Arnold. *Equations Différentielles Ordinaires*. Editions Mir, 5 ème edition, 1996.

[Bab73] A. Babakhanian. Exponentials in differentially algebraic extension fields. *Duke Math. J.*, 40:455–458, 1973.

basis functions	operators	recursive classes
$0, 1, U_i^n$	$\circ, \int, x \cdot 0 = 0, (4\mathbf{j}+3).\mu$	$M_{4j+3} \supseteq \Sigma_j^1, \Pi_j^1$
$0, 1, U_i^n$	$\circ, \int, x \cdot 0 = 0, 3.\mu$	$M_3 \supseteq \Sigma_1^0$
$0, 1, U_i^n$	$\circ, \int, 2.\mu$	$M_2 \supseteq \Sigma_0^0$
$0, 1, U_i^n, \theta_k$	$\circ, \int_{\text{linear}}, 1.\mu$	$\mathcal{L} + \theta_k + \mu \supseteq \Sigma_1^0$
$0, 1, -1, U_i^n, \theta_k$	$\circ, \int$	$\mathcal{G} + \theta_k \supseteq \mathcal{PR}$
$0, 1, -1, U_i^n, \theta_k$	$\circ, \int_{\text{linear}}, (\mathbf{n}-3).\int_2$	$\mathcal{G}_n + \theta_k = \mathcal{E}^n, n \geq 3$
$0, 1, -1, \pi, U_i^n, \theta_k$	$\circ, \int_{\text{linear}}$	$\mathcal{L} + \theta_k = \mathcal{E}$
$0, 1, -1, U_i^n$	$\circ, \int$	$\mathcal{G} =$ d.a. functions

Table 1. Summary of the main results of the paper. The operations in the definitions of the recursive classes on the reals are denoted by: $\circ$ for composition, $\int_{\text{linear}}$ for linear integration, $\int_2$ for integrating equations of the form (2) as in definition 23, $\int$ for unrestricted integration and μ for zero-finding on the reals. A number before an operation, as in n.μ, means that the operator can be applied at most n times.

[Ban75] S. Bank. Some results on analytic and meromorphic solutions of algebraic differential equations. *Advances in mathematics*, 15:41–62, 1975.

[BBV37] S. Bose, N. Basu and T. Vijayaraghavan. A simple example for a theorem of Vijayaraghavan. *J. London Math. Soc.*, 12:250–252, 1937.

[Bou99] O. Bournez. Achilles and the tortoise climbing up the hyper-arithmetical hierarchy. *Theoretical Computer Science*, 210(1):21–71, 1999.

[Bou99b] O. Bournez. *Complexité algorithmique des systèmes dynamiques co hybrides*. PhD thesis, École Normale Supérieure de Lyon, 1999.

[Bow96] M.D. Bowles. U.S. technological enthusiasm and the British technological skepticism in the age of the analog brain. *IEEE Annals of the History of Computing*, 18(4):5–15, 1996.

[Bra95] M. S. Branicky. Universal computation and other capabilities of hybrid and continuous dynamical systems. *Theoretical Computer Science*, 138(1), 1995.

[BSF00] A. Ben-Hur, H. Siegelmann, and S. Fishman. A theory of complexity for continuous time systems. To appear in *Journal of Complexity*.

[BSS89] L. Blum, M. Shub, and S. Smale. On a theory of computation and complexity over the real numbers: NP-completnes, recursive functions and universal machines. *Bull. Amer. Math. Soc.*, 21:1–46, 1989.

[CMC99] M.L. Campagnolo, C. Moore, and J.F. Costa. Iteration, inequalities, and differentiability in analog computers. To appear in *Journal of Complexity*.

[CMC00] M.L. Campagnolo, C. Moore, and J.F. Costa. An analog characterization of the subrecursive functions. In P. Kornerup, editor, *Proc. of the 4th Conference on Real Numbers and Computers*, pages 91–109. Odense University, 2000.

[Cut80] N. J. Cutland. *Computability: an introduction to recursive function theory*. Cambridge University Press, 1980.

[Grz53] A. Grzegorczyk. Some classes of recursive functions. *Rosprawy Matematyzne*, 4, 1953. Math. Inst. of the Polish Academy of Sciences.

[Grz55] A. Grzegorczyk. Computable functionals. *Fund. Math.*, 42:168–202, 1955.

[Grz57] A. Grzegorczyk. On the definition of computable real continuous functions. *Fund. Math.*, 44:61–71, 1957.

[Hay96] H.K. Hayman. The growth of solutions of algebraic differential equations. *Rend. Mat. Acc. Lincei.*, 7:67–73, 1996.

[Kál43] L. Kálmar. Egyzzerü példa eldönthetetlen aritmetikai problémára. *Mate és Fizikai Lapok*, 50:1–23, 1943.

[KCG94] P. Koiran, M. Cosnard, and M. Garzon. Computability with low-dimensional dynamical systems. *Theoretical Computer Science*, 132:113–128, 1994.

[Kel76] W. Thomson (Lord Kelvin). On an instrument for calculating the integral of the product of two given functions. *Proc. Royal Society of London*, 24:266–268, 1876.

[KM99] P. Koiran and C. Moore. Closed-form analytic maps in one or two dimensions can simulate Turing machines. *Theoretical Computer Science*, 210:217–223, 1999.

[Lac55] D. Lacombe. Extension de la notion de fonction récursive aux fonctions d'une ou plusieurs variables réelles I. *C. R. Acad. Sci. Paris*, 240:2478–2480, 1955.

[LR87] L. Lipshitz and L. A. Rubel. A differentially algebraic replacement theorem, and analog computation. *Proceedings of the A.M.S.*, 99(2):367–372, 1987.

[Mee93] K. Meer. Real number models under various sets of operations. *Journal of Complexity*, 9:366–372, 1993.

[Moo90] C. Moore. Unpredictability and undecidability in dynamical systems. *Physical Review Letters*, 64:2354–2357, 1990.

[Moo96] C. Moore. Recursion theory on the reals and continuous-time computation. *Theoretical Computer Science*, 162:23–44, 1996.

[Moo98] C. Moore. Dynamical recognizers: real-time language recognition by analog computers. *Theoretical Computer Science*, 201:99–136, 1998.

[Odi89] P. Odifreddi. *Classical Recursion Theory*. Elsevier, 1989.

[Orp97a] P. Orponen. On the computational power of continuous time neural networks. In *Proc. SOFSEM'97, the 24th Seminar on Current Trends in Theory and Practice of Informatics*, Lecture Notes in Computer Science, pages 86–103. Springer-Verlag, 1997.

[Orp97b] P. Orponen. A survey of continuous-time computation theory. In D.-Z. Du and K.-I Ko, editors, *Advances in Algorithms, Languages, and Complexity*, pages 209–224. Kluwer Academic Publishers, Dordrecht, 1997.

[PE74] M. B. Pour-El. Abtract computability and its relation to the general purpose analog computer. *Trans. Amer. Math. Soc.*, 199:1–28, 1974.

[PR89] M. B. Pour-El and J. I. Richards. *Computability in Analysis and Physics*. Springer-Verlag, 1989.

[Ros84] H. E. Rose. *Subrecursion: functions and hierarchies*. Clarendon Press, 1984.

[Rub89] L. A. Rubel. Digital simulation of analog computation and Church's thesis. *The Journal of Symbolic Logic*, 54(3):1011–1017, 1989.

[Rub89b] L. A. Rubel. A survey of transcendentally transcendental functions. *Amer. Math. Monthly*, 96:777–788, 1989.

[SF98] H. T. Siegelmann and S. Fishman. Analog computation with dynamical systems. *Physica D*, 120:214–235, 1998.

[Sha41] C. Shannon. Mathematical theory of the differential analyser. *J. Math. Phys. MIT*, 20:337–354, 1941.

[Sie98] H. Siegelmann. *Neural Netwoks and Analog Computation: Beyond the Turing Limit*. Birkhauser, 1998.

[Vij32] T. Vijayaraghavan. Sur la croissance des fonctions définies par les équations différentielles. *C. R. Acad. Sci. Paris*, 194:827–829, 1932.

[VSD86] A. Vergis, K. Steiglitz, and B. Dickinson. The complexity of analog computation. *Mathematics and computers in simulation*, 28:91–113, 1986.

[Zho97] Q. Zhou. Subclasses of computable real functions. In T. Jiang and D. T. Lee, editors, *Computing and Combinatorics*, Lecture Notes in Computer Science, pages 156–165. Springer-Verlag, 1997.

P Systems with Valuations

Carlos Martín-Vide[1] and Victor Mitrana[2]*

[1] Research Group in Mathematical Linguistics, Rovira i Virgili University, Tarragona, Spain
[2] Faculty of Mathematics, University of Bucharest, Bucharest, Romania

Abstract. We propose a new variant of P systems having as simple as possible evolution rules in which the communication of objects and the membrane dissolving are regulated by a valuation mapping (a morphism which assigns to each word an integer value). Two NP-complete problems are solved by P systems with valuation in linear time.

1 Introduction

A new computing model inspired by the hierarchical and modularized cell structure has been recently proposed in [9]. This model - simply called P system - consists of a hierarchical architecture, the so-called *membrane structure*, composed of membranes recurrently distinguished in a unique outermost membrane. Each membrane is meant as the border of a *cell* containing certain *objects* and *evolution rules* acting on these objects. The system might be viewed as a computing device by permitting the objects to evolve, in parallel, in accordance with the evolution rules from the cells where the objects are situated, and to pass the adjacent membranes. The membranes, excepting the outermost membrane, can be dissolved, the content (objects and rules) of a cell delimited by a dissolved membrane being lost. Thus, the model has the potential to supply massive parallel and distributed computations. For more motivations we refer to the aforementioned seminal paper [9].

The large number of possibilities for choosing the objects (strings, symbols, etc.), the evolution rules (rewriting rules, splicing rules, etc.), and the ways of passing membranes, dissolving and creating new cells has led to many variants of P systems; see [7, 11–15] and the early survey [10].

P systems have been investigated as language or integer vector set generating devices [3, 17] and as potential tools for solving various problems, especially problems which require resources exceeding those of contemporary electronic computers, see [7, 15]. The latter offers a different way of performing and looking at computations: data could be encoded in objects and the P system cells execute computational operations on their own objects. Thus, when discussing the potential of a given variant of P system to solve a problem one may say that one designs a P system algorithm, or a P algorithm.

---

* Research supported by the Dirección General de Enseñanza Superior e Investigación Científica, SB 97–00110508.

This work proposes a new variant of P systems and discusses the potential of this variant for solving hard computational problems. To this aim, we propose two P algorithms for two NP-complete problems, namely the 3-colorability problem and the Bounded Post Correspondence Problem. Both algorithms require polynomial time in the size of the given instances.

In our variant, the objects are strings and the evolution rules are as simple as possible (point mutations: substitutions, deletions, insertions) without any information regarding the target cell for the new objects as in the most of the variants considered so far. Furthermore, any cell contains rules of one type only, namely either substitution or deletion or insertion rules. Each cell is electrically charged (as in [7, 15]). There are some important differences between the variant presented here and the other ones considered so far, which could make this variant more adequate to be biologically implemented. The main and completely new feature of this variant is the valuation mapping which assigns to each string an integer value, depending on the values assigned to its symbols. Actually, we are not interested in computing the value of a string (as it is done, for instance, in valence grammars [8]) but the sign of this value. By means of this valuation, one may say that the strings are also electrically polarized. Thus, the strings migration from one cell to another through the membrane which separates the two cells seems to be more natural. Another more natural assumption concerns the dissolving way of a membrane. In this model, a membrane is automatically dissolved when it contains no string, excepting the outermost membrane which is never dissolved.

In the next section we define formally our variant and then propose two P algorithms for solving the two NP-complete problems mentioned above. Some conclusions and final remarks end the paper.

2 P Systems: Basic Definitions

We start by summarizing the notions used throughout the paper. An *alphabet* is a finite and nonempty set of symbols. Any sequence of symbols from an alphabet V is called *string (word)* over V. The set of all strings over V is denoted by V^* and the empty string is denoted by ε. The length of a string x is denoted by $|x|$. For additional formal language definitions and notations we refer to [18].

A *multiset* over a set X is a mappingg $M : X \longrightarrow \mathbf{N} \cup \{\infty\}$. The number $M(x)$ expresses the number of copies of $x \in X$ in the multiset M. When $M(x) = \infty$, then x appears arbitrarily many times in M. The set $supp(M)$ is the support of M, i.e., $supp(M) = \{x \in X \mid M(x) > 0\}$. For two multisets M_1 and M_2 over X we define their union by $(M_1 \cup M_2)(x) = M_1(x) + M_2(x)$. For other operations on multisets the reader may consult [1].

A homomorphism from the monoid V^* into the monoid (group) of additive integers $\mathbf{Z}$ is called *valuation* of V^* in $\mathbf{Z}$.

An *elementary cell* is meant as a construct $[C]$, where C is the content of the cell, represented by a triple $C = (S, R, p)$ placed in a membrane, represented by the pair of square brackets [,]. The parameters appearing in $C = (S, R, p)$ are:

- S is a multiset of objects, of a finite support,
- R is a finite set of evolution rules,
- $p \in \{-, 0, +\}$ is the polarity of the cell.

For an elementary cell $[C]$ as above we denote $C(1) = S, C(2) = R$, and $C(3) = p$.

A *super-cell* (SC for short) is defined as follows:

(i) Every elementary cell is an SC.
(ii) If $\rho_1, \rho_2, \ldots, \rho_n$ are SC and $[C]$ is an elementary cell, then $\rho = [C\rho_1\rho_2 \ldots \rho_n]$ is an SC. The polarity of ρ is $C(3)$.
(iii) Every SC is obtained by applying iteratively the rules (i) and (ii).

For a given SC $\rho = [C\rho_1\rho_2 \ldots \rho_n]$ as above, $\rho_i, 1 \leq i \leq n$, are called the direct SC-components of ρ. In a recursive way, one defines the SC-components of an SC; more precisely, each direct SC-component is an SC-component and if τ is a direct SC-component of ρ, then τ is an SC-component of any SC having ρ as an SC-component. Moreover, let $cont(\rho) = C$.

A *P system with valuation* is a triple

$$\Pi = (V, \mu, \varphi),$$

where:

- V is an alphabet.
- μ is the initial SC such that the objects of all its SC-components are strings over V of length at most one. For each SC-component ρ of μ, the following conditions are satisfied
  (i) $cont(\rho)(2)$ is a finite set of evolution rules of just one of the following forms:
    - $a \to b, a, b \in V$ (substitution rules),
    - $a \to \varepsilon, a \in V$ (deletion rules),
    - $\varepsilon \to a, a \in V$ (insertion rules),
  (ii) The sign of $\varphi(x)$ is exactly $cont(\rho)(3)$ for all strings x in $cont(\rho)(1)$.
- φ is a valuation of V^* in $\mathbf{Z}$.

In what follows, we consider that each string appearing in any SC has an arbitrarily large number of copies, so that we shall identify multisets by their supports. A substitution rule $a \to b$ is applied to a string (one of its copies) in the usual way (one occurrence of a, nondeterministically chosen, is replaced by b). The same rule may replace different occurrences of the same symbol in different copies of the same string. If more than one rule applies to a string, one of them is chosen nondeterministically for each copy. Unlike their common use, deletion and insertion rules are applied only to the end of the string.

Thus, a deletion rule $a \to \varepsilon$ can be applied only to a string which ends by a, say wa, leading to the string w, and an insertion rule $\varepsilon \to a$ applied to a string x consists of adding the symbol a to the end of x, obtaining xa.

A P system with valuation Π works as follows. Assume that ρ is the current SC of Π. This SC changes into another SC η - we write $\rho \Longrightarrow \eta$ - by performing the following sequence of actions:

1. In parallel, for each SC-component γ of ρ, one applies a rule from $cont(\gamma)(2)$ to all copies of the strings in $cont(\gamma)(1)$.
2. One computes the integer value assigned to each obtained string by means of the valuation mapping φ. Then, the obtained strings are distributed as follows:
    - Those strings whose assigned value sign is the same with the polarity of γ remain in $cont(\gamma)(1)$.
    - Those strings whose assigned value sign is the same with the polarity of a direct SC-component of γ or of a SC having γ as a direct SC-component go to the content of the respective SC. If several such SCs have the same polarity, then this distribution is done nondeterministically.
    - The other strings are completely removed.
3. If $cont(\gamma)(1) = \emptyset$ after the above step, then the membrane of γ is *dissolved*. More formally, if $\tau = [C_1 \tau_1 \tau_2 \ldots \tau_k]$, with $\tau_i = \gamma$ for some $1 \leq i \leq k$, and $\gamma = [C_2 \gamma_1 \gamma_2 \ldots \gamma_p]$ then, by dissolving the membrane of γ one gets the new SC

$$[C_1 \tau_1 \ldots \tau_{i-1} \gamma_1 \gamma_2 \ldots \gamma_p \tau_{i+1} \ldots \tau_k].$$

By definition, the outermost membrane is never dissolved.

A computation of a given P system $\Pi = (V, \mu, \varphi)$ is a finite sequence of SCs

$$\mu = \mu_0 \Longrightarrow \mu_1 \Longrightarrow \mu_2 \Longrightarrow \ldots \Longrightarrow \mu_n$$

for some $n \geq 0$ such that μ_n cannot change anymore into another SC. The output of the above computation is represented by the multiset of strings in the outermost membrane of μ_n, that is $cont(\mu_n)(1)$. If μ_n is an elementary cell, we say that the computation is *pure* and Π purely computes $cont(\mu_n)(1)$. Note that there could exist computations which do not end in an elementary cell. The time complexity of the above computation is the number of steps, that is n.

3 Solving NP-complete Problems

In this section we attack two problems known to be NP-complete and construct P systems with valuations for solving them. Furthermore, the algorithms based on P systems presented in this section compute all solutions.

First we need an auxiliary result which will be used in the two announced algorithms.

Lemma 1 *Let M be a multiset of finite support, $supp(M) = \{x_1, x_2, \ldots, x_n\}$, every string appearing arbitrarily many times in M. One can construct a P system which purely computes M in $O(p)$ time, where $p = max\{|x_i| \mid 1 \leq i \leq n\}$.*

Proof. Assume that $x_i = a_1^{(i)} a_2^{(i)} \ldots a_{p_i}^{(i)}$, $a_j^{(i)} \in V$, $1 \leq i \leq n$, $1 \leq j \leq p_i$; furthermore $|x_i| > 1$, $1 \leq i \leq t$, and $|x_i| \leq 1$, $t+1 \leq i \leq n$. We construct the P system $\Pi = (U, \mu, \varphi)$ with

$$U = V \cup \{b_j^{(i)} \mid 1 \leq i \leq n, 1 \leq j \leq p_i\} \cup \{q_j^{(i)} \mid 1 \leq i \leq n, 1 \leq j \leq p_i - 2\} \cup$$
$$\{[r_j^{(i)}, k] \mid 1 \leq i \leq n, 1 \leq j \leq p_i, 1 \leq k \leq 2p\} \cup \{c, \bar{c}, \hat{c}\},$$
$$\mu = [T \rho_{x_1} \rho_{x_2} \ldots \rho_{x_t}].$$

We set
$$T(1) = \{x_i \mid t+1 \leq i \leq n\}, \quad T(2) = \emptyset, \quad T(3) = 0.$$

For each $1 \leq i \leq t$, the SC-component ρ_{x_i} is defined as follows, assuming that p_i is odd (the case when p_i is even is left to the reader):

$$\rho_{x_i} = [S_1^i [S_2^i [\ldots S_{p_i}^i [A_{p_i}^i [A_{p_i-1}^i [\ldots A_2^i] \ldots]]] \ldots]],$$

where

$$A_2^i = (\{b_1^{(i)}\}, \{\varepsilon \to b_2^{(i)}\}, +),$$
$$A_j^i = (\{q_{j-2}^{(i)}\}, \{\varepsilon \to b_j^{(i)}\}, sgn((-1)^j)), 3 \leq j \leq p_i$$
$$S_j^i = (\{[r_j^{(i)}, 1]\}, \{b_j^{(i)} \to a_j^{(i)}\} \cup \{[r_j^{(i)}, k] \to [r_j^{(i)}, k+1] \mid 1 \leq k \leq 2p_i - j - 1\}$$
$$\cup \{[r_j^{(i)}, 2p_i - j] \to c\}, sgn((-1)^{j+1})), 2 \leq j \leq p_i,$$
$$S_1^i = (\{[r_1^{(i)}, 1]\}, \{b_1^{(i)} \to a_1^{(i)}\} \cup \{[r_1^{(i)}, k] \to [r_1^{(i)}, k+1] \mid 1 \leq k \leq 2p_i - 2\} \cup$$
$$\{[r_1^{(i)}, 2p_i - 1] \to \bar{c}\}, +).$$

Here $sgn((-1)^m)$ delivers the sign of $(-1)^m$.

The valuation mapping is defined by

$$\varphi(a_j^{(i)}) = \varphi(c) = 0, \ 1 \leq i \leq n, \ 1 \leq j \leq p_i,$$
$$\varphi(b_1^{(i)}) = 1, \ 1 \leq i \leq n,$$
$$\varphi(b_j^{(i)}) = (-1)^{j+1} \cdot 2, \ 1 \leq i \leq n, \ 2 \leq j \leq p_i,$$
$$\varphi(q_j^{(i)}) = (-1)^j \cdot 2j, \ 1 \leq i \leq n, \ 1 \leq j \leq p_i - 2,$$
$$\varphi([r_j^{(i)}, k]) = (-1)^{j+1}, \ 1 \leq i \leq n, \ 1 \leq j \leq p_i, \ 1 \leq k \leq 2p,$$
$$\varphi(\bar{c}) = -1,$$
$$\varphi(\hat{c}) = 1.$$

It is relatively easy to observe that each SC-component ρ_{x_i} produces and sends to the content of T an arbitrarily large number of copies of the string

x_i and no other strings. In the first phase, copies of the string $b_1^{(i)} b_2^{(i)} \ldots b_{p_i}^{(i)}$ are produced, all cells directly involved in this generating process being lost at the end of the process. Then, step by step, the symbols $b_{p_i}^{(i)}, b_{p_i-1}^{(i)}, \ldots, b_1^{(i)}$ are replaced by $a_{p_i}^{(i)}, a_{p_i-1}^{(i)}, \ldots, a_1^{(i)}$, respectively, at every step the most inner membrane being dissolved.

As soon as such a component finishes this task - the copies of x_i have been sent to the content of T - it is dissolved. Consequently, the computation ends in an elementary cell which is $[T]$ and the result is exactly the given multiset M. Clearly, the time complexity of this computation is $2p$, which concludes the proof.

Remark. It is worth mentioning here that the output of the P system from the previous lemma is the content of an elementary cell having a neutral charge and no evolution rule. Thus, this system may be used as an SC-component of another P system in the aim of generating cells having a certain content. By this remark, without loss of generality, in the next two constructions we may start with P systems having the initial SC with some SC-components containing multisets of arbitrary finite support of strings of arbitrary length.

The first problem we deal with is the well known "3-colorability problem". This problem is to decide whether each vertex in an undirected graph can be colored by using three colors (say red, blue, and green) in such a way that after coloring, no two vertices which are connected by an edge have the same color.

Proposition 1. *The 3-colorability problem is solvable by a P system with valuation in time linearly bounded by the sum of the numbers of vertices and edges of the graph.*

Proof. Consider the undirected graph $\Gamma = (\{1, 2, \ldots, n\}, U)$, where the edges of U are $e_1, e_2, \ldots, e_m$ for some m. We define the P system $\Pi = (V, \mu, \varphi)$, where

- $V = \{c, c''\} \cup \bigcup_{i=1}^{n} (\{r_i, b_i, g_i\} \cup \{r_i^{(j,i)}, \bar{r}_i^{(j,i)}, b_i^{(j,i)}, \bar{b}_i^{(j,i)}, g_i^{(j,i)}, \bar{g}_i^{(j,i)} \mid (j,i) \in U\}) \cup \{p_j \mid 0 \leq j \leq 4(m-1)\} \cup \{q_j \mid 0 \leq j \leq 4(m-1)+1\} \cup \{t_j \mid 0 \leq j \leq 4(m-1)+2\} \cup \{c_j' \mid 0 \leq 4(m-1)+3\}$,

- $\mu = [T[F(e_1)[E(e_1)[D(e_1)[C(e_1)[F(e_2)[E(e_2)[D(e_2)[C(e_2)[\ldots[F(e_m) [E(e_m)[D(e_m)[C(e_m)]]]]]\ldots]]]]]]]]]]$, with

$$C(i, j) = (A(i, j), R_1(i, j), 0),$$
$$D(i, j) = (\{q_0\}, R_2(i, j), -),$$
$$E(i, j) = (\{t_0\}, R_3(i, j), 0),$$
$$F(i, j) = (\{c_0\}, R_4(i, j), +),$$

for all edges $(i,j) \in U$, and
$$T = (\emptyset, \emptyset, 0),$$

$$A(i,j) = \begin{cases} \{a_1 a_2 \ldots a_n \mid a_t \in \{r_t, b_t, g_t\}, 1 \leq t \leq n\}, & \text{if } e_m = (i,j), \\ \{p_0\}, & \text{otherwise,} \end{cases}$$

$$R_1(i,j) = \begin{cases} \{b_i \to b'_i, r_i \to r'_i, g_i \to g'_i\}, & \text{if } e_m = (i,j), \\ \{b_i \to b'_i, r_i \to r'_i, g_i \to g'_i\} \cup \{p_j \to p_{j+1} \mid 0 \leq j \leq 4(m-k)\} \cup \\ \{p_{4(m-k)} \to c\}, & \text{if } e_k = (i,j), \end{cases}$$

$$R_2(i,j) = \{b_j \to b_j^{(i,j)}, b_j \to \bar{b}_j^{(i,j)}, r_j \to r_j^{(i,j)}, r_j \to \bar{r}_j^{(i,j)}, g_j \to g_j^{(i,j)},$$
$$g_j \to \bar{g}_j^{(i,j)}, b_j^{(i,j)} \to c, \bar{b}_j^{(i,j)} \to c, r_j^{(i,j)} \to c, \bar{r}_j^{(i,j)} \to c, g_j^{(i,j)} \to c,$$
$$\bar{g}_j^{(i,j)} \to c\} \cup \{q_j \to q_{j+1} \mid 0 \leq j \leq 4(m-k)+1\} \cup$$
$$\{q_{4(m-k)+1} \to c\}, \text{ if } e_k = (i,j),$$

$$R_3(i,j) = \{b'_i \to b_i, r'_i \to r_i, g'_i \to g_i\} \cup \{t_j \to t_{j+1} \mid 0 \leq j \leq 4(m-k)+2\}$$
$$\cup \{t_{4(m-k)+2} \to c\}, \text{ if } e_k = (i,j),$$

$$R_4(i,j) = \{b_j^{(i,j)} \to b_j, \bar{b}_j^{(i,j)} \to b_j, r_j^{(i,j)} \to r_j, \bar{r}_j^{(i,j)} \to r_j, g_j^{(i,j)} \to g_j,$$
$$\bar{g}_j^{(i,j)} \to g_j\} \cup \{c'_j \to c'_{j+1} \mid 0 \leq j \leq 4(m-k)+3\} \cup$$
$$\{c \to c'', c'_{4(m-k)+3} \to c''\}, \text{ if } e_k = (i,j).$$

- $\varphi : V^* \longrightarrow \mathbf{Z}$, defined by

$$\varphi(b_i) = \varphi(r_i) = \varphi(g_i) = 0, \ 1 \leq i \leq n,$$
$$\varphi(c) = \varphi(c'_j) = 4, \ 0 \leq j \leq 4(m-1)+3,$$
$$\varphi(b'_i) = -1, \ \varphi(r'_i) = -2, \ \varphi(g'_i) = -3, \ 1 \leq i \leq n,$$
$$\varphi(t_j) = 0, \ 0 \leq j \leq 4(m-1)+2,$$
$$\varphi(p_j) = 0, \ 0 \leq j \leq 4(m-1),$$
$$\varphi(c'') = \varphi(q_j) = -4, \ 0 \leq j \leq 4(m-1)+1,$$
$$\varphi(r_j^{(i,j)}) = \varphi(g_j^{(i,j)}) = 1, \ 1 \leq j \leq n, \ (i,j) \in U,$$
$$\varphi(b_j^{(i,j)}) = \varphi(\bar{g}_j^{(i,j)}) = 2, \ 1 \leq j \leq n, \ (i,j) \in U,$$
$$\varphi(\bar{r}_j^{(i,j)}) = \varphi(\bar{b}_j^{(i,j)}) = 3, \ 1 \leq j \leq n, \ (i,j) \in U.$$

Some very informal description of the roles of the C, D, E, and F components might help in the further reasoning: C switches symbols on the current edge's i vertex, D switches symbols to set up possible colors for that edge's j

vertex, and E and F undo these switches in the strings that survive the transition from D to E, i.e., represent valid vertex-color assignments relative to the current edge.

Now, let us examine a computation in this P system. First we note that $C(e_m)(1)$ contains strings corresponding to all possible ways of coloring the vertices of Γ, not necessarily fulfilling the requirements of the 3-colorability problem. One starts with μ and after the first step the following situation appear:

- all strings of the content of $[C(e_m)]$ are changed into new strings by using the substitution rules from $C(e_m)(2)$,
- the values assigned to these strings by the valuation mapping φ are negative, so that all of them go to the content of $D(e_m)$ whose polarity is negative,
- the membrane of $C(e_m)$ is dissolved, because its content is now empty; this membrane is the only one dissolved in this step,
- the strings in all SC-components contents, excepting $[C(e_m)]$, are modified according to substitution rules and remain in the same components.

In the next step, after applying the evolution rules, some strings from $D(e_m)(1)$ go to $E(e_m)$, because their value sign by φ is 0, and the others remain in $D(e_m)$, because their value sign by φ is still negative. The only exception is the string c produced by q_1 which is removed because $\varphi(c) > 0$. Those strings which have arrived in $E(e_m)$ were originally those strings in $A(e_m)$ for which the symbols a_i and a_j were different, providing that $e_m = (i,j)$.

In the next step, all obtained strings from those remained in $D(e_m)$ are removed, because their value sign by φ is positive, and the membrane of the cell $[D(e_m)]$ is dissolved. The membrane of the cell $E(e_m)$ is also dissolved, all strings produced by its former strings being now in $F(e_m)$. Now the symbols modified in the previous steps are restored, the membrane of $F(e_m)$ is dissolved and the current SC is now:

$$\mu = [T[F'(e_1)[E'(e_1)[D'(e_1)[C'(e_1)[F'(e_2)[E'(e_2)[D'(e_2)[C'(e_2)[\ldots \\ [F'(e_{m-1})[E'(e_{m-1})[D'(e_{m-1})[C'(e_{m-1})]]]]\ldots]]]]]]]]],$$

with

$$C'(i,j) = (A'(i,j), R_1(i,j), 0),$$
$$D'(i,j) = (\{q_4\}, R_2(i,j), -),$$
$$E'(i,j) = (\{t_4\}, R_3(i,j), 0),$$
$$F'(i,j) = (\{c_4\}, R_4(i,j), +),$$

for all edges $(i,j) \in U \setminus \{e_m\}$, and

$$A'(i,j) = \begin{cases} \{a_1 a_2 \ldots a_n \mid a_t \in \{r_t, b_t, g_t\}, 1 \leq t \leq n,\ a_r \neq a_s, e_m = (r,s)\}, \\ \quad \text{if } e_{m-1} = (i,j), \\ \{p_4\}, \text{ otherwise} \end{cases}$$

By the above explanations one may easily conclude that any computation ends by an elementary cell $[(T', \emptyset, 0)]$. Thus, when the P system halts, we check whether there is a string in the outermost membrane (in $supp(T')$); this instance of the "3-colorability problem" has a solution if and only if the outermost membrane contains at least one string. Furthermore, all strings in the outermost membrane are solutions of the given instance. Clearly, the number of steps in this computation is four times the number of edges of Γ. By Lemma 1, the generating process of all strings in $C(e_m)(1)$ requires time linearly bounded by the number of vertices of the graph, hence the overall claimed complexity follows.

We consider now the Bounded Post Correspondence Problem [2, 4] which is a variant of a much celebrated computer science problem, the Post Correspondence Problem (PCP) known to be unsolvable [5] in the unbounded case. An instance of the PCP consists of an alphabet V and two lists of strings over V

$$u = (u_1, u_2, \ldots, u_n) \quad \text{and} \quad v = (v_1, v_2, \ldots, v_n).$$

The problem asks whether or not a sequence $i_1, i_2, \ldots, i_k$ of positive integers exists, each between 1 and n, such that

$$u_{i_1} u_{i_2} \ldots u_{i_k} = v_{i_1} v_{i_2} \ldots v_{i_k}.$$

The problem is undecidable when no upper bound is given for k and NP-complete when k is bounded by a constant $K \leq n$. A DNA-based solution to the bounded PCP is proposed in [6].

Proposition 2. *The bounded PCP can be solved by a P system with valuation in time linearly bounded by the product of K and the length of the longest string of the two Post lists.*

Proof. Let $u = (u_1, u_2, \ldots, u_n)$ and $v = (v_1, v_2, \ldots, v_n)$ be two Post lists over the alphabet V and $K \geq n$. Consider a new alphabet U and a one-to-one mapping $\alpha : V^* \longrightarrow U^*$ such that for each sequence $i_1, i_2, \ldots, i_j, j \leq K$, of integers between 1 and n, $\alpha(u_{i_1} u_{i_2} \ldots u_{i_j})$ and $\alpha(v_{i_1} v_{i_2} \ldots v_{i_j})$ are two strings, none of them containing two occurrences of the same symbol from U.

We construct the P system $\Pi = (W, \mu, \varphi)$, where

- $W = U \cup \bar{U} \cup \hat{U} \cup U' \cup \{b_1, b_2, \ldots, b_n\} \cup Q$ with
  $\bar{U} = \{\bar{a} \mid a \in U\}$,
  $\hat{U} = \{\hat{a} \mid a \in U\}$,
  $U' = \{a' \mid a \in U\}$,
  and Q will be specified later.
- $\mu = [T(\rho_x)_{x \in \{1,2,\ldots,n\}^{\leq K}}]$, where
  - $T = (\emptyset, \emptyset, 0)$,
  - $\{1, 2, \ldots, n\}^{\leq K}$ is the set of all sequences of integers between 1 and n of length at most K.

- For each sequence of integers between 1 and n, $x = i_1 i_2 \ldots i_j$, $j \leq K$, we assume that $\alpha(u_{i_1} u_{i_2} \ldots u_{i_j}) = c_1 c_2 \ldots c_{p_x}$ and $\alpha(v_{i_1} v_{i_2} \ldots v_{i_j}) = d_1 d_2 \ldots d_{q_x}$. Now, the SC ρ_x is defined as follows (for simplicity we consider that q_x is odd and p_x is even):

$$\rho_x = [\bar{D}(d_1)[\bar{C}(d_1)[\hat{D}(d_2)[\hat{C}(d_2)[\ldots[\bar{D}(d_{q_x})[\bar{C}(d_{q_x})$$
$$[\hat{D}(c_1)[\hat{C}(c_1')[\bar{D}(c_2)[\bar{C}(c_2')[\ldots[\bar{D}(c_{p_x})[\bar{C}(c_{p_x}')]]]\ldots]]].$$

with

$$\bar{D}(c_t) = (\{\varepsilon\}, \{\bar{c}_t \to \varepsilon\}, +), 1 \leq t \leq p_x,$$
$$\hat{D}(c_t) = (\{\varepsilon\}, \{\hat{c}_t \to \varepsilon\}, -), 1 \leq t \leq p_x,$$
$$\bar{D}(d_t) = (\{\varepsilon\}, \{\bar{d}_t \to \varepsilon\}, +), 1 \leq t \leq q_x,$$
$$\hat{D}(d_t) = (\{\varepsilon\}, \{\hat{d}_t \to \varepsilon\}, -), 1 \leq t \leq q_x,$$

$$\bar{C}(c_t') = \begin{cases} (\{b_{i_1} b_{i_2} \ldots b_{i_j} c_1 c_2 \ldots c_{p_x} d_1' d_2' \ldots d_{q_x}'\}, \\ \{c_{p_x}' \to \bar{c}_{p_x}\}, 0), \text{ if } t = p_x, \\ (\{z_0\}, \{c_t' \to \bar{c}_t, z_{2(p_x-t)} \to \bar{c}_t\} \cup \\ \{z_r \to z_{r+1} \mid 0 \leq r \leq 2(p_x - t) - 1\}, 0), \text{ otherwise,} \end{cases}$$

$$\hat{C}(c_t') = (\{z_0\}, \{c_t' \to \hat{c}_t, z_{2(p_x-t)} \to \hat{c}_t\} \cup$$
$$\{z_r \to z_{r+1} \mid 0 \leq r \leq 2(p_x - t) - 1\}, 0),$$
$$\bar{C}(d_s) = (\{z_0\}, \{d_s \to \bar{d}_s, z_{2p_x+2(q_x-t)} \to \bar{d}_s\} \cup$$
$$\{z_r \to z_{r+1} \mid 0 \leq r \leq 2p_x + 2(q_x - t) - 1\}, 0),$$
$$\hat{C}(d_s) = (\{z_0\}, \{d_s \to \hat{d}_s, z_{2p_x+2(q_x-t)} \to \hat{d}_s\} \cup$$
$$\{z_r \to z_{r+1} \mid 0 \leq r \leq 2p_x + 2(q_x - t) - 1\}, 0).$$

All symbols $z_0, z_1, \ldots, z_{2p_x+2(q_x-1)}$ are gathered in Q.
- The valuation φ is defined by

$$\varphi(y) = 0, \, y \in U \cup U' \cup Q \cup \{b_1, b_2, \ldots, b_n\},$$
$$\varphi(y) = 1, \, y \in \bar{U},$$
$$\varphi(y) = -1, \, y \in \hat{U}.$$

Here are some informal considerations about the computing mode of this P system. For a better understanding, we shall restrict our discussion to the transformations of the SC-component ρ_x during the computation process. Each SC-component q_y, $y \in \{1, 2, \ldots n\}^{\leq K}$, works in the same way. Initially, the cell $[\bar{C}(c_{p_x}')]$ contains an arbitrarily large number of copies of the string $w = b_{i_1} b_{i_2} \ldots b_{i_j} c_1 c_2 \ldots c_{p_x} d_1' d_2' \ldots d_{q_x}'$ with $x = i_1 i_2 \ldots i_j$, $\alpha(u_{i_1} u_{i_2} \ldots u_{i_j}) = c_1 c_2 \ldots c_{p_x}$ and $\alpha(v_{i_1} v_{i_2} \ldots v_{i_j}) = d_1 d_2 \ldots d_{q_x}$. An occurrence of c_{p_x}' in all copies of w is replaced by $\bar{c}_{p_x}$ and all obtained strings (now their value sign by φ is +) go through the membrane of $[\bar{C}(c_{p_x}')]$ to the cell $[\bar{D}(c_{p_x})]$, the membrane of $[\bar{C}(c_{p_x}')]$ being dissolved. Thus, the cell $[\bar{D}(c_{p_x})]$ has copies of a string other than the empty one, if and only if there exists $1 \leq i \leq q_x$ such that $d_i = c_{p_x}$. If $i \neq q_x$, then all these copies remain in $[\bar{D}(c_{p_x})]$ for the rest of the computation, hence the computation ends after at least $2(p_x + q_x)$ steps but no copy of a string coming from ρ_x is observed in the outermost membrane. If $i \neq q_x$, then $u_{i_1} u_{i_2} \ldots u_{i_j} \neq v_{i_1} v_{i_2} \ldots v_{i_j}$.

If $i = q_x$, then $\bar{d}_{q_x}$ is deleted in $[\bar{D}(c_{p_x})]$ from all strings, and all new strings go to the cell $[\hat{C}(c'_{p_x-1})]$ because their value sign by φ is again 0. Thus, the system has just checked whether or not $c_{p_x} = d_{q_x}$. Now, the checking process applies to the pair of symbols c_{p_x-1} and d_{q_x-1} for the next two computation steps, and so forth. Therefore, after $2(p_x + q_x)$ steps the cell $[T]$ contains a string coming from the SC-component ρ_x, more precisely the string $b_{i_1} b_{i_2} \ldots b_{i_j}$, if and only if $c_1 c_2 \ldots c_{p_x} = d_1 d_2 \ldots d_{q_x}$ which is equivalent, by the definition of α, to $u_{i_1} u_{i_2} \ldots u_{i_j} = v_{i_1} v_{i_2} \ldots v_{i_j}$.

In conclusion, when the computation is finished, $[T]$ contains a string if and only if the given instance of the Bounded PCP has a solution. Furthermore, all solutions are found in $[T]$. Obviously, one can choose the encoding mapping α in such a way that the time complexity required is two times the product between K and the length of the longest string in the two given Post lists. Furthermore, the generating phase is of the same complexity, which concludes the proof.

Remark. All the P systems presented in the previous section have cells that behave like counters (they contain single symbols which evolve in a sequence). They prevent those membranes of the cells which do not contain any string but only other cells from dissolving spontaneously. In order to simplify the constructions, the strings in these cells could be eliminated by allowing cells to exist that only contain rules and other cells.

4 Conclusions

We have proposed a new variant of P systems having a different computational mode than that of the other P systems considered in the literature. A natural question arises: How far is this model from the biological reality? More precisely, is it possible to have adjacent membranes of the same polarity? Can the sum-model valuation of strings be biologically implemented? Is there any biological mechanism able to destroy those strings of a cell which have a bad polarity? We hope that at least some answers to these questions is affirmative.

With respect to this model, we have presented two linear P algorithms which provide all solutions of two NP-complete problems.

Further, one can go to still simpler variants. A natural question concerns the computational power of this variant (it is known that most of the variants introduced so far are computationally complete). Our belief is that those variants which are "specialized" in solving a few classes of problems have better chances to get implemented, at least in the near future.

Obviously, the P system model of computation is still in its incipient stage of finding a suitable formalism in the aim of identifying molecular biology techniques appropriate for its implementation. Nevertheless, the existence of many different variants with complementing features suggests the versatility of the model and increases the probability of a practical implementation.

Acknowledgements

We express our gratitude to the anonymous referees for their valuable comments and suggestions.

References

1. J. P. Banâtre, A. Coutant, D. Le Metayer, A parallel machine for multiset transformation and its programming style, *Future Generation Computer Systems*, 4 (1988), 133–144.
2. R. Constable, H. Hunt, S. Sahni, On the computational complexity of scheme equivalence, *Technical Report* No. 74-201, Dept. of Computer Science, Cornell University, Ithaca, NY, 1974.
3. J. Dassow, Gh. Păun, On the power of membrane computing, *J. Univ. Computer Sci.*, 5, 2 (1999), 33–49.
4. M. Garey, D. Johnson, *Computers and Intractability. A Guide to the Theory of NP-completeness*, Freeman, San Francisco, CA, 1979.
5. J. Hopcroft, J. Ulmann, *Formal Languages and Their Relation to Automata*, Addison-Wesley, Reading, MA, 1969.
6. L. Kari, G. Gloor, S. Yu, Using DNA to solve the Bounded Correspondence Problem, *Theoret. Comput. Sci.*, 231 (2000), 193–203.
7. S. N. Krishna, R. Rama, A variant of P systems with active membranes: Solving NP-complete problems, *Romanian Journal of Information Science and Technology*, 2, 4(1999), 357–367.
8. Gh. Păun, A new type of generative devices: valence grammars, *Rev. Roum. Math. Pures Appl.* 25, 6(1980), 911–924.
9. Gh. Păun, Computing with membranes, *J. Comput. Syst. Sci.* 61(2000) to apper, (see also *TUCS Research Report* No. 208, November 1998, http://www.tucs.fi.)
10. Gh. Păun, Computing with membranes. An introduction, *Bulletin of the EATCS*, 67 (Febr. 1999), 139–152.
11. Gh. Păun, Y. Sakakibara, T. Yokomori, P systems on graphs of restricted forms, submitted 1999.
12. Gh. Păun, T. Yokomori, Membrane computing based on splicing, *Preliminary Proc. of Fifth Inter. Meeting on DNA Based Computers* (E. Winfree, D. Gifford, eds.), MIT, June 1999, 213–227.
13. Gh. Păun, T. Yokomori, Simulating H systems by P systems, *Journal of Universal Computer Science*, 6, 1(2000), 178–193.
14. Gh. Păun, S. Yu, On synchronization in P systems, *Fundamenta Inform.*, 38, 4(1999), 397–410.
15. Gh. Păun, P systems with active membranes: Attacking NP-complete problems, *Journal of Automata, Languages and Combinatorics*, in press. (See also *CDMTCS Research Report* No.102, 1999, Auckland Univ., New Zealand, www.cs.auckland.ac.nz/CDMTCS.)

16. Gh. Păun, G. Rozenberg, A. Salomaa, *DNA Computing. New Computing Paradigms*, Springer-Verlag, Berlin, 1998.
17. Gh. Păun, G. Rozenberg, A. Salomaa, Membrane computing with external output, *Fundamenta Inform.*, 41, 3(2000), 259–266.
18. G. Rozenberg, A. Salomaa, eds. *Handbook of Formal Languages*, 3 volumes, Springer, Berlin, 1997.

The Quantum Domain As a Triadic Relay

Anna B. Mikhaylova[1] and Boris S. Pavlov[1,2]

[1] Laboratory of Complex Systems Theory, Institute for Physics, St. Petersburg State University, Russia
[2] Department of Mathematics, University of Auckland, Auckland, New Zealand

Abstract. A solvable model of the three-positions quantum switch - triadic relay - is constructed as a circular domain with 4 one-dimensional wires attached to it. In resonance case, when Fermi level in the wires coincides with some energy level in the domain, the magnitude of the governing constant electric field is specified such that manipulation of the quantum current through the relay may be done via the change of the direction of the field in a plane parallel to the circle.

1 Introduction: Low-Dimensional Structures and Hybrid Schrödinger Operators

In our previous papers [1–6] we considered mathematical models for quantum electronic devices designed to manipulate quantum current in physical networks constructed of quantum wires and quantum domains on the interface of an electrolyte and a *narrow-gap semiconductor*. The wires are created usually by itching as narrow channels where the molecules of an electrolyte form a conducting chain. Quantum domains may be formed on the surface of the semiconductor by epitaxy. Due to the narrow-gap property [2] the effective mass m_e of electrons in the semiconductor is small compared with conventional electron's mass, $\frac{m_e}{m_0} \approx 0.01$, and hence the De-Broghlie wave-length Λ calculated from the effective Schrödinger equation

$$-\frac{\hbar^2}{2m_e}\Delta\Psi + V\Psi = E_g\Psi$$

at the Fermi level E_g may be large if compared with the width of the wires or the thickness of the domain (approximately $100 \div 200\text{Å}$)). This may be illustrated by the following Table [1] (see [1], Section 3 and [2]):

---

[1] The partial valency of compounds is denoted by x.

Semiconductor	E_g (300 K) eV	$\frac{m_e^*}{m_0}$	$\frac{m_p^*}{m_0}$	Λ (300 K) Å	Λ (77 K) Å
$GaAs$	1.430	0.070	0.40	290	580
$InAs$	0.360	0.022	0.40	515	1020
$Cd_xHg_{1-x}Te$					
$x = 0.20$	0.150	0.0130	0.45	670	1320
$x = 0.25$	0.220	0.0165	0.45	590	1160
$x = 0.27$	0.250	0.0180	0.45	570	1125
$HgTe$	-0.117	0.012	0.50	700	1380
$Zn_{0.15}Hg_{0.85}Te$	0.190	0.015	0.45	620	1220

This means that in narrow-gap case the wires are effectively one-dimensional and the domains are two-dimensional objects, so that the physical network of wires and domains becomes a *hybrid system* and the effective Schrödinger operator becomes effectively *hybrid Schrödinger operator* - a coupling of ordinary differential operators on one-dimensional wires and partial differential operators on domains, connected by proper boundary conditions at the points of contact. An approach to hybrid differential equations based on Operator Extensions Theory was proposed in [7], [8]. Specific properties of differential equations on graphs were first discussed in [9]. Scattering theory on discrete graphs based on simplectic geometry is developed in [10]. The above mentioned "narrow-gap properties" imply also important physical phenomena such as high-temperature quantum behavior, in particular the possibility of simulated Peierls transition, which may be also used for manipulating quantum current, [1].

Basing on the mentioned above low-dimensional properties of narrow-gap semiconductors and following general ideas described in [5] we construct below [2] a solvable model of a *triadic relay* as a two-dimensional circular domain Ω with a few (eventually $N = 4$) one-dimensional wires Γ_s attached to it. The corresponding hybrid Hamiltonian is formed as a common selfadjoint extension of a couple of Schrödinger operators acting in the orthogonal sum $\mathcal{H} = \oplus \sum_s L_2(\Gamma_s) \oplus L_2(\Omega)$ of Hilbert spaces of all square-integrable functions on wires and domain. At the points of contact of the domain and wires we connect the functions via special boundary conditions which characterize the strength of the contact between wires and domain. To avoid awkward notations we assume for the beginning that the dimensionless variables are used, so that the underlying differential equations become

$$-\frac{d^2\Psi}{dx^2} = \lambda\Psi, \quad -\Delta\Psi + V\Psi = \lambda\Psi, \qquad (1)$$

in wires and on the domain, respectively. The exact potential V for the manipulation of transmission across the domain will be specified later. The simplest but still convenient option is a linear potential of some constant electric field

---
[2] See also the preprint [11].

intensity ε directed along the vector **e**:

$$V = V_0 + V_g(\mathbf{X}) = V_0 + \langle \mathbf{x}, \mathbf{e} \rangle \, \varepsilon, \quad \nabla \mathbf{V_g} = \varepsilon \, \mathbf{e}.$$

The intensity ε and the shift V_0 will be defined from the conditions for the transmission coefficients, see Section 3.

To introduce the boundary conditions at the points of contact we begin with operators defined on the wires Γ_i $1 = 1, 2, 3, 4$ and on the domain Ω with homogeneous Neumann boundary conditions: $\frac{du_s}{dx}|_{x=0} = 0$ $\frac{\partial u_\Omega}{\partial n}|_{\partial \Omega} = 0$, respectively. The third boundary condition $\frac{\partial u_\Omega}{\partial n} - \sigma' u|_{\partial \Omega} = 0$ at the boundary $\partial \Omega$ of the domain gives even more interesting option permitting modeling a large scale of boundary behavior including the loose ($\sigma' = 0$) and hard ($\sigma' = \infty$) connection between the domain and the environment. The magnitude of σ may be interpreted as a common characteristics of the potential barrier on the border of the domain and the density of the electrical double layer on the interface of the domain and the environment, and formally represented as an additional generalized potential. Then the corresponding Schrödinger equation may be written in form

$$-\Delta \Psi + [V_{bulk} + V_g + \sigma' \delta(dist(x, \partial \Omega))] \Psi = \lambda \Psi,$$

where V_{bulk} is a step-vise function, $V_{bulk} = V_0$ inside the domain and $V_{bulk} = V_1 \gg 1$ outside of it, V_g is the *governing* potential of a constant electric field $V_g = \langle x, e \rangle \varepsilon$ used for manipulating of electron's current across the domain, and the singular term $\sigma' \delta$ attached to the boundary corresponds to the additional polarization of the electrolyte near the boundary of the domain. The straightforward integration by parts shows that the singular term gives the boundary condition

$$\left[\frac{\partial u}{\partial n} \right] = \frac{\partial u}{\partial n}_{out} - \frac{\partial u}{\partial n}_{in} = \sigma' u|_{\partial \Omega}.$$

For the high barrier $V_1 \gg E_g$ we may assume that for electrons at Fermi level at low temperatures the approximate equation is fulfilled:

$$\frac{\partial u}{\partial n}_{out} \cong -(V_1 - E_g)^{1/2} u|_{\partial \Omega}$$

which gives due to continuity of u at the boundary

$$\frac{\partial u}{\partial n}_{in} + (\sigma' + [V_1 - E_g]^{1/2}) u|_{\partial \Omega} = 0.$$

If we denote $\sigma' + (V_1 - E_g)^{1/2}$ by σ then the homogeneous Neumann boundary condition corresponds to the special case $\sigma' + (V_1 - E_g)^{1/2} = 0$. In what follows we assume that this is exactly the case, though the analysis based on a more general third boundary condition may add in fact only minor technical complications. The main objective of our paper is optimization of the design of the triadic relay. Basing on a solvable model of the relay we specify the value ε of the covering electric field and the shift V_0 such that the pattern of zero-lines of the resonance eigenfunction ϕ_λ, $\lambda = E_g$ are compatible with positions of contact points to ensure the manipulation of the current by the rotation of the macroscopic electric field $\varepsilon \mathbf{e}$.

2 Solvable Model of a Triadic Relay

Following ideas of our previous papers we construct the solvable model of a triadic relay as a "mixing" selfadjoint extension of the underlying operators

$$l_s = -\frac{d^2}{dx^2} \text{ in } L_2(\Gamma_s), \quad L = -\Delta + V \text{ in } L_2(\Omega),$$

on the wires Γ_s, $s = 1, 2, \ldots N$ and on the domain with Neumann boundary condition, both initially reduced onto the linear set of elements from the corresponding domains vanishing near the points of contact $a_1, a_2, \ldots$. To describe the extension procedure we need a description of the deficiency elements and the connecting boundary conditions. Note that the Green function $G_\lambda(x, y)$ of the selfadjoint operator $L = -\Delta + V$ in $L_2(\Omega)$ with homogeneous Neumann boundary condition is a singular solution of the homogeneous equation

$$-\Delta G + VG = \lambda G, \qquad (2)$$

with a special behavior at the inner pole y, $x \to y$:

$$G_\lambda(x, y) = \frac{1}{2\pi} \log \frac{1}{|x-y|} + O(1).$$

For zero-potential the main singular solution is just a sum Hankel-function of the first kind and a smooth correcting term $G_\lambda^0(x, y) = \frac{i}{4} H_0^1(\sqrt{\lambda}|x-y|) + c_\lambda^0(x, y)$ selected to fulfill the boundary conditions. The Green-function of the Schrödinger equation (2) may be found as a solution of the Lippman-Schwinger equation

$$G_\lambda(x, y) = G_\lambda^0(x, y) + \int_\Omega G_\lambda^0(x, s) V(s) G_\lambda^0(s, y) ds^2 \qquad (3)$$

from which one can deduce that it has the same singularity at the pole as the Green-function of G_λ^0 the Laplace equation. If the boundary $\partial \Omega$ is smooth, then one derives that the Green function of Laplace equation has the logarithmic singularity at the boundary pole $y \in \partial \Omega$

$$G_\lambda^0(x, y) \approx \frac{1}{\pi} \log \frac{1}{|x-y|}, \quad x \to y. \qquad (4)$$

From the above Lippman-Schwinger equation we deduce that the Green-function of the Schrödinger equation has the singularity at the boundary points described by (4), which implies the square integrability of the Green-function for any position of the pole in Ω. This means that the deficiency indices of the operator L reduced $L \to L_0$ onto the linear set of all elements from the domain D_L of L vanishing near the points $a_1, a_2, a_3, \ldots a_N$ are equal (N, N), and the Green-functions $G_\lambda(x, a_s)$, $s = 1, 2, \ldots N$ play the role of deficiency elements.

Planning to use the simplectic version of the operator extension procedure, see [10], we introduce the asymptotic boundary values, *singular amplitudes* $A_s(u)$ and *regularized values* $B_s(u)$ for elements u of the domain of the adjoint operator L_0^+ at points a_s. We assume now that the nonperturbed Schrödinger operator $L + I$ on the domain Ω with Neumann boundary condition is positive $L + I > 0$. Then its resolvent $[L + 1]^{-1}$ is a bounded integral operator with the kernel $G_{-1}(x, y)$. We use this kernel as a unit of growing rate for elements from the domain of adjoint operator at the poles:

$$u(x) = A_s^1 G_{-1}(x, a_s) + B_s^1 + o(1), \quad x \to a_s. \tag{5}$$

Lemma 1 *For any regular point λ from the complement of the spectrum $\sigma(L)$ of the operator L and any $a \in \{a_s\}_{s=1}^N$ the following representation is true:*

$$G_\lambda(x, a) = G_{-1}(x, a) + (\lambda + 1)G_{-1} * G_\lambda (x, a),$$

where the second addend $(\lambda + 1)G_{-1} * G_\lambda (x, a) \equiv g_\lambda(x, a)$ is a continuous function of x and its spectral series is absolutely and uniformly convergent in Ω. The separation of the singularity at each eigenvalue λ_0 is possible:

$$g_\lambda(x, a) = (\lambda + 1) \sum_l \frac{\varphi_l(x)\varphi_l(a)}{(\lambda_l + 1)(\lambda_l - \lambda)} = \frac{\varphi_0(x)\varphi_0(a)}{\lambda_0 - \lambda} + g_\lambda^0(x, a) \tag{6}$$

with uniformly and absolutely convergent spatial series for $g_\lambda^0(x, a)$ in a neighborhood of λ_0.

The proof follows the same pattern as the three–dimensional given in [5] and is based on the classical Mercer Theorem.

The previous lemma implies existence of *regularized boundary values*

$$\lim_{x \to a_s} [G_\lambda(x, a_s) - \Re G_i(x, a_s)],$$

for each deficiency element $G_i(x, a_s)$ and hence for all linear combinations of them. In particular, each element $u = u_0 + \sum_s A_s^+ G_i(x, a_s) + \sum_s A_s^- G_{-i}(x, a_s)$ of the domain of the adjoint operator D_0^+ also has *regularized boundary values* defined as

$$\lim_{x \to a_s} [u(x) - A_s \Re G_i(x, a_s)] = B_s, \quad A_s = A_s^+ + A_s^-.$$

Together with *singular amplitudes* A_s they form a set of simplectic coordinates which may be used to represent the boundary form of L_0^+ as a hermitian simplectic form

$$\mathcal{J}_L(u,v) = \langle L_0^+ u, v\rangle - \langle u, L_0^+ v\rangle = \sum_{s=1}^{N} B_s^u \bar{A}_s^v - A_s^u \bar{B}_s^v = \langle \boldsymbol{B}^u, \boldsymbol{A}^v\rangle - \langle \boldsymbol{A}^u, \boldsymbol{B}^v\rangle. \quad (7)$$

Using the above representation for elements of the domain of adjoint operator (1) we derive the explicit expression for regularized boundary values of elements of $\bar{\lambda}$ − deficiency subspace $u = \sum_{s=1}^{N} A_s G_\lambda(x, a_s)$. They serve as solutions of the adjoint homogeneous equation

$$L_0^+ u = \lambda u. \quad (8)$$

It follows from Lemma 1 that the singular amplitudes A_s^u of u coincide with A_s. An important step is to express the regularized boundary values B_s in terms of singular amplitudes for any solutions u of the *adjoint homogeneous equation* (8). The operator Q which transforms the vector of singular amplitudes $\boldsymbol{A}$ into the vector of corresponding regularized boundary values $\boldsymbol{B}$ for the solutions of (8) is a simplest analog of so-called *Dirichlet to Neumann map* for elliptic partial differential equations and at the same time it is a multi-dimensional version of Titchmarsh-Weyl function. In our scheme it appears in the form of Krein's Q-function, see [12]. Its dependence on λ contains actually all essential spectral information about the nonperturbed operator $\mathcal{A}$.

To recover this information note first that the deficiency element $e_{\bar\lambda}$ for the spectral point $\bar\lambda$ is a solution of the adjoint equation $(L_0^+ - \lambda)e_{\bar\lambda} = 0$ and is expressed through the deficiency element e_i for the spectral point i as

$$e_{\bar\lambda} = \frac{L+iI}{L-\lambda I} e_i = G_\lambda(x, a_s)$$

and

$$G_\lambda(x, a_s) - \Re G(x, a_s, i) = \frac{L+iI}{L-\lambda I} e_i^s - \frac{L}{L^2+1}(L+iI)e_i^s.$$

Then denoting $\delta(x - a^s)$ by δ^s we obtain:

$$\frac{I}{L-\lambda I}\delta^s - \frac{L}{L^2+1}\delta^s = \frac{1}{L-iI}\frac{1+\lambda L}{L-\lambda I}\frac{1}{L+iI}\delta^s.$$

One can easily see from the above lemma that $\frac{I}{L-\lambda I}\delta^s - \frac{L}{L^2+1}\delta^s$ is a continuous function of x near the point a_s and

$$\lim_{x\to a_s}[G_\lambda(x, a_s) - \Re G_i(x, a_s)] = \langle \frac{I+\lambda L}{L-\lambda I} G_i(a_s), G_i(a_s)\rangle = g^s(\lambda). \quad (9)$$

is just the regularized boundary value of $G_\lambda^L(x, a_s)$ at the point a_s. In particular, for the sum of deficiency elements $u = \sum_s A_s G_\lambda(x, a_s)$ we have the following asymptotic $x \to a_s$:

$$u = A_s \Re G_i(x, a_s) + g^s(\lambda) A_s + \sum_{t\neq s} A_t G_\lambda(a_s, a_t) + O(1). \quad (10)$$

where A_s, $B_s = g^s(\lambda) A_s + \sum_{t \neq s} A_t G_\lambda(a_s, a_t)$ define the simplectic variables of the element u.

Summarizing

Lemma 2 *The singular amplitudes A and the regularized boundary values B of solutions of the homogeneous equation $L_0^+ u = \lambda u$ are connected by the formula:*
$$B^u = QA^u,$$
where

$$Q(\lambda) = \begin{pmatrix} g^1(\lambda) & G_\lambda(a_1, a_2) & \cdots & \cdots & G_\lambda(a_1, a_N) \\ G_\lambda(a_2, a_1) & g^2(\lambda) & \cdots & \cdots & G_\lambda(a_2, a_N, \lambda) \\ \cdots & \cdots & g^3(\lambda) \cdots & & \cdots \\ \cdots & & \cdots & \cdots & \cdots \\ G_\lambda(a_N, a_1) & \cdots & \cdots & \cdots & g^N(\lambda) \end{pmatrix}, \quad (11)$$

and $g_s(\lambda) = \sum_l \frac{1+\lambda_l \lambda}{\lambda_l - \lambda} |\varphi_l(a_s)|^2$. Near the simple eigenvalue λ_0 the Q-matrix is decomposed as

$$Q_{st}(\lambda) = \frac{\bar\varphi_0(a_s) \varphi_0(a_t)}{\lambda_0 - \lambda} + (K_0)_{st}(\lambda), \quad (12)$$

where K_0 is an analytic matrix-function in a neighborhood of λ_0:

$$(K_0)_{st}(\lambda_0) = \begin{cases} -\lambda_0 |\varphi_0(a_s)|^2 + \sum_{l \neq 0} \frac{1+\lambda_l^2}{(1+\lambda_l^2)(\lambda_l - \lambda_0)} |\varphi_l(a_s)|^2, & s = t, \\ \left[G_\lambda(x, y) - \frac{\bar\varphi_0(x) \varphi_0(y)}{\lambda_0 - \lambda} \right]|_{x=a_s, y=a_t}, & s \neq t. \end{cases}$$

Note that the last statement is just a version of the general formula connecting the boundary values of abstract Hermitian operators, and Q plays the role of Krein's Q-function.

The practical construction of selfadjoint extensions of operators in the domain can be based on the following statement:

Theorem 1 *The restriction of the operator*
$$L = -\Delta + V \to L_0,$$
onto the class of all W_2^2- smooth functions which satisfy homogeneous Neumann boundary conditions and vanish near the boundary points a_s, $s = 1, 2, \ldots N$ is a symmetric operator with deficiency indices equal (N, N). The corresponding deficiency elements are Green-functions $G_\lambda(x, a_s)$. The domain of the adjoint operator L_0^+ consists of all locally W_2^2-smooth functions u defined on $\Omega \setminus \{a_1, a_2, \ldots, N\}$ which have the asymptotic boundary values - singular amplitudes and regularized boundary values - at the boundary points $\{a_1, a_2, \ldots\} \in \partial\Omega$:

$$u(x) \approx A_s^u \Re G_i(x, a_s) + B_s^u, \quad s = 1, 2, \ldots, \quad (13)$$

connected via Q-matrix: $B^u = QA^u$. The boundary form of the adjoint operator may be represented in terms of asymptotic boundary values as a Hermitian simplectic form:

$$\mathcal{J}_L(u,v) = \langle L_0 u, v\rangle - \langle u, L_0 v\rangle = \sum_s \{B^u_s \bar{A}^v_s - A^u_s \bar{B}^v_s\}. \tag{14}$$

All selfadjoint extensions of the operator L_0 in $L_2(\Omega)$ are parameterized by Lagrangian planes $\mathcal{L}$ of the boundary form (15) in the complex $2N$-dimensional space of variables A, B:

$$\mathcal{J}_L(u,v) = 0, \text{ if } \{A^u, B^u\}, \{A^v, B^v\} \in \mathcal{L}.$$

A proof of this statement can be obtained following the technique used in the three-dimensional case in ([5]). Theorem 1 will be used as a base for calculating the resolvents of our hybrid Hamiltonians.

Together with the operator L_0 in $L_2(\Omega)$ we consider also the ordinary differential operators $l_s = -\frac{d^2}{dx^2}$ on the wires in $L_2(\Gamma_s)$, $s = 1, 2, \ldots$. Their restrictions $l_{s,0}$ onto classes of all W_2^2-smooth functions on the wires vanishing at the origin $x_s = 0$ give Hermitian operators with deficiency indices $(1,1)$, and adjoint operators $l^+_{s,0}$ with boundary forms represented in terms of boundary values $u_s(0)$, $u'_s(0)$ of elements from the corresponding domains D^+_s. Denoting by l^+ the orthogonal sum $\oplus \sum_s l^+_{s,0}$ we represent the boundary form of l^+ in $\oplus \sum_s L_2(\Gamma_s)$ on elements $\boldsymbol{u} = \{u_1, u_2 \ldots\}$, $\boldsymbol{v} = \{v_1, v_2, \ldots\}$ as

$$\mathcal{J}_l(\boldsymbol{u}, \boldsymbol{v}) = \langle l^+ \boldsymbol{u}, \boldsymbol{v}\rangle - \langle \boldsymbol{u}, l^+ \boldsymbol{v}\rangle = \sum_s \{u'_s \bar{v}_s - u_s \bar{v}'_s\}. \tag{15}$$

We construct now the selfadjoint extensions of the total reduced operator $L_0 \oplus l_0$ in $L_2(\Omega_0) \oplus \sum_{s=1}^N L_2(\Gamma_s)$ based on the analysis of the boundary form of the corresponding adjoint operator $L_0^+ \oplus l_0^+$ on elements $\{u, \boldsymbol{u}\}$, $\{v, \boldsymbol{v}\}$

$$\mathcal{J}_{L_0^+ \oplus l_0^+} = \mathcal{J}_L(u,v) + \mathcal{J}_l(\boldsymbol{u},\boldsymbol{v}) = \sum_s \{B^u_s \bar{A}^v_s - A^u_s \bar{B}^v_s\} + \sum_s \{u'_s \bar{v}_s - u_s \bar{v}'_s\}. \tag{16}$$

All selfadjoint extensions of $L_0 \oplus l_0$ may be parameterized by Lagrangian planes of the form (16). Imposing *special* boundary conditions at the boundary points of contact a_s, $s = 1, 2, \ldots$:

$$\begin{pmatrix} A_s \\ u_s(0) \end{pmatrix} = \begin{pmatrix} 0 & \beta \\ \beta & 0 \end{pmatrix} \begin{pmatrix} B_s \\ u'_s(0) \end{pmatrix}, \quad \beta > 0, \ s = 1, 2, \ldots. \tag{17}$$

we obtain, after closure, the selfadjoint operator $\mathbf{L}_\beta$ in $L_2(\Omega) \oplus \sum_{s=1}^N L_2(\Gamma_s)$ which will serve, subject to proper choice of the field ε, a as solvable Hamiltonian of the triadic relay. The parameter β may be interpreted as an exponential of the "strength" of the potential barrier high H and width d separating the domain from the wires: $\beta = e^{-\sqrt{H}d}$.

Following [5] we describe the spectral properties of the hybrid Schrödinger operator $\mathbf{L}_\beta$ based on Krein formula connecting the resolvents of the underlying operator $\mathbf{L}_\infty = L \oplus \sum_{s=1}^{N} l_s$ with the resolvent of $\mathbf{L}_\beta$.

We suggest below an explicit formula for the component of the resolvent kernel (Green function) $G_\lambda^\beta(x,y)$ of the operator $\mathbf{L}_\beta$ on the domain Ω and complete formulae for scattered waves of the operator — the eigenfunctions of absolutely-continuous spectrum $\sigma(\mathbf{L}_\beta)$. Again as in [5] we consider the *generic case* when neither eigenfunction of the nonperturbed operator L in $L_2(\Omega)$ vanishes at all contact points a_s, $s = 1, 2, \ldots, N$ simultaneously. In what follows we assume that the singular amplitudes $A = \{A_1, A_2, \ldots\}$ and regularized boundary values are elements of an auxiliary finite-dimensional Hilbert space E, $\dim E = N$. We call this space the *channel space*, having in mind the role it plays in scattering problem. In particular we use vectors combined of the values of the nonperturbed Green functions attached to the points $\{a_s\} : \{G_\lambda(x, a_s)\} \equiv \boldsymbol{G}_\lambda(x)$. We assume that the metric form of E is standard, for instance the dot product of two vectors above is given by the formula:

$$\langle \boldsymbol{A}, \boldsymbol{G}_\lambda(x) \rangle = \sum_s \bar{A}_s G_\lambda(x, a_s).$$

The resolvents of the hybrid Hamiltonians are described by the following statement which is in fact a version of the Krein formula and is similar to the corresponding $3-d$ statement proven in [5]:

Theorem 2 *The component $G_\lambda^\beta(x,y)$ of the resolvent kernel of the operator $\mathbf{L}_\beta$ on the domain, $x, y \in \Omega$, is represented in terms of Green function of the nonperturbed operator L in the following way:*

$$G_\lambda^\beta(x,y) = G_\lambda(x,y) + < \overline{\boldsymbol{G}_\lambda(y)}, \left[\frac{1}{ik\beta^2} - Q\right]^{-1} \boldsymbol{G}_\lambda(x) >, \quad k^2 = \lambda, \ \Im k \geq 0. \tag{18}$$

The spectrum $\sigma(\mathbf{L}_\beta)$ of the perturbed operator $\mathbf{L}_\beta$ consists of all singularities of the matrix

$$\left[\frac{1}{ik\beta^2} - Q\right]^{-1}$$

in the complex plane of the spectral parameter $\lambda = k^2$, where Q is the Krein's Q-matrix described above, see (11). In particular the absolutely continuous spectrum of $\mathbf{L}_\beta$ fills the positive half-axis $\lambda \geq 0$ with the constant multiplicity N. The eigenvalues $\lambda_r = k_r^2$, $\Im k_r > 0$ and the resonances $\lambda_r = k_r^2$, $\Im k_r < 0$ of the operator $\mathbf{L}_\beta$ are defined, counting multiplicity, respectively: by the poles of the matrix $\left[\frac{1}{ik\beta^2} - Q\right]^{-1}$ on the positive imaginary half-axis (for eigenvalues $\lambda = k^2 < 0$) or in lower half-plane $\Im k < 0$ (for resonances). They are roots of the corresponding dispersion equation in upper $\Im k > 0$ and lower $\Im k < 0$ half-planes (physical and nonphysical sheets) respectively:

$$\det\left[\frac{1}{ik\beta^2} - Q\right] = 0. \tag{19}$$

The eigenfunctions of the absolutely-continuous spectrum of the operator $\mathbf{L}_{\beta,0}$ are presented by the scattered waves which form a complete orthogonal system of eigenfunctions in the absolutely-continuous subspace of $\mathbf{L}$. In particular, the scattered wave

$$\Psi = \left(\Psi^1(x), \psi^1(x_1), \psi^1(x_2), \ldots \psi^1(x_N)\right),$$

initiated by the incoming plane wave e^{-ikx_1} in the first channel (on the wire attached to a_1) has the form:

$$\psi_s^1(x_s) = \delta_{1,s} e^{-ikx_s} + S_{s,1} e^{ikx_s}, \quad x_s \in \Gamma_s,$$

$$\Psi^1(x) = 2\beta^{-1} < \bar{G}_\lambda(x), \left[\frac{1}{ik\beta^2} - Q\right]^{-1} \delta_1, \quad x \in \Omega,$$

where $S_{s,t}$ is the scattering matrix calculated by the formula

$$S = \frac{\frac{I}{ik\beta^2} + Q}{-\frac{I}{ik\beta^2} + Q}. \tag{20}$$

The proof closely follows [5]. In particular, the explicit expression for the scattered waves $\{\Psi_1, \psi_{1,1}, \psi_{1,2}, \ldots \psi_{1,N}\}$ initiated by the plane wave in the first channel Γ_1 and the formulae for the elements $S_{s,1}$ of the scattering matrix is derived from the ansatz

$$\psi_s^1(x_s) = \delta_{1,s} e^{-ikx_s} + S_{s,1} e^{ikx_s}, \quad x_s \in \Gamma_s,$$

$$\Psi^1(x) = \sum_s A_{1,s} G_\lambda(x, a_s),$$

using the boundary conditions and the connection between the regularized boundary values and singular amplitudes defined by Q-matrix, see (11).

3 Manipulation of Electron Current in Resonance Case

The expression for effective conductance ρ_s^{-1} of an elementary scatterer characterized by the transmission coefficient T was suggested by Landauer [13] in the form $\rho_s^{-1} = (e^2/h)|T|^2/(1-|T|^2)$ which corresponds to the connection between the quantum-mechanical conductance $\rho^{-1} = (e^2/h)|T|^2$ and the basic conductance of the quantum wire $\rho_0^{-1} = (e^2/h)$ represented by the standard Matiessen rule for corresponding resistance $\rho = \rho_0 + \rho_s$. This means, that for all but two quantum blocked wires we may obtain the conductance of the quantum domain from Landauer formula, if the corresponding transmission coefficient is already calculated. Thus the problem of conductance is reduced to the calculation of the scattering matrix.

We shall do all spectral calculations for the dimensionless equation in the unit disc and then choose physical parameters using a proper change of variables. In [5] the scattering matrix is calculated in the resonance case when the

Fermi-level E_g in the wires Γ_s coincides with a simple *resonance* eigenvalue λ_0 of the nonperturbed operator L on the domain. Assuming that the interaction between wires and the domain is weak, $\beta \ll 1$, we calculate approximately the transmission coefficients $S_{s,1}$ from the wire Γ_1 to the wire Γ_s in terms of values of the resonance eigenfunction at the points of contact. Let us denote by φ_0 a vector in the auxiliary channel space E of values of the normalized resonance eigenfunction at the points of contact

$$\varphi_0 = (\varphi_0(a_1), \varphi_0(a_2), \ldots, \varphi_0(a_N)),$$

and by P_0 the orthogonal projection in E onto φ_0

$$P_0 h = \frac{\langle \varphi_0, h \rangle_E}{|\varphi_0|^2}, \quad P_0^\perp = I_E - P_0.$$

In view of Lemma 1 and (12) the separation of singularity is possible:

$$[Q(\lambda) - \frac{1}{ik\beta^2}] = \frac{|\varphi_0|^2}{\lambda_0 - \lambda} P_0 - \frac{1}{ik\beta^2} + P_0 K_0 P_0 + P_0 K_0 P_0^\perp$$

$$+ P_0^\perp K_0 P_0 + P_0^\perp K_0 P_0^\perp. \tag{21}$$

Theorem 3 *The scattering matrix S^β of the operator $\mathbf{L}_\beta$ for the weakening boundary condition $\beta \to 0$ has in generic case the following asymptotic at the simple resonance eigenvalue:*

$$S(k) = -I - 2ik\beta^2 K_0 + (I + ik\beta^2 K_0)\frac{2k\beta^2|\varphi_0|^2}{k\beta^2|\varphi_0|^2 + i(\lambda_0 - \lambda)} P_0 + O(|\beta|^4) \tag{22}$$

$$\approx -I + \frac{2k\beta^2|\varphi_0|^2}{k\beta^2|\varphi_0|^2 + i(\lambda_0 - \lambda)} P_0 + o(|\beta|^2).$$

This result can be derived from (20) using the technique in [5]. It follows from the formula (22) that the transmission coefficient $S_{1s}(k)$ from the wire Γ_1 to the wire Γ_s for the resonance energy $k^2 = \lambda_0$ is approximately equal to

$$2 \frac{\bar{\varphi}_0(a_s)\varphi_0(a_1)}{\sum_r |\varphi_0(a_r)|^2}, \tag{23}$$

for $\beta \ll 1$, hence it vanishes for all s such that $\varphi_0(a_s) = 0$.

The main objective of actual paper is to choose the contact points $\{a_s\}$ on the unit circle and the intensity of the homogeneous field ε such that the switching of electron current from one direction to another may be manipulated by the direction of the vector e only. This can be done if, according to (23), zeroes of the eigenfunction φ_0 coincide with two of these contact points.

It is clear that for a constant electric field oriented along the vector e all eigenfunctions of the nonperturbed operator may be represented as functions of the polar coordinates r, θ, where θ may be chosen as an azimuth with respect to the vector e, so that each eigenfunction is rotated by the angle $\delta\theta$ together with the vector e. This observation shows that the problem of switching the electron's current will be solved if we may find the intensity ε such that some zeroes b_s of the resonance eigenfunction on the boundary of the unit disc Ω satisfy the condition $\frac{|b_s-b_t|}{|b_s-b_r|} = 2$. In the simplest case there may be only 2 zeroes on the unit circle b_1 and b_2 dividing it in the ratio $2 \div 1$, that is, $b_1 b_2 = \frac{2\pi}{3}, b_2 b_1 = \frac{4\pi}{3}$. Then attaching the wires to the unit circle at the points a_2, a_3, a_4 such that $|a_2 - a_3| = |a_3 - a_4| = \frac{2\pi}{3}$ (see Fig. 1) we may choose the direction of the vector e such that zeroes of the corresponding resonance eigenfunction coincide either with a_2, a_3 or with a_3, a_4 or with a_2, a_4 and at the same time $\varphi_0(a_1)$, $\varphi_0(a_4) \neq 0$ in the first case and $\varphi_0(a_1)$, $\varphi_0(a_2) \neq 0$ in the second case, and $\varphi_0(a_1)$, $\varphi_0(a_3) \neq 0$ in the third case. Then the only nonzero transmission coefficient for resonance energy will be S_{14} in the first case, or S_{13} in the second case, or S_{13} in the last case. We fix now the number of points of contact a_s, $N = 4$ and reserve the convenient symbol N for other needs in the last section.

A simulation using Mathematica in which the intensity $\varepsilon = 3.558$ constructed the corresponding dimensionless eigenfunction of the Schrödinger operator in the unit disc with Neumann boundary conditions such that this eigenfunction has only two zeroes on the unit circle which divide the circle in ratio $2 \div 1$. Then attaching the wires at the points with azimuth $0, \pm\frac{2\pi}{3}, \pi$ we obtained the pattern of contact points which has the required properties, see Fig. 1, so that we can control the flow of electrons from the ray Γ_1 to only one of the rays Γ_2, Γ_3 or Γ_4; we can block two complementary rays by adjusting the direction of the field V in such a way that both zeroes of the resonance eigenfunction φ_0 of the operator L coincide with a_2 and a_3 or a_2 and a_4 or a_3 and a_4. In this way, only one ray (Γ_4 or Γ_3 or Γ_2 respectively) will be allowed for the flow of electron (see Fig. 2). The formula (23) shows that we can manipulate in this way the quantum current across the domain if we choose the direction of the vector e such that the transmission coefficients S_{1k} are equal to zero for $k = 2, 3$ and $S_{14} \neq 0$ and so on. Now we discuss the change of variables which can reduce the real problem to the problem for the dimensionless equation. In reality we should consider dynamical processes described by the two-dimensional Schrödinger equation written in polar coordinates with real potential $V = \mathcal{E}r\cos(\theta) + \mathcal{V}_0$:

$$\left(-\frac{\hbar^2}{2m_e}\Delta\Psi + (\mathcal{E}r\cos(\theta) + \mathcal{V}_l)\right)\Psi = \mathcal{E}_g\Psi, \tag{24}$$

with the Neumann boundary condition

$$\frac{\partial \Psi}{\partial n}|_R = 0,$$

in a circular domain with radius R on the surface of the narrow-gap semiconductor with effective electron-mass m_e, Fermi level $\mathcal{E}_g$ and the potential shift $\mathcal{V}_0$. The radius R should be chosen such that $2R$ is less or equal to the free path of the electron, and $\mathcal{V}$, should be chosen such that $\mathcal{E}_g$ is exactly equal to the resonance energy. The typical values of these parameters are given in the Table in Section 1.

To reduce the spectral analysis of the real equation (24) to the spectral analysis of the corresponding dimensionless equation on the disc radius $r = 1$, we transform the real equation just by division through $\dfrac{\hbar^2}{2m_e}$, followed by the change of variable $r \to \xi = r/R$, and then one more division by R_2. This results in dimensionless equation for properly transformed resonance eigenfunction $\varphi_0(\xi, \theta) = \Psi(x, \theta)$

$$-\Delta_\xi\ \varphi_0 + \mathcal{E}\frac{2m_e R^3}{\hbar^2}|\xi|\cos\theta\ \varphi_0 + V_0\frac{2m_e R^2}{\hbar^2} = \mathcal{E}_g\frac{2m_e R^2}{\hbar^2}\ \varphi_0 \qquad (25)$$

in the unit disc with the Neumann boundary condition at the boundary

$$\lambda = \mathcal{E}_g\frac{2m_e R^2}{\hbar^2}, \quad V_0 = \frac{2m_e R^2}{\hbar^2}\mathcal{V}_0, \quad \varepsilon = \mathcal{E}\frac{2m_e R^3}{\hbar^2}.$$

4 Perturbation Procedure and Error Estimation

In this section we reduce the spectral problem for the Schrödinger operator to the spectral problem for some specially arranged finite matrix and obtain explicit expressions for its eigenvectors and numerical values of its eigenvalues. We also estimate the error arising from this simplification.

In order to obtain a solution of dimensionless equation we develop the perturbation procedure for the dimensionless operator

$$L\varphi = -\Delta_\xi \varphi + \varepsilon \xi \varphi, \quad \xi = R\cos\varphi$$

in the unit disc with Neumann boundary condition using the representation of it in form of infinite matrix in l_2 with respect to the basis of normalized eigenvectors Φ_{ns} of the *nonperturbed operator*

$$-\Delta_\xi \Phi_{ns}^{c,s} = (k_n^s)^2 \Phi_{ns}^{c,s},$$

which may be constructed in explicit form:

$$\Phi_{nc}^c = \frac{J_n(k_n^s r)\cos(n\phi)}{(\pi \int_0^1 |J_n(k_n^s r)|^2 r dr)^{1/2}}, \quad n = 0, 1, 2, \ldots,$$

$$\Phi_{ns}^s = \frac{J_n(k_n^s r)\sin(n\phi)}{(\pi \int_0^1 |J_n(k_n^s r)|^2 r dr)^{1/2}}, \quad n = 1, 2, \ldots$$

Here J_n is n-th Bessel function, and k_n^s is s-th root of the first derivative J_n' of J_n. In this basis the perturbed operator is an orthogonal sum of two block-symmetric matrices $\mathcal{A}^c \oplus \mathcal{A}^s$, the first addend corresponding to the cosine-part of the basis (including $n = 0$), the second one corresponding to the sine-part of the basis ($n > 0$). The second addend $\mathcal{A}^s$ coincides with the submatrix of the first addend for $n \geq 1$. For this reason we calculate now the first addend $\mathcal{A}^c$ only. It may be represented as a sum of the diagonal matrix $\mathcal{A}^{diag} = \mathrm{diag}(k_n^s)^2$ which corresponds to the nonperturbed operator and the perturbation V caused by the homogeneous field:

$$\mathcal{A} = \mathcal{A}^{diag} + V = \begin{pmatrix} A_{00} + V_{00} & V_{01} & 0 & 0 & \cdots \\ V_{10} & A_{11} + V_{11} & V_{12} & 0 & \cdots \\ 0 & V_{21} & A_{22} + V_{22} & V_{23} & \cdots \\ \cdot & \cdot & \cdot & \cdot & \cdots \end{pmatrix}.$$

Here A_{ii} are infinite diagonal blocks with elements $A_{ii}^{jj} = (k_i^j)^2$ and $V_{ik} = V_{ki}^+$,

$$V_{01}^{st} = \frac{\varepsilon}{\sqrt{(2)}} \frac{\int_0^1 J_1(k_1^s r) J_0(k_0^t r) r^2 dr}{\left(\int_0^1 [J_0(k_0^t r)]^2 r dr\right)^{1/2} \left(\int_0^1 [J_1(k_1^t r)]^2 r dr\right)^{1/2}}$$

$$V_{nm}^{st} = \frac{\varepsilon}{2} \frac{\int_0^1 J_n(k_n^s r) J_m(k_m^t r) r^2 dr}{\left(\int_0^1 [J_n(k_n^t r)]^2 r dr\right)^{1/2} \left(\int_0^1 [J_m(k_m^t r)]^2 r dr\right)^{1/2}}, \quad m = n \pm 1.$$

We know that all eigenvalues of the nonperturbed operator have multiplicity 2 except one which corresponds to the constant angular factor, because for any $\lambda = (k_n^s)^2$, $n > 0$ there are 2 eigenfunctions:

$$\frac{J_n(k_n^s r) \cos(n\phi)}{(\pi \int_0^1 |J_n(k_n^s r)|^2 r dr)^{1/2}}, \quad \frac{J_n(k_n^s r) \sin(n\phi)}{(\pi \int_0^1 |J_n(k_n^s r)|^2 r dr)^{1/2}}.$$

Having in mind future estimates of errors of the perturbation procedure it is convenient to rearrange the matrix A in the following way. We consider the matrix A as a set of infinite rows, each one containing one "diagonal" element $(k_n^s)^2$. From the properties of Bessel functions the roots k_n^s depend monotonically of both indices: $k_n^s < k_n^t$, if $s < t$, and $k_n^s < k_m^s$, if $n < m$. So, for any fixed positive number M we may find a curve $\{n, s\}_M$ which divides the root lattice Z_2 such that $(k_n^s)^2 \leq M$ and for any $m, t, m > n$ and $t > s$ inequality holds $(k_m^t)^2 > M$. Now we rearrange all rows and columns in the order of increasing (non decreasing) "diagonal" elements $(k_n^s)^2$. We denote the rearranged matrix by $\mathbf{A}$ and its "diagonal" submatrix by $\mathbf{A}^{diag} = \mathrm{diag}(k_n^s)^2$. Now we have $\mathbf{A}_{ll}^{diag} < \mathbf{A}_{l+1\,l+1}^{diag}$. So for any fixed M we may find some finite-dimensional subspace E_N, $N = N_M$ characterized by the corresponding orthogonal projection P_N such that the "diagonal part" $\mathbf{A}^{diag}$ of the corresponding block $\mathbf{A}_{NN} = P_N \mathbf{A} P_N$ is not greater than than M. The diagonal part $\mathbf{A}^{diag}_{N^\perp N^\perp} = P_{N^\perp} \mathbf{A}^{diag} P_{N^\perp}$ of the restriction of the whole operator $\mathbf{A}$ onto the orthogonal complement $E_{N^\perp}$ is not less than M, and the whole rearranged matrix becomes

$$A = \begin{pmatrix} \mathbf{A}_{NN} & 0 & 0 \\ & V_{NN^\perp} & 0 \\ 0 & V_{N^\perp N} & \mathbf{A}_{N^\perp N^\perp} \\ 0 & 0 & \end{pmatrix}$$

We write $\mathbf{A}_{N^\perp N^\perp} = \mathbf{A}^{diag}_{N^\perp N^\perp} + V_{N^\perp N^\perp}$ where $V_{N^\perp N^\perp} = P_{N^\perp} V P_{N^\perp}$. In our case the operator V is bounded and hence the symmetric matrix $\mathbf{A}$ is *asymptotically diagonal - dominated* - ADD. That is for each real λ,

$$\sup_f \frac{\|P_{N^\perp} V P_{N^\perp} f\|}{\left\|\mathbf{A}^{diag}_{N^\perp N^\perp} f - \lambda f\right\|} \longrightarrow 0, \quad \sup_f \frac{\|P_N V P_{N^\perp} f\|}{\left\|\left(\mathbf{A}^{diag}_{N^\perp N^\perp} - \lambda I\right)^{1/2} f\right\|} \longrightarrow 0,$$

when $M \to \infty$. Note that symmetric ADD-matrices have discrete spectrum and the following estimates are true:

$$\left\|\left(\mathbf{A}_{N^\perp N^\perp} - \lambda I\right)^{-1}\right\| \longrightarrow 0, \quad \left\|V_{NN^\perp}\left(\mathbf{A}_{N^\perp N^\perp} - \lambda I\right)^{-1} V_{NN^\perp}\right\| \longrightarrow 0.$$

The explicit expression for the resolvent $\Psi = [\mathbf{A} - \lambda I]^{-1} f$ of the operator $\mathbf{A}$ is a solution of the the nonhomogeneous equation

$$[\mathbf{A} - \lambda I]\Psi = f,$$

or in decomposed form $\Psi = \Psi^s_N + \Psi^s_{N^\perp}$ for $f = f_N \in E_N$ with respect to the splitting of the space: $E_N \oplus E_{N^\perp}$

$$\begin{cases} \mathbf{A}_{NN} \Psi^s_N + V_{NN^\perp} \Psi^s_{N^\perp} - \lambda \Psi^s_N = f_N \\ V_{N^\perp N} \Psi^s_N + \mathbf{A}^{diag}_{N^\perp N^\perp} \Psi^s_{N^\perp} + V_{N^\perp N^\perp} \Psi^s_{N^\perp} - \lambda \Psi^s_{N^\perp} = 0. \end{cases} \quad (26)$$

Solving this equation as in [14], [15], [16] we obtain

$$f = (\mathbf{A}_{NN} - \lambda I)\Psi^s_N - V_{NN^\perp} \times (\mathbf{A}^{diag}_{N^\perp N^\perp} - \lambda I_{N^\perp} + V_{N^\perp N^\perp})^{-1} V_{N^\perp N} \Psi^s_N =$$

$$\left(\mathbf{A}_{NN} - \lambda I\right) \Psi^s_N + \mathbf{R}_{NN} \Psi^s_N = \mathbf{A}_{NN} \Psi^s_N \quad (27)$$

which means that the operator $[\mathbf{A}]_N^{-1}(\lambda)$ is exactly the block of the resolvent of $[\mathbf{A} - \lambda I]^{-1}$ in E_N:

$$P_N [\mathbf{A} - \lambda I]^{-1} P_N := [\mathbf{A}]_N^{-1}(\lambda).$$

Then from the general fact for selfadjoint operators we obtain the estimate of the inverse via the distance of λ from the spectrum $\sigma(\mathbf{A})$ of the operator A:

$$\|[\mathbf{A}]_N^{-1}(\lambda)\| \leq \frac{1}{\text{dist}(\lambda, \ \sigma(\mathbf{A}))}. \quad (28)$$

In particular this estimate is true for λ sitting on the circle $C_s^\alpha = C_{\lambda_s}^\alpha = \{\lambda : |\lambda - \lambda_s| = \alpha \delta_s$, so that

$$\|\mathbf{A}_N^{-1} \mathbf{R}_N(\lambda)\| \to 0 \qquad (29)$$

for s, ϵ fixed, $M \to \infty$.

Theorem 4 *For every ADD-matrix $\mathbf{A}$ in l_2 and for any isolated and simple eigenvalue λ^s of $\mathbf{A}$*

$$\mathbf{A}\Psi^s = \lambda^s \Psi^s, \ \text{dist}(\lambda_s, \lambda_t) = \delta_s, s \neq t > 0,$$

there exists, for large M, a block $\mathbf{A}_{NN}$ with all "diagonal" elements less than M, such that the eigenvalue μ_N^s of the block $\mathbf{A}_{NN}$ is also isolated, simple,

$$lim_{N \to \infty} \mu_N^s = \lambda^s,$$

and the corresponding eigenvector ψ_N^s of the block-matrix $\mathbf{A}_{NN}$ tends to the eigenfunction of $\mathbf{A}$

$$lim_{N \to \infty} \psi_N^s = \Psi^s$$

Given M consider the orthogonal decomposition of l_2 into the orthogonal sum $E_N \oplus E_{N\perp}$:

$$\mathbf{A} = P_N \mathbf{A} P_N + P_N V P_{N\perp} + P_{N\perp} V P_N + P_{N\perp} \mathbf{A}^{diag} P_{N\perp} + P_{N\perp} V P_{N\perp}. \qquad (30)$$

For given eigenvalue λ^s of $\mathbf{A}$ choose $M \gg \lambda^s$, so that $\|\mathbf{R}_N\| = r \ll 1$. We decompose the eigenfunction of $\mathcal{A}$ into the sum of two components

$$\Psi^s = \Psi_N^s + \Psi_{N\perp}^s$$

in E_N and $E_N^\perp$ respectively. Hence Ψ_N^s satisfies the equation:

$$(\mathbf{A}_{NN} - \lambda^s I)\Psi_N^s + \mathbf{R}_N \Psi_N^s = 0. \qquad (31)$$

Consider now the spectral problem in E_N for the operator $\mathbf{A}_{NN}$ as a perturbation of the spectral problem for $\mathbf{A}$. Comparing the finite dimensional operator-function $[\mathbf{A}_N](\lambda)$ from (31) and the reduced operator $\left(\mathbf{A}_{NN} - \lambda I\right)$ we see that their ratio

$$\left(\mathbf{A}_{NN} - \lambda I\right)[\mathbf{A}_N]^{-1}(\lambda) = I - \frac{I}{[\mathbf{A}_N](\lambda)} R_N(\lambda),$$

is an invertible operator for large M (since the second term in the last formula is small on the circle C_s). Then, from the operator version of the Rouchet Theorem [17] we see that for M large enough both operator functions have equal numbers of vector zeroes in C_s. Hence the reduced operator $\mathbf{A}_N$ for large M has exactly *one simple* eigenvalue μ inside C_s.

A similar statement remains true for the neighboring eigenvalues of λ_{s-1}, λ_{s+1} **A**: the corresponding eigenvalues of the operator $\mathbf{A}_{NN}$ sit in some small neighborhoods $|\lambda - \lambda_{s\pm 1}| \leq \alpha \delta_{s\pm 1}$, for large M, though they may not be simple, if λ_{s-1}, λ_{s+1} are not simple. Still, their distance to μ is not less than $\delta_s - \alpha \delta_s - \alpha \delta_{s\pm 1} := \delta_\alpha$.

Now we estimate the error we make using the nearest eigenvalues and eigenfunctions μ, ψ of the reduced operator $\mathbf{A}_{NN}$ instead of corresponding eigenvalues and eigenfunctions of the operator $\mathbf{A}$. From the homogeneous equations ($f = 0$) similar to (26, 27) follows, that the complementary component $\Psi^s_{N^\perp}$ of the eigenfunction is small for large M. Hence to derive the announced result we should prove that $\delta\psi = \Psi^s_N - \psi$ is small. Then we assume that $|\psi| = 1$ and $\delta\psi$ is orthogonal to it. We decompose the operator $\mathbf{A}_{NN}$ as an orthogonal sum

$$\mathbf{A}_{NN} = \mu P_\psi + \hat{\mathbf{A}}_{NN}, \qquad (32)$$

where $\hat{\mathbf{A}}_{NN} = \sum_{\mu^t \neq \mu} \mu^t P_t$ is the spectral representation of the component of the operator $\mathbf{A}_{NN}$ in the orthogonal complement of ψ. Then due to the separation of the eigenvalues of $\mathbf{A}_{NN}$ proven before we have

$$\|(\hat{\mathbf{A}}_{NN} - \mu)^{-1}\| << max_{\mu \neq \mu^t} \frac{1}{|\mu - \mu^t|} \leq \frac{1}{\delta_\alpha}. \qquad (33)$$

Using this representation of Ψ^s_N we can rewrite (31) as:

$$(\mathbf{A}_{NN} - \lambda^s I)(\psi_N + \delta\psi) - R(\psi_N + \delta\psi) = 0.$$

Then (32) implies

$$(\mu - \lambda^s)\psi + (\hat{\mathbf{A}}_{NN} - \lambda^s)\delta\psi - \mathbf{R}_N(\lambda_s)\psi - \mathbf{R}_N \delta\psi = 0. \qquad (34)$$

Now we look after $(\mu - \lambda^s) = \delta\lambda$ and $\delta\psi$. Applying $P_\psi = \psi >< \psi$ and $P^\perp = P_N - P_\psi$ to (34) we obtain the first order approximations:

$$\delta\lambda \approx \langle \mathbf{R}_N(\lambda^s)\psi, \psi \rangle, \quad \delta\psi = [\hat{\mathbf{A}}_{NN} - \lambda^s I - P^\perp R(\lambda^s) P^\perp]^{-1} P^\perp R(\lambda^s) \psi_N \approx \qquad (35)$$

$$[\hat{\mathbf{A}}_{NN} - \lambda^s I]^{-1} P^\perp \mathbf{R}_N(\lambda^s)\psi. \qquad (36)$$

Here $[\hat{\mathbf{A}}_{NN} - \lambda^s I - P^\perp \mathbf{R}_N(\lambda^s) P^\perp]$ is also invertible if $\mathbf{R}_N(\lambda^s)$ is small, because the operator $[\hat{\mathbf{A}}_{NN} - \lambda^s I]^{-1}$ is estimated by $max_{s \neq t} \frac{1}{|\lambda^s - \lambda_t|} = \frac{1}{\delta_\alpha}$, due to above separation results. Then we obtain for large M,

$$\delta\lambda \approx \langle \mathbf{R}_N(\lambda^s)\psi, \psi \rangle \leq \|\mathbf{R}_N\| = r,$$

$$\delta\psi \approx [\hat{\mathbf{A}}_{NN} - \lambda^s I]^{-1} P^\perp_s R(\lambda^s) \psi_N \leq \frac{\|\mathbf{R}_N\|}{\delta_\alpha} = \frac{\epsilon}{\delta_\alpha}.$$

Note that in our case the operator V is bounded. Hence assuming $\delta_\alpha \approx \delta_s$ we obtain $r \leq \frac{\|V\|^2}{M - \lambda_s - \|V\|}$. Using Mathematica we have found eigenvalues and eigenfunctions of $\mathbf{A}_{NN}$. We chose $M = 100$, then $N = 17$; that is, there are 17 diagonal elements k_n^s which satisfy the requirement $(k_n^s)^2 \leq 100$. The matrix $\mathbf{A}_{NN}$ and its eigenvalues for the required intensity of the electric field $\varepsilon = 3.558$ were found. To obtain explicit expressions for the eigenfunctions and numerical values of the eigenvalues we applied two criteria for choosing one of these 17 eigenfunctions as a resonance one. First, it has only two zeroes on the unit circle, and these zeroes divide the circle in ratio $2 \div 1$. Actually, their positions can be controlled to certain extent by appropriate choice of ε. Secondly, the corresponding eigenvalue should not be too close to its neighbors for the reason of stability of the perturbation procedure. In our situation this condition is fulfilled since the error arising from replacement of the operator A by the finite matrix is defined by the norm of

$$\mathbf{R} = V_{NN^\perp} V_{N^\perp N^\perp} \frac{1}{A_{N^\perp N^\perp} - \lambda I_{N^\perp} + V_{N^\perp N^\perp}} V_{N^\perp N}:$$

$$\Delta = inf_t |\mu^s - \mu^t| = 4.6 \gg \frac{\|V\|^2}{100 - 3.6 - 3.8} \approx 0.14 \gg \|R\|.$$

The eigenfunctions which correspond to the lowest three eigenvalues including the resonance eigenvalue μ_{17}^2 can be calculated. Eigenvalues and eigenfunctions were found for different values of ε. Then by interpolation the value of ε which gives the desired disposition of zeroes of chosen eigenfunction was obtained with a precision of 4 significant digits. So, $\varepsilon = 3.558$ (in dimensionless case) the eigenfunction that corresponds to the second smallest eigenvalue $\mu_{17}^2 = 3.79$ has 2 zeroes on the unit circle which divide it in the ratio $2 \div 1$. This gives the possibility to choose the contact points a_s on the circle such that the switching of electron current from one ray to another may be manipulated by the direction of electric field, as was described above. It can be seen that $\mu_{17}^2 = 3.79$ is not really too close to the neighboring eigenvalues $\mu_{17}^1 = -0.79$ and $\mu_{17}^3 = 9.39$. The condition $\kappa T < E_g \Delta = 4.16 * E_g$ is fulfilled for sufficiently high temperatures. It means that the relay exhibits a staple high-temperature behavior: the parallel modes associated with neighboring eigenvalues μ_{17}^1 and μ_{17}^3 are not excited unless the critical value of the temperature $T = E_g \frac{\Delta}{\kappa} = 4.16 \frac{E_g}{\kappa}$ is reached.

It should be noted that the error $\frac{\|R\|}{\Delta}$ arising from using the finite matrix $\mathbf{A}_{NN}$ instead of the operator A can be made as small as desired by increasing M and, correspondingly by increasing the order N of $\mathbf{A}_{NN}$. For the case of $M = 100$ it is estimated by 3% only.

Similar results are obtained for Dirichlet boundary condition and may be easily derived for the third boundary condition with the boundary parameter defined by the technical implementation and the choice of materials used. The constructed relay may be used as an element of multi-valued logic systems. Note, that the construction of corresponding two-position switch does not require special conditions on distribution of zeroes of the resonance eigenfunction.

Acknowledgments

We are grateful to the Commission of the European Communities for financial support in the frame of EC-Russia Exploratory Collaborative Activity under EU ESPRIT Project 28890 NTCONGS and partial support from the Russian Academy of Sciences (Grant RFFI 97 - 01 - 01149).

References

1. B. Pavlov, G. Miroshnichenko. *Quantum Interference Transistor*, Patent of Russian Federation RU (11) 2062530 (13) C1, registered 20 June 1996, priority date 12 March 1992, in Russian, 10pp.
2. I. Antoniou, B. Pavlov, Y. Yafyasov. Quantum electronic Devices, based on Metal-Dielectric Transition in Low-Dimensional Quantum Structures, In: D. S. Bridges, C. Calude, J. Gibbons, S. Reeves, I. Witten (eds.). *Combinatorics, Complexity, Logic, Proceedings of DMTCS'96*, Springer-Verlag, Singapore, 1996, 90–104.
3. B. Pavlov, G. Roach, A. Yafyasov. Resonance scattering and design of quantum gates, In: C.S. Calude, J. Casti, M.J. Dinneen (eds.). *Unconventional Models of Computation*, Springer Verlag, Singapore, 1998, 336–351.
4. V. Bogevolnov, A. Mikhaylova, B. Pavlov, A. Yafyasov. About scattering on the ring, *Department of Mathematics, Report Series* No.413, University of Auckland, 16pp. Accepted for publication in *Operator Theory, Advances and Applications* (Gohberg's Volume).
5. A. Mikhaylova, B. Pavlov, I. Popov, T. Rudakova, A. Yafyasov. Scattering on a compact domain with few semi-infinite wires attached: resonance case, *Department of Mathematics, Report Series* No 420, University of Auckland, 17pp. Accepted for publication in *Mathematishe Nachrichten.*
6. B. Gejler, B. Pavlov, I. Popov. *Possible construction of a quantum triple logic device. New technologies for narrow-gap semiconductors*, Esprit project N 28890 ESPRIT NTCONGS Progress Reports (July 1, 1998 - June 30 1999).
7. B.S. Pavlov. The theory of extensions and explicitly-solvable models, *Russian Math. Surveys* 42: 6 (1987), 127–168.
8. S. Albeverio, P. Kurasov. *Singular Perturbations of Differential Operators*, London Mathematical Society lecture Notes Series 271, Cambridge University Series, 2000.
9. N. Gerasimenko, B. Pavlov. Scattering problem on noncompact graphs, *Theor. Math. Physics*, 75 (1988), 230–240.
10. S. Novikov. Schrödinger operators on graphs and simplectic geometry, In: E. Bierstone, B. Khesin, A. Khovanskij and J. Marsden (eds.). *Proceedings of the Fields Institute Conference in Honour of 60 th Birthday of Vladimir I. Arnold*, Communications of Fields Institute, AMS 1999.
11. A. Mikhaylova, B. Pavlov. Quantum domain as a triadic relay, *Department of Mathematics, Report Series* No.439, University of Auckland, Auckland, 200, 24pp.
12. M. Krein. Concerning the resolvents of an Hermitian operator with the deficiency-index (m, m), *Comptes Rendues (Doklady) Acad. Sci. URSS (N.S.)*, 52 (1946), 651–654.
13. C.W.J. Beenakker, H. van Houten. Quantum transport in semiconductor nanostructures. In: H. Ehrenreich, D. Turnbull (eds.). *Solid State Physics. Advances in Research and Applications*, 44 (1991), 1–228.

14. V. Adamjan, H. Langer. Spectral properties of a class of rational operator-valued functions, *Journ. Operator Theory*, 33 (1995), 259–277.
15. F.V. Atkinson, H. Langer, R. Mennicken, A. Shkalikov. Essential spectrum of some matrix operators, *Mathem. Nachrichten*, 167 (1994), 5–20.
16. R. Mennicken, A. Shkalikov. Spectral decomposition of symmetric operator matrices, *Math. Nachrichten*, 179 (1996), 259–273.
17. I.S. Gohberg, E.I. Sigal. Operator extension of the theorem about logarithms residue and Rouchet theorem, *Mat. Sbornik*. 84 (1971), 607.

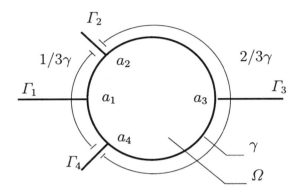

Fig. 1. Design of the triadic relay with required properties.

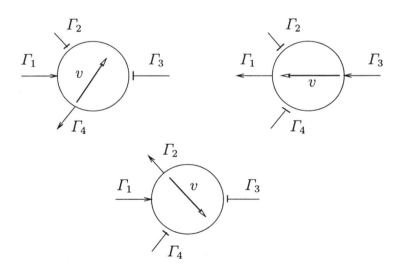

Fig. 2. The flow of electrons in dependence of the applied electrostatic field V from the ray Γ_1: **1** - to the ray Γ_4, **2** - to the ray Γ_3, **3** - to the ray Γ_2.

On P Systems with Active Membranes

Andrei Păun*

Department of Computer Science, University of Western Ontario, London, Ontario, Canada

Abstract. The paper deals with the vivid area of computing with membranes (P systems). We improve here two recent results about the so-called P systems with active membranes. First, we show that the Hamiltonian Path Problem can be solved in polynomial time by P systems with active membranes where the membranes are only divided into two new membranes (a result of this type was obtained by Krishna and Rama, [4], but making use of the possibility of dividing a membrane in an arbitrary number of new membranes). We also show that HPP can be solved in polynomial time also by a variant of P systems, with the possibility of dividing non-elementary membranes under the influence of objects present in them. Then, we show that membrane division (and even membrane dissolving) is not necessary in order to show that such systems are computationally complete.

1 Introduction

In the framework of Molecular Computing, a new model for distributed parallel computing has been recently introduced in [6], cell-like devices called P systems, which try to model the work of the alive cells.

Each cell is delimited by its membrane and can contain objects and other cells (the objects and the other cells are floating in this membrane). The objects are represented by symbols, their multiplicity is taken into account (that is, we work with multisets), and they can evolve by using given evolution rules (which are specific to each membrane).

The objects can pass from one cell to another one (if the cells are adjacent), they can dissolve a membrane (and in this case all other objects go into the immediately higher cell), or leave the biggest membrane (the skin membrane), which means that that specific object leaves the system.

A sequence of transitions among configurations of the system is interpreted as a computation. The result of a halting computation is the vector describing the multiplicities of objects which have left the system during the computation.

Many variants of P systems have been considered and investigated in [2], [3], [5], [8], etc. In [7] one considers also the possibility of letting the number

---

* This work has been partially supported by the Natural Sciences and Engineering Research Council of Canada grants OGP0041630 and a graduate scholarship.

of membranes increase during a computation by using some division rules (we will define them later as rules of types (e), (e'), (e_I), (f) and (f_I)). That is modeling the division of a cell from nature.

In the initial variant ([7]) the division rules were defined such that from the division of a membrane we obtain always two membranes (we call it a 2-bounded division); a generalization was defined in [4] where a membrane can divide in several (but a finite number) of membranes (we call it a k-bounded division).

Using the 2-bounded division it was shown that the NP-complete problem SAT can be solved in linear time [7]. By using the k-bounded division, in [4] was shown that also the Hamiltonian path problem (HPP) can be solved in linear time [4]. Remember that HPP was the problem dealt with in the first experiment in DNA Computing, [1].

In [4], the open problem is formulated whether or not a system with k-bounded division can be simulated by a system with 2-bounded division, with a reasonable slowdown (polynomial, at most). We do not attack here this problem, but we directly shown that HPP can also be solved in polynomial time by using P systems with a 2-bounded division. Thus, the extension from 2-bounded to k-bounded division is not necessary from this point of view.

Actually, we give two results of this type, one in the framework of the definition in [7] (which is generalized in [4]), and one with the variant of evolution rules as considered in [9] (elementary or non-elementary membranes are divided under the influence of objects, but not under the influence of lower membranes).

The P systems with active membrane were shown in [7] to be computationally complete: any set of vectors of natural numbers which can be computed by a Turing machine can be computed by such a system. The result was improved in [9]: rules of type (f) (non-elementary membrane division under the influence of lower membranes) are not necessary. We improve here this result: no membrane division operation is necessary in order to obtain the computability completeness.

2 Basic Definitions

We refer the reader to [10] for basics of formal language theory and to [6], [7] for basic notions, notations and results about P systems; here we directly introduce the class of systems we will investigate in the subsequent sections. We only mention that the membrane structures are represented as strings of matching parentheses, with the parentheses labeled by elements in a given set, and with each membrane electrically polarized, that is, marked with one of the symbols $+, -, 0$. For example, the string $[_1[_2\ ]_2^+[_3[_4\ ]_4^-]_3^0]_1^0$ is a membrane structure as illustrated in Figure 1, with four membranes arranged in a hierarchy of depth three with membrane 1 neutral, membrane 2 positive, membrane 3 neutral and membrane 4 negative.

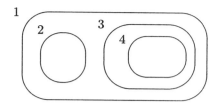

Figure 1: A membrane structure.

A *P system with active membranes and 2-bounded membrane division* is a construct
$$\Pi = (V, T, H, \mu, w_1, \ldots, w_m, R),$$
where: $m \geq 1$; V is an alphabet (the *total alphabet* of the system); $T \subseteq V$ (the *terminal* alphabet); H is a finite set of *labels* for membranes; μ is a *membrane structure*, consisting of m membranes, labeled with elements of H, and having a neutral charge (initially, all membranes are marked with 0); $w_1, \ldots, w_m$ are strings over V, describing the *multisets of objects* placed in the m regions of μ; R is a finite set of *rules*, of the following forms:

(a) $[_h a \to v]_h^\alpha$, for $h \in H$, $a \in V$, $v \in V^*$, $\alpha \in \{+, -, 0\}$ (object evolution rules),
(b) $a[_h\,]_h^{\alpha_1} \longrightarrow [_h b]_h^{\alpha_2}$, where $a, b \in V$, $h \in H$, $\alpha_1, \alpha_2 \in \{+, -, 0\}$ (an object is introduced in the membrane h),
(c) $[_h a]_h^{\alpha_1} \to b[_h\,]_h^{\alpha_2}$, for $h \in H$, $\alpha_1, \alpha_2 \in \{+, -, 0\}$, $a, b \in V$ (an object is sent out of the membrane h),
(d) $[_h a]_h^\alpha \to b$, for $h \in H$, $\alpha \in \{+, -, 0\}$, $a, b \in V$ (the membrane h was dissolved),
(e) $[_h a]_h^{\alpha_1} \to [_h b]_h^{\alpha_2} [_h c]_h^{\alpha_3}$,
for $h \in H$, $\alpha_1, \alpha_2, \alpha_3 \in \{+, -, 0\}$, $a, b, c \in V$
(2-division rules for elementary membranes; in reaction with an object, the membrane is divided into two membranes with the same label, maybe of different polarizations; the object specified in the rule is replaced in the two new membranes by possibly new objects, all other objects are duplicated in the two new copies of the membrane);
(f) $[_{h_0} [_{h_1}\,]_{h_1}^{\alpha_1} \cdots [_{h_k}\,]_{h_k}^{\alpha_1} [_{h_{k+1}}\,]_{h_{k+1}}^{\alpha_2} \cdots [_{h_n}\,]_{h_n}^{\alpha_2}]_{h_0}^{\alpha_0}$
$\to [_{h_0} [_{h_1}\,]_{h_1}^{\alpha_3} \cdots [_{h_k}\,]_{h_k}^{\alpha_3}]_{h_0}^{\alpha_5} [_{h_0} [_{h_{k+1}}\,]_{h_{k+1}}^{\alpha_4} \cdots [_{h_n}\,]_{h_n}^{\alpha_4}]_{h_0}^{\alpha_6}$,
for $k \geq 1, n > k, h_i \in H, 0 \leq i \leq n$, and $\alpha_0, \ldots, \alpha_6 \in \{+, -, 0\}$ with $\{\alpha_1, \alpha_2\} = \{+, -\}$; if this membrane with the label h_0 contains other membranes than those with the labels $h_1, \ldots, h_n$ specified above, then they must have neutral charges in order to make this rule applicable; these membranes are duplicated and then are part of the contents of both copies of the membrane h_0;
(2-division of non-elementary membranes; this is possible only if a membrane directly contains two membranes of opposite polarization, + and −; the membranes of opposite polarizations are separated in the two new membranes, but their polarization can change; always, all membranes of opposite polarizations are separated by applying this rule).

We refer to [7] and [9] for a more precise definition of the way a P system as above works and we only describe here informally the way of passing from a configuration of the system to the next one.

The rules of types (a) to (d) are applied in a maximally parallel manner: in each time unit, any objects which can evolve, have to evolve. If a membrane with label h is divided by a rule of type (e), which involves an object a, then all other objects in membrane h which do not evolve are introduced in each of the resulting membranes h. Similarly, when using a rule of type (f), the membranes which are not specified in the rule, (that is, they are neutral), are reproduced in each of the resulting membranes with the label h, unchanged if no rule is applied to them (the contents of these membranes are reproduced unchanged in these copies, providing that no rule is applied to their objects).

When applying a rule of type (f) or (e) to a membrane, if there are objects in this membrane which evolve by a rule of type (a), changing the objects, then in the new copies of the membrane we introduce the results of evolution. The rules are applied "from bottom up", in one step, but first the rules of the innermost region and, then, level by level until the region of the skin membrane. The rules associated with a membrane h are used for all copies of this membrane; it doesn't matter whether or not the membrane is an initial one or it was obtained by division. At one step, a membrane h can be the subject of only one rule of types (b) – (f). The skin membrane can never divide (or dissolve). As any other membrane, the skin membrane can be "electrically charged". During a computation, objects can leave the skin membrane (using rules of type (c)).

The result of a halting computation (no rule can be used in the last configuration) consists of the vector of natural numbers describing the multiplicities of terminal objects which have left the system during the computation. Because of the non-determinism of choosing the rules and the objects which evolve at each step, several results are obtained starting from an initial configuration. We denote by $N(\Pi)$ the set of all vectors of natural numbers computed as above by a P system Π.

The symbols not in T which leave the skin membrane as well as all the symbols from T which remain in the system at the end of a halting computation are not contributing to the generated vectors; if a computation goes forever, then it provides no output, it does not contribute to the set $N(\Pi)$.

Further variants for these systems have been defined. For example, in [9] one allows the division of non-elementary membranes by using rule $[_i a]_i^{\alpha_1} \to [_i b]_i^{\alpha_2} [_i c]_i^{\alpha_3}$ of type (e); we say that this is a rule of type (e').

Moreover, in [4] one considers rules (e_I) and (f_I) that generalize the rules (e) and (f) allowing one membrane to divide in a number of other membranes (in the initial variant they were allowed to divide in two membranes):

(e_I) $[_h a]_h^\alpha \to [_h a_1]_h^{\alpha_1} [_h a_2]_h^{\alpha_2} \ldots [_h a_n]_h^{\alpha_n}$,
for $\alpha, \alpha_i \in \{+, -, 0\}, a \in V, a_i \in V^*, i = 1, \ldots, n, h \in H$ (n-division rules for elementary membranes),

(f_I) $[_{h_0}[_{h_1}]_{h_1}^{\alpha_1} \cdots [_{h_k}]_{h_k}^{\alpha_k} [_{h_{k+1}}]_{h_{k+1}}^{\alpha_{k+1}} \cdots [_{h_n}]_{h_n}^{\alpha_n}]_{h_0}^{\alpha_0}$
$\to [_{h_0}[_{h_1}]_{h_1}^{\beta_1}]_{h_0}^{\beta_0} \cdots [_{h_0}[_{h_k}]_{h_k}^{\beta_k}]_{h_0}^{\beta_0} [_{h_0}[_{h_{k+1}}]_{h_{k+1}}^{\beta_{k+1}}]_{h_0}^{\beta_0} \cdots [_{h_0}[_{h_n}]_{h_n}^{\beta_n}]_{h_0}^{\beta_0}$,
for $k \geq 1$, $n > k$, $h_i \in H$, $0 \leq i \leq n$, and there is i, $1 \leq i \leq n-1$, such that $\{\alpha_i, \alpha_{i+1}\} = \{+, -\}$; moreover, $\beta_j \in \{+, -, 0\}$, $1 \leq j \leq n$ (n-division of non-elementary membranes).

One can notice that in the rules of type (e_I) we have a double generalization, one dealing with the number of membranes produced (from 2 to n) and also in each membrane produced we have a string $a_i \in V^*$ (in the initial model we had only one symbol).

By using P systems with rules of types (e_I) and (f_I) instead of types (e) and (f) (it is immediate to see that such systems are more powerful than the initial ones), in [4] one shows that the Hamiltonian path problem can be solved in linear time. In that paper, it is also conjectured that the systems with rules (a)...(d), (e_I), and (f_I) can be simulated with systems with rules (a)...(f). We do not have an answer to this problem, but we directly show that HPP can be solved in polynomial time by using systems with rules of the forms (a)...(f).

3 Solving HPP by P Systems with Active Membranes

We will continue in this section the work from [7], namely, we will solve the Hamiltonian Path Problem in polynomial time (specifically, in $O(n^3)$ steps).

Theorem 1. *HPP can be solved in polynomial time by P systems with rules of types $(a) \ldots (f)$.*

Proof. Consider a graph $G = (V, E)$, where $V = \{0, 1, \ldots, n-2, n-1\}$, $2 \leq n$, is the set of vertices and E is the set of edges (without loss of the generality we may assume that no edge of the form $\{i, i\}$ exists). We will construct a P system $\Pi = (V, H, \mu, w_0, w_1, \ldots, w_{n-1}, w_n, R)$, taking G as input and producing as the output the letter t at the moment $2n^3 + 2n$ if and only if there is a Hamiltonian path in G starting in node 0.

The components of the P system Π are as follows:

$V = \{i, b_i, d_i, u_i \mid 0 \leq i \leq n-1\} \cup \{c_i \mid 0 \leq i \leq 2n-2\} \cup \{t_i \mid 0 \leq i \leq 2n^2 - 2\}$
$\quad \cup \{u_{j_1 \ldots j_k} \mid 2 \leq k \leq n-2,\ 0 \leq j_1, \ldots, j_k \leq n-1\}$
$\quad \cup \{z, z_0, z_1, z_2, a, b, d, e, e', u', x, y, t, \#\}$,
$H = \{0, 1, \ldots, n-1, n\}$,
$\mu = [_n[_{n-1} \cdots [_1[_0\ ]_0^0\ ]_1^0 \cdots]_{n-1}^0\ ]_n^0$,
$w_0 = u_0 c_0, w_n = \lambda$,
$w_i = e$, for all $i = 1, 2, \ldots, n-1$,

and the set R contains the following rules:

0) $[_i e]_i^0 \to [_i]_i^+ e'$, for $i = 1, \ldots, n-1$,
1) $[_0 c_i \to c_{i+1}]_0^0$, for $0 \leq i < 2n-2$,
1') $[_0 c_{2n-2}]_0^0 \to t$,
2) $[_0 u_i \to i b_i d]_0^0$, for $1 \leq i \leq n-1$,
2') $[_0 u_0 \to b_0 d]_0^0$,
3) $[_0 d]_0^0 \to d_0 [_0]_0^+$,
4) $[_1 d_i \to d_{i+1}]_1^+$, for $0 \leq i \leq n-2$,
5) $[_0 b_i]_0^+ \to [_0 u_{j_1}]_0^+ [_0 u_{j_2\ldots j_k}]_0^+$,
   $[_0 u_{j_2\ldots j_k}]_0^+ \to [_0 u_{j_2}]_0^+ [_0 u_{j_3\ldots j_k}]_0^+$,
   ......
   $[_0 u_{j_{k-1} j_k}]_0^+ \to [_0 u_{j_{k-1}}]_0^+ [_0 u_{j_k}]_0^+$,
   where $j_1, \ldots, j_k$ are the nodes adjacent with the node i in the graph,
5') $[_0 b_i]_0^+ \to [_0 u_j]_0^+$, if j is the only node adjacent with i in G,
6) $[_{j+1} [_j]_j^+ \ldots [_j]_j^+ [_j]_j^-]_{j+1}^\alpha \to [_{j+1} [_j]_j^+ \ldots [_j]_j^+]_{j+1}^+ [_{j+1} [_j]_j^+]_{j+1}^-$,
   for $0 \leq j \leq n-2, \alpha \in \{0, +\}$,
7) $d_{n-1} [_0]_0^+ \to [_0 u']_0^-$,
8) $[_0 u' \to a t_0]_0^-$,
9) $[_0 t_i \to t_{i+1}]_0^-$, for $0 \leq i \leq 2n^2 - 2$,
10) $[_0 a]_0^+ \to b [_0]_0^-$,
10') $[_j b]_j^+ \to b [_j]_j^0$, for $1 \leq j \leq n-2$,
11) $[_{n-1} b]_{n-1}^- \to a [_{n-1}]_{n-1}^0$,
12) $a [_j]_j^+ \to [_j a]_j^0$, for $1 \leq j \leq n-1$,
13) $a [_0]_0^+ \to [_0 a]_0^-$,
14) $[_0 t_{2n^2-2}]_0^- \to x [_0]_0^-$,
15) $[_j x]_j^0 \to x [_j]_j^0$, for $1 \leq j \leq n-1$,
16) $[_n x \to y^{n-1} z_0 z]_n^0$,
17) $y [_j]_j^0 \to [_j y]_j^+$, for $1 \leq j \leq n-1$,
18) $y [_0]_0^- \to [_0 y]_0^0$,
19) $[_n z_0 \to z_1]_n^+$,
20) $[_n z_1 \to z_2]_n^+$,
21) $[_n z]_n^0 \to \# [_n]_n^+$,
22) $[_n \alpha \to \#]_n^+$, for $\alpha \in \{y, z\}$

23) $[_n z_2]_n^+ \to \#[_n]_n^0$,

23') $[_n z \to \#]_n^+$,

24) $[_i i]_i^0 \to i$, for $1 \leq i \leq n-1$,

25) $[_n t]_n^0 \to t[_n]_n^+$.

We will explain now how this system works. We start with the symbols u_0 and c_0 in the membrane 0. At the first step, these symbols are changed into db_0 and c_1, respectively, by rules 2' and 1; in parallel, all membranes $1, 2, \ldots, n-1$ will get a positive charge by means of rules of type 0. At the next step we will change the polarization of the membrane 0 using a rule of type 3; in parallel with this rule we apply again a rule of type 1.

In this moment membrane 0 has the polarization $+$, so no other rules of types 1, 2, 2', 3 can be applied.

We start by using the rules of types 4 and 5 in parallel; notice that the rules of type 4 simulate a timer (it will finish the work after the last rule of type 5 was applied). The rules of type 5 create many 0 membranes, in each one having (in the end) an object u_j, where $\{i, j\}$ is an edge in the initial graph. A node can have at most $n-1$ different nodes adjacent with it, so the number of rules of type 5 that can be used for a specific node i is at most $n-1$. One can observe also that using the special characters $u_{j_1 \ldots j_k}$ forces us to follow only one path. The computation can only continue using the rule of type 7, d_{n-1} enters one of the 0 membranes and changes the polarization of that particular membrane from $+$ to $-$ (and it is introduced in the membrane as u'). So, the membrane structure is now:

$$\mu = [_n[_{n-1} \cdots [_2[_1[_0 \,]_0^+[_0 \,]_0^+ \cdots [_0 \,]_0^+[_0 \,]_0^-]_1^0]_2^0 \cdots]_{n-1}^0]_n^0.$$

Now, u' generates two symbols, a and t_0 (rule 8), t_0 is a timer and a will change the sign of another membrane of type 0; in parallel, also a rule of type 6 is applied at this step (because we have a negative membrane). Because of the form of rule 6 we will produce two membranes, one positive and one negative, so another rule of type 6 will be applied at the next step and so on. Because the sign of membrane 0 is positive again we can only apply a rule of type 10 and the membrane is negative again (we cannot apply a rule of type 6 to this negative membrane because this is the only membrane of type 0 into membrane 1). Because membrane 0 is negatively charged, the timer starts working (rule 9). Now the rules of type 10') are applied in parallel with the rules of type 6, so after $n-1$ steps the membrane structure will be as follows (the symbol b changes the membranes to neutral polarization when exiting them):

$$\mu = [_n[_{n-1} \cdots [_2[_1[_0 \,]_0^+[_0 \,]_0^+ \cdots [_0 \,]_0^+]_1^+]_2^+ \cdots]_{n-1}^+[_{n-1}[_{n-2} \cdots [_2[_1[_0 \,]_0^-]_1^0]_2^0$$
$$\cdots]_{n-2}^0]_{n-1}^-]_n^0$$

and in this moment the symbol b is in the membrane $n-1$ that is negative. So next we apply the rule of type 11 (notice that rules of type 6 cannot apply because the membrane n cannot divide), so we get:

$$\mu = [_n[_{n-1}\cdots[_2[_1[_0]_0^+[_0]_0^+\cdots[_0]_0^+]_1^+]_2^+\cdots]_{n-1}^+[_{n-1}[_{n-2}\cdots[_2[_1[_0]_0^-]_1^0]_2^0$$
$$\cdots]_{n-2}^0]_{n-1}^0]_n^0$$

(the only change is in the polarization of the membrane $n-1$ and we have now the symbol a in the membrane n). We continue with rules of type 12; that is, the symbol a passes through membranes $n-1$ to 1 changing their polarization from + to 0. In $n-1$ steps we get:

$$\mu = [_n[_{n-1}\cdots[_2[_1[_0]_0^+[_0]_0^+\cdots[_0]_0^+]_1^0]_2^0\cdots]_{n-1}^0[_{n-1}[_{n-2}\cdots[_2[_1[_0]_0^-]_1^0]_2^0$$
$$\cdots]_{n-2}^0]_{n-1}^0]_n^0$$

at the next step the symbol a enters one of the 0 membranes (rule 13) and changes the polarization of that membrane to $-$. Then the rule of type 6 is applied again; the symbol a exits the membranes $0, 1, \ldots, n-1$ and so on until we will have in each membrane 1 only one membrane 0; in that moment we cannot apply a rule of type 6, so the polarization of the 0 membrane remains $-$, so the symbol a remains in that membrane and this "branch" of the computation halts. We continue only after the timer t_i reaches the value t_{2n^2-2}, and at that moment we can apply the rule of type 14 that produces an x in membrane 1, then, by using rules of type 15, x exits the membranes $1, 2, \ldots, n-1$. In the membrane n, x produces $n-1$ copies of y, a copy of z_0 and a copy of z (rule 16). Next the y-s enter the membranes $n-1$ changing their polarities using rules of type 16). Notice that only one copy of y can enter one membrane, so there may be some number of y-s left in the membrane n at the next step, but they cannot enter any membrane $n-1$ because these membranes had their polarities changed at the previous step (and y can enter a membrane only if it is neutral). At the same time, z leaves the membrane n, changing its polarity in positive (rule 21), so at the next step all the remaining y-s will transform in the trap symbol # (rule 22) and at the next step z_2 leaves the membrane n making it neutral again (rule 23). After $n-1$ steps the y-s are present in all the membranes 1, so at the next step rule 18 will be applied (y enters the membranes 0 changing their polarizations to neutral). This change of polarization is very important, because we can apply again the rules of types 1, 2 and/or 2', so another step begins and the computation continues as described previously.

In the end, c_i will reach the value c_{2n-2}, which means that we cycled $n-1$ times, so we have paths of length n in the graph, so at the next step, when membranes 0 becomes neutral, we apply rules 2 and/or 2' and, at the same time, the rule 1'; the rules of types 2, 2' introduce i instead of u_i and rules 1' dissolve all the 0 membranes at the same time (producing the symbol t). Then we will try to apply $n-1$ times the rules of type 24 (we dissolve the membrane i only if the symbol i is present; this means that we reached somewhere in that path the node i). If all the membranes $1, \ldots, n-1$ are dissolved somewhere then the symbol t reaches the membrane n and then can exit the system using rule 25 (we have found a path starting with node 0 that passed through all the nodes of the graph). If no copy of t reaches the membrane n, then we know that there is no Hamiltonian path starting with 0 in the graph G.

Following the steps of computation one can easily see that the symbol t exits the system at the moment $(n+3+2n^2-2+n+1)(n-1)+n+1+1+n-1+1 = 2n^3+2n$.

So, exactly after $2n^3 + 2n$ computation steps, the symbol t will exit the system if and only if there is a Hamiltonian path in the graph starting with the node 0.

4 Solving HPP by P Systems of a Restricted Form

In this section we will show that the Hamiltonian path problem can be also solved using a restricted variant of the P systems with active membranes. Specifically, instead of using all types of rules: (a), ..., (f), we will prove that we can use only rules (a), ..., (d), (e').

Theorem 2. *HPP can be solved in polynomial time by P systems with rules of types (a), ..., (d), (e').*

Proof. We will use the same idea as that from the proof of Theorem 1: we will construct all the paths in the graph starting with the node 0 and of length n, and then we will verify that they go through all the nodes of the graph (in other words, they pass through a node exactly once, because the length of the path is n). The system outputs the symbol t if and only if a Hamiltonian path has been found.

As input we will have the graph $G = (V, E)$, where $V = \{0, 1, \ldots, n-2, n-1\}$, $2 \leq n$, is the set of vertices and E is the set of edges. The output of the P system will be the letter t at the moment $4n^2 + 2n - 3$ if and only if a Hamiltonian path starting with the node 0 exists in G.

The P system is $\Pi = (V, H, \mu, w_0, w_1, \ldots, w_{n-1}, w_n, R)$, with the following components:

$$V = \{i, b_i, b'_i, b''_i, u_i \mid 0 \leq i \leq n-1\} \cup \{c_i \mid 0 \leq i \leq 2n-2\}$$
$$\cup \{u_{j_1 \ldots j_k} \mid 2 \leq k \leq n-2, 0 \leq j_1, \ldots, j_k \leq n-1\} \cup \{d_i \mid 0 \leq i \leq 2n+1\}$$
$$\cup \{y, z, z_0, z_1, t, \#\},$$

$H = \{0, 1, \ldots, n-1, n, n+1\}$,

$\mu = [_{n+1}[_n[_{n-1} \cdots [_1[_0\ ]_0^0]_1^0 \cdots]_{n-1}^0]_n^0]_{n+1}^0$,

$w_0 = u_0 c_0$,

$w_i = \lambda$, for all $i = 1, 2, \ldots, n+1$,

and the set R contains the following rules:

0) $[_0 c_i \to c_{i+1}]_0^0$, for $0 \leq i < 2n-2$,

0') $[_0 c_{2n-2}]_0^0 \to t$,

1) $[_0 u_i \to i b_i'']_0^0$, for $1 \leq i \leq n-1$,

1') $[_0 u_0 \to b_0'']_0^0$,

2) $[_0 b_i'']_0^0 \to b_i'[_0]_0^+$, for $0 \leq i \leq n-1$,

2') $[_j b_i']_j^0 \to b_i'[_j]_j^0$, for $1 \leq j \leq n-1$, $0 \leq i \leq n-1$,

3) $[_n b_i' \to b_i d_0]_n^0$, for $0 \leq i \leq n-1$,

4) $[_n d_0]_n^0 \to [_n]_n^+$,

5) $[_{n+1} d_i \to d_{i+1}]_{n+1}^0$, for $0 \leq i \leq 2n$,

6) $[_n b_i]_n^+ \to [_n u_{j_1}]_n^+ [_n u_{j_2 \ldots j_k}]_n^+$,

   $[_n u_{j_2 \ldots j_k}]_n^+ \to [_n u_{j_2}]_n^+ [_n u_{j_3 \ldots j_k}]_n^+$,

   $\ldots \ldots$

   $[_n u_{j_{k-1} j_k}]_n^+ \to [_n u_{j_{k-1}}]_n^+ [_n u_{j_k}]_n^+$,

   where $j_1, \ldots, j_k$ are the nodes adjacent with the node i in the graph G,

6') $[_n b_i \to u_j]_n^+$, if j is the only node adjacent with i in G,

7) $u_i[_j]_j^\alpha \to [_j u_i]_j^\alpha$, for $0 \leq i, j \leq n-1, \alpha \in \{0, +\}$,

8) $[_{n+1} d_{2n+1} \to y^{n-1} z_0 z]_{n+1}^0$,

9) $y[_i]_i^\alpha \to [_i y]_i^0$, for $\alpha \in \{+, 0\}, 0 \leq i \leq n$,

10) $[_{n+1} z]_{n+1}^0 \to \#[_{n+1}]_{n+1}^+$,

11) $[_{n+1} y \to \#]_{n+1}^+$,

11') $[_{n+1} z \to \#]_{n+1}^+$,

12) $[_{n+1} z_0 \to z_1]_{n+1}^+$,

13) $[_{n+1} z_1]_{n+1}^+ \to \#[_{n+1}]_{n+1}^0$,

13') $[_{n+1} z_1 \to \#]_{n+1}^0$,

14) $[_i i]_i^0 \to i$, for $1 \leq i \leq n-1$,

15) $[_n t]_n^0 \to t[_n]_n^+$,

16) $[_{n+1} t]_{n+1}^0 \to t[_{n+1}]_{n+1}^+$.

The computation starts in membrane 0 with a rule of type 1 (or 1' for u_0) and a rule of type 0 (which increments the counter c_i) applied in parallel. At the next step b_i'' exits the membrane 0 making it positive (rule 2) and again rule 0 is applied. Because the membrane changed the polarization, no other rule can be applied in the membrane 0 while b_i' exits the membranes $1, 2, \ldots, n-1$ (rule 2').

Reaching membrane n, b_i' in transformed in $b_i d_0$ (rule 3), the next rule applied is rule 4: d_0 exits the membrane n making it positive. The membrane structure at this moment is:

$$\mu = [_{n+1}[_n[_{n-1} \cdots [_1[_0]_0^{+0}]_1^0 \cdots]_{n-1}^0]_n^{+0}]_{n+1}.$$

Now the rules of type 6 can start to be applied in the same time with rules of type 5 (the timer d_i), so in one step we get the membrane structure:

$$\mu = [_{n+1}[_n[_{n-1} \cdots [_1[_0]_0^{+0}]_1^0 \cdots]_{n-1}^0]_n^{+} [_n[_{n-1} \cdots [_1[_0]_0^{+0}]_1^0 \cdots]_{n-1}^0]_n^{+0}]_{n+1}.$$

Using rules of type 7 the u_i-s go "deeper" into the membrane structure until each of them arrives in a membrane of type 0; because of the way they are produced, no two u_i-s can be found in the same membrane. So, after $2n - 2$ steps this "branch" of the computation stops (u_i-s have reached membranes 0), and after $2n+1$ steps we reach d_{2n+1}, so we can apply rule 8, generating z, z_0 in the membrane $n + 1$ together with $n - 1$ copies of y. The next step the y-s go into the membranes n changing their polarity from + to 0 (rule 9); at the same time, z exits the membrane $n + 1$ and change its polarity from 0 to +. Thus, at the next step all the remaining y-s and z-s will transform into the trap symbol # (rules 11, 11'). Then, z_1 exits the membrane $n + 1$ making it neutral again (rule 13) and in the end we apply the rule 13' to get rid of the remaining z_1-s. In $n + 1$ steps the y-s will reach all the 0 membranes, making them neutral again, so another cycle of computation can start with rules 0 and 1. We perform $n - 1$ such cycles; when c_i reaches the value c_{2n-2}, rule 0' is applied and it dissolves all the 0 membranes. In this way, only the rules of type 14 can be applied the next $n - 1$ steps, and after that if t reaches membrane n we can apply rules 15 and 16.

One can see that the time required by the system to output t is $(4n + 5)(n - 1) + n + 2 = 4n^2 + 2n - 3$.

5 Universality without Membrane Division

In this section we will show that P systems with active membranes using only rules of types (a), (b), and (c) are computationally universal. This improves the best result known in the literature, Theorem 1 in [9], where rules of types (a), (b), (c), and (e) were shown to be sufficient.

We denote with NPA_r the family of vectors of natural numbers $N(\Pi)$ computed by systems Π which use only rules of types (a), (b) and (c); in other words, we don't use dissolving or division rules. By $PsRE$ we denote the set family of recursively enumerable sets of vectors of natural numbers; clearly, this is the same with the family of Parikh images of recursively enumerable languages.

Theorem 3. $PsRE = NPA_r$.

Proof. We will only prove the inclusion $PsRE \subseteq NPA_r$ and to this aim we use the fact that each recursively enumerable language can be generated by a matrix grammar with appearance checking in the binary normal form.

Let $G = (N, T, S, M, F)$ be such a grammar, that is, with $N = N_1 \cup N_2 \cup \{S, \#\}$ and with matrices in M either of the form $(X \to \alpha, A \to x)$, for $X \in N_1, \alpha \in N_1 \cup \{\lambda\}, A \in N_2, x \in (N_2 \cup T)^*$, or of the form $(X \to Y, A \to \#)$, for $X, Y \in N_1, A \in N_2$; moreover, a unique matrix of the form $(S \to XA)$ exists, with $X \in N_1, A \in N_2$. Each matrix of the form $(X \to \lambda, A \to x)$ is replaced by $(X \to Z, A \to x)$, where Z is a new symbol. Assume that we have n_1 matrices of the form $(X \to Y, A \to x)$, with $X \in N_1, Y \in N_1 \cup \{Z\}, x \in (N_2 \cup T)^*$, and n_2 matrices of the form $(X \to Y, A \to \#), X, Y \in N_1, A \in N_2$.

We construct the P system (with $p = n_1 + n_2 + 1$ membranes) $\Pi = (V, H, \mu, w_1, \ldots, w_p, R)$, with

$$V = N_1 \cup N_2 \cup T \cup \{Z, \#\} \cup \{X', X'' \mid X \in N_1\}$$
$$\cup \{\langle x \rangle \mid x \in (N_2 \cup T)^*, (X \to Y, A \to x) \text{ is a matrix in } G'\},$$
$$H = \{1, 2, \ldots, p\},$$
$$\mu = [_p[_1\ ]_1^0[_2\ ]_2^0 \cdots [_{n_1}\ ]_{n_1}^0[_{n_1+1}\ ]_{n_1+1}^0 \cdots [_{n_1+n_2}\ ]_{n_1+n_2}^0]_p^0,$$
$$w_i = \lambda, \text{ for all } i \in H - \{p\},$$
$$w_p = XA, \text{ for } (S \to XA) \text{ being the initial matrix of } G,$$

and the following set R of rules:

1. For each matrix $m_i = (X \to Y, A \to x), 1 \leq i \leq n_1$, we introduce the rules:
$X[_i\ ]_i^0 \to [_iY]_i^+,$
$[_iY \to Y]_i^+,$
$A[_i\ ]_i^+ \to [_iA]_i^0,$
$[_iA]_i^0 \to \langle x \rangle[_i\ ]_i^0,$
$[_iY]_i^0 \to Y'[_i\ ]_i^0,$
$[_p\langle x \rangle \to x]_p^0,$
$[_pY' \to Y]_p^0.$

2. For each matrix $m_i = (X \to Y, A \to \#), n_1 + 1 \leq i \leq n_1 + n_2$, we introduce the rules:
$X[_i\ ]_i^0 \to [_iY]_i^+,$
$[_iY \to Y']_i^+,$
$[_iY' \to Y'']_i^+,$
$A[_i\ ]_i^+ \to [_iA]_i^0,$
$[_iA]_i^0 \to \#[_i\ ]_i^0,$
$[_iY'']_i^+ \to Y[_i\ ]_i^0.$

3. To finish the simulation we need the following rules:
$[_p Z]_p^0 \to Z[_p]_p^+$,
$[_p a]_p^+ \to a[_p]_p^+$, for $a \in T$,
as well as the following "trap" rules:
$[_p A \to \#]_p^+$, for $A \in N_2$
$[_p \# \to \#]_p^\alpha$, for $\alpha \in \{0, +\}$.

The system works as follows.

Assume that at a given moment in membrane p we have a multiset Xw, for $X \in N_1$ and $w \in (N_2 \cup T)^*$. Initially, we have $w = A$, for $(S \to XA)$ being the starting matrix of G.

As long as symbols X from N_1 are present, the computation is not finished, a rule $X[_i]_i^0 \to [_i Y]_i^+$ can be used. That is, we have to remove all nonterminal symbols from membrane p that are from N_1 and this is done by simulating a derivation in G, in the following way.

The unique copy of X will go to a membrane i. Assume that we have $1 \leq i \leq n_1$, hence corresponding to a matrix $m_i : (X \to Y, A \to x)$ from G. Inside membrane i we have Y and the membrane gets a positive charge. The rule $[_i Y \to Y]_i^+$ can be used forever. The way to avoid this is to also send to membrane i the symbol A. The arriving of the right symbol A changes the polarity of membrane i to 0, which allows both the symbols Y and A to be sent out of this membrane as the symbols Y' and $\langle x \rangle$, respectively. In membrane p, the first one is replaced by Y and the latter one is replaced by the multiset x. In this way, the matrix m_i was simulated. Note that even if Y' is sent out before $\langle x \rangle$, the symbol Y is available only at the next step, hence we can start simulating a new matrix only after completing the simulation of m_i.

Assume now that the symbol X was sent into a membrane i with $n_1 + 1 \leq i \leq n_1 + n_2$, that is, corresponding to a matrix $m_i : (X \to Y, A \to \#)$. The polarity of the membrane is changed (and x is changed into Y). If there is at least a copy of the nonterminal A in the membrane p, then it will enter the membrane i using the rule $A[_i]_i^+ \to [_i A]_i^0$ at the same time as Y becomes Y' using the rule $[_i Y \to Y']_i^+$. Notice that if the symbol A enters the membrane i, then the polarity of that membrane is changed to neutral. If this doesn't happen (the polarity of the membrane remains positive), then we can transform Y' in Y'' (using the rule $[_i Y' \to Y'']_i^+$) and Y'' can exit the membrane, by using the rule $[_i Y'']_i^+ \to Y[_i]_i^0$ (the membrane returns to the neutral polarity while outside it we have the symbol Y). If A has entered membrane i, then it will send out the trap-symbol $\#$ and the computation will continue forever by using the rule $[_p \# \to \#]_p^0$. Thus, we can continue correctly only if the appearance checking application of the matrix is correctly simulated.

These operations can be iterated, hence any derivation in G can be simulated by a computation in Π and, conversely, the computations in Π correspond to correct derivations in G.

When the object Z has been produced, the process stops – providing that no element of N_2 is still present. This means that the derivation in G is terminal. This is done because the symbol Z exits the system and changes the polarization of the p membrane to positive (rule $[_pZ]_p^0 \to Z[_p]_p^+$). In this moment the nonterminals A from N_2 are transformed in the trap symbol #: $[_pA \to \#]_p^+$, for $A \in N_2$, and then the computation never stops because of the rule $[_p\# \to \#]_p^\alpha$, for $\alpha \in \{0, +\}$.

The terminal symbols $a \in T$ can exit the system using the rule $[_pa]_p^+ \to a[_p]_p^+$, for $a \in T$ (the membrane p will become positive only after the symbol Z exited the system).

One can see that in our construction we used only two (out of three possible) polarities for the membranes (neutral and positive).

It is worth noticing that a system can be constructed that outputs only terminal symbols and only when the computation was successful. To this aim, we can use one further membrane, $p+1$, which will contain the membrane p, and which, in the presence of the symbol Z will allow to leave the system only the terminal symbols. The details are left to the reader.

6 Final Remarks

The question whether or not the systems with n-bounded division can be simulated in linear/polynomial time by systems with 2-bounded division still remains *open*. It is also *open* whether or not HPP can be solved in linear time by P systems with 2-bounded division. It would be also of interest to bound the number of membranes in the universality result; in our proof, the number of membranes depends on the number of matrices in the matrix grammar we start with, but we believe that systems with a small number of membranes are still computationally complete.

References

1. L. M. Adleman, Molecular computation of solutions to combinatorial problems, *Science*, 226 (1994), 1021–1024.
2. J. Dassow, Gh. Păun, On the power of membrane computing, *Journal of Universal Computer Science*, **5**, 2 (1999), 33–49.
3. R. Freund, Generalized P systems, *Fundamentals of Computation Theory, FCT'99*, Iaşi, 1999, (G. Ciobanu, Gh. Păun, eds.), *LNCS* 1684, Springer, 1999, 281–292.
4. S. N. Krishna, R. Rama, A variant of P systems with active membranes: Solving NP-complete problems, *Romanian J. of Information Science and Technology*, 2, 4 (1999), 357–367.
5. C. Martin-Vide, V. Mitrana, P systems with valuation, submitted, 2000.
6. Gh. Păun, Computing with membranes, *J. of Computer and System Sciences*, 61 (2000).
7. Gh. Păun, P systems with active membranes: Attacking NP complete problems, *J. of Automata, Languages and Combinatorics*, to appear.
8. Gh. Păun, G. Rozenberg, A. Salomaa, Membrane computing with external output, *Fundamenta Informaticae*, 41, 3 (2000), 259–266.

9. Gh. Păun, Y. Suzuki, H. Tanaka, T. Yokomori, On the power of membrane division in P systems, *Proc. Conf. on Words, Languages, and Combinatorics*, Kyoto, 2000.
10. G. Rozenberg, A. Salomaa, eds., *Handbook of Formal Languages*, Springer-Verlag, Berlin, 1997.
11. Y. Suzuki, H. Tanaka, On a LISP implementation of a class of P systems, *Romanian J. of Information Science and Technology*, 3, 2 (2000).

Spatial Computing on Self-Timed Cellular Automata

Ferdinand Peper

Communications Research Laboratory, Ministry of Posts and Telecommunications, Kobe, Japan

Abstract. The huge computational power required to support paradigms of brain-like and evolvable computers in the 21st century will likely be delivered by computers with very homogeneous structures like cellular automata. It is expected that such computers can be manufactured cost-effectively in nanotechnology and will be at least ten orders of magnitude more powerful than current computers. Attempts to do general-purpose computations on cellular automata, however, have not gone beyond simulating Turing Machines on them. This paper proposes more effective ways to exploit the universal computing power of cellular automata. First, by using self-timing of cells, as opposed to synchronous timing, different parts of a cellular automaton can work more independently from each other, so is important to program them in a modular way. Secondly, by laying out programs for cellular automata spatially, the von Neumann communication bottleneck between processing elements and memory elements virtually disappears. Thirdly, by organizing programs as self-timed hierarchical modules, the ease of programming is greatly enhanced. Experiments with hierarchical modular programming are shortly reported.

1 Introduction

While it is generally believed that computers in the 21st century will move away from being the static human-programmed machines they are nowadays, it is less agreed upon how to provide the huge computational power needed to support alternative paradigms, like brain-like and evolutionary computers. The trend towards processor chips with further shrinking feature sizes that are based on the von Neumann architecture is expected to continue to the year 2015 [28]. Further in the future, progress in nanotechnology promises computers composed of the order of Avogadro's number (6×10^{23}) of transistor-like logical devices [11], possibly realized as molecular electronics. This number is larger than the number of transistors of all the computers together in the world today. Apart from the lack of manpower needed to design computers with nonhomogeneous structures on such scales, the shortcomings of von Neumann computer architectures will impose severe restrictions on performance. Important aspects responsible for this are the large amount of wiring needed in such architectures, the timing being centralized and synchronous, and the communication bottleneck between memory and processor(s) through which instructions and data are transported. To reap the benefits of the progress in nanotechnology, alternatives that are radically different from the von Neumann computer architecture are required, for example [5], [29].

One such alternative is *Cellular Computing* [24], [1], a paradigm bordering on neural networks, evolutionary computing, artificial life, reconfigurable hardware [14], and massively parallel computing. Cellular Computing is characterized by a vast number of identical computing cells that are very simple and interconnect locally only with a small fixed neighborhood of cells. This paradigm seeks to answer the question how an integrated way of processing emerges from the simple computations of individual cells. It is expected that computers organized in a cellular way can be designed and manufactured much more cost-effectively than current computers. Promising in this respect are self-assembling chemical techniques like those in [31] that use the growth of crystals, structured as two-dimensional arrays of DNA molecules, to perform computations. Using similar techniques, manufacturing of highly regular structures with the order of 10^{20} simple cells might be possible in the future. This number of cells represents a huge computational power if run in parallel with frequencies around 1 GHz, and might result in a performance gain of ten orders of magnitude as compared to current computers (see also [22]).

A well-known class of computing models, Cellular Automata (CA), fits particularly well in the Cellular Computing paradigm. CA can operate with high speed, since in each transition cycle signals only have to be propagated over the neighborhoods of the cells. Though the standard CA model assumes synchronous updating of the cells, in physical implementations it will be hard to distribute a central clock signal that arrives within a narrow time window in all the cells, certainly if there are many of them. The clock signal itself suffering from wire delays will cause different parts of the device to experience different delays, and this problem is aggravated at higher levels of integration and higher clock-frequencies. Alternatively, the locality of CA can be taken a step further, such that timing of every cell is done locally, i.e. asynchronously [17]. A particular mode of asynchronous timing, *self-timing*, is increasingly employed in computer architectures [6], [26], [27], but it has never been used for CA.

With self-timing applied to CA, the topic of this paper, an action in a cell is triggered by an action in a neighboring cell. Unlike with synchronous timing, self-timed cells are completely autonomous: each cell only operates when there is input for it, without regard of the timing of cells outside their neighborhood. Self-timing frees configurations in the cellular space from timing-restrictions, thus keeping interactions between different parts of CA local. This allows the configurations to be organized as modules that can work independently from each other, which is helpful for designing programs for

CA. As discussed in this paper, laying out the modules spatially on a cellular space greatly enhances the efficiency by which the parallelism of CA can be exploited. Spatial computing, nowadays employed in reconfigurable hardware [9], has its origins in systolic architectures. Finally, this paper discusses the programmability of CA. It is argued that programming is eased substantially by organizing programs hierarchically, as self-timed modules that contain sub-modules, and so on, until the level of atomic units is reached, which constitute the cells of the CA.

Section 2 discusses self-timed CA in detail, followed by an explanation of spatial computing in section 3. Then, section 4 focuses on self-timed hierarchical modules and shortly reports on experiments with organizing programs in this way.

2 Self-Timed Cellular Automata

A Self-Timed Cellular Automaton (STCA) consists of an array of cells, each of which is in a state that is a member of a finite state set Q [15]. Each cell has a finite neighborhood consisting of, say, n cells, and undergoes transitions in accordance to a set of indexed transition functions, each of which is associated with a different neighbor of the cell. Accompanied by a state transition of a cell are signals that the cell sends to a (possibly empty) subset of its set of neighbors. There are two types of signals by which cells trigger each other and reply to each other: Request signals and Acknowledge signals. Request signals are sent by a cell to a neighboring cell it wishes to activate. Once the neighboring cell is activated it does its transition and replies with an Acknowledge signal.

Formally, let N be the neighborhood index set that uniquely identifies the neighbors of each cell. Then the set of indexed transition functions is defined as

$$\{f_i : Q^{n+1} \to Q \times 2^N \mid i \in N\},$$

where $n = |N|$. Transition function f_i is applied to a cell c when c receives a Request signal from neighbor i, and determines the next state of c based on the current states of c and its n neighbors. In addition, f_i selects a subset among the neighbors of c to which Request signals are to be sent to activate them. Cells execute their actions in accordance with the steps in the following cycle:

1. Cell receives Request signal from a neighbor (let this be neighbor i). If there are Request signals from more than one neighbor, the cell selects only one of the signals to be received in accordance to some selection/scheduling mechanism (for example, at random, clockwise, etc.), and will put the remaining Request signals on hold to be received later.
2. Cell does transition and sends Request signal(s) to its neighbor(s) in accordance with the transition function f_i.

3. Cell sends an Acknowledge signal to the neighbor from which it received a Request signal in step 1.
4. If cell has pending Request signals to its neighbors (i.e., it has sent Request signals in step 2 but did not receive back the corresponding Acknowledge signals) then it waits until all these signals are acknowledged.
5. Go to step 1.

Processing of an STCA starts when at least one cell receives a Request signal from its neighborhood, and stops when all Request signals die out.

The signaling mechanism of the STCA is represented by so-called *signal bits* in each cell, one bit for each neighbor. When a cell sends a Request signal to a neighbor, the cell's bit that corresponds to the neighbor is set to 1. When the neighbor acknowledges the signal, the cell's bit is returned to 0. The state of a cell is depicted in the center of the cell, and the signal bits of the cell are denoted by small circles near the cell's borders with its neighbors. A bit with the value 1 is denoted by a filled circle, and a bit with the value 0 by an open circle. The examples in this paper deal with 2-dimensional STCA, in which each cell has a neighborhood consisting of the four non-diagonal neighboring cells (von Neumann neighborhood). Fig. 1 shows a cell in state S which sends a Request signal to its eastern neighbor. The neighbor acknowledges the signal, clearing the eastern signal bit of the cell.

Fig. 1. Cell sends a Request signal to its eastern neighbor (left); The neighbor acknowledges the signal, clearing the corresponding signal bit of the cell (right).

The following example concerns a wire transferring binary numbers eastward, implemented by cells having states denoted by $\rightarrow_0$ and $\rightarrow_1$, the 0 or 1 index indicating the number carried by the cell. When the state of a neighbor does not make a difference as to the result, it is denoted by an underscore. Used as an index, an underscore indicates that any number may be carried by the cell. A transition function implementing this functionality is

$$f_{\text{West}} : \begin{array}{c} \overline{} \\ \rightarrow_a \rightarrow_{\_} \rightarrow_{\_} \\ \underline{} \end{array} \implies (\rightarrow_a, \{\text{East}\}), \quad \text{with } a \in \{0, 1\},$$

where the neighborhood index set is $N = \{\text{North}, \text{South}, \text{East}, \text{West}\}$. This transition function gives rise to the following functionality: when a cell receives a Request signal from its western neighbor, it copies the state of its western neighbor, and sends a Request signal to its eastern neighbor. Fig. 2 shows the working of this rule on a wire configuration in a 2-dimensional STCA carrying the sequence of numbers $\langle 1, 0 \rangle$. The timing of signals is nondeterministic, giving an arbitrarily narrowing or widening gap between consecutive active cells at times. The ordering on a wire of two consecutive numbers in the sequence, however, will never change: a number can not surpass its successor in the sequence, because a cell will only process incoming Request signals when all its outgoing Request signals are acknowledged. Obviously, the STCA model also allows different cell types to be defined, like NOT gates, AND gates, etc. It can be proven that the STCA model is computational universal [20].

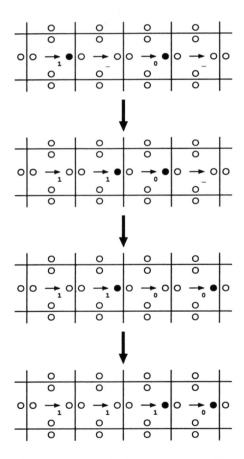

Fig. 2. Transportation of the sequence $\langle 1, 0 \rangle$ along a wire. The index of a cell's state indicates the number carried by the cell. An underscore in the index indicates that any number may be carried by the cell.

3 Spatial Computing on Cellular Automata

Simulating Turing Machines (TMs) on CA fails to deliver efficient computation: apart from their inefficiency as compared to conventional computers, TMs are von Neumann machines, which do not optimally exploit the parallelism of CA. Better fit to the structure of CA is *Spatial Computing* [9]. In spatial computing, a program is specified as a computation graph, which is a scheme of nodes connected by arrows (see Fig. 3 for an example of a simple computation graph). Each node specifies an operation that is executed when data enters the node via its inward arrows. The result of this computation leaves the node via its outward arrows, and then enters other nodes, and so on. This uniquely specifies the way at which data flows along the nodes and operations are executed. The computation graph can be laid out spatially on the hardware elements of a computer, which in the case of CA are the cells. Data flowing along the cells is executed upon by the cells in accordance with the computation graph (see Fig. 4). The preferred mode of timing in spatial computing is self-timing, since it prevents "bubbles" in pipelined structures [27], thus allowing a simple and local control of computations.

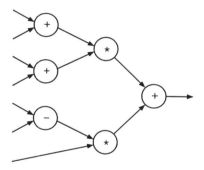

Fig. 3. Computation graph. Data flows through the graph following the arrows, and in the process is operated upon by the nodes. Here, the nodes execute arithmetic operations.

Unlike processing in von Neumann computers, no instructions are involved in spatial computing, so there is no need to fetch instructions from memory to the processor. Rather, the spatial encoding of operations in the computation graph determines an appropriate order and level of parallelism in which they can be executed. Furthermore, transport of data to and from memory is unnecessary, since calculations are done locally on data passing by in accordance with the computation graph. In other words, spatial computing suffers much less from the communication bottleneck between processing elements and memory elements in von Neumann computers.

CA are particularly suited to be programmed spatially for the following reasons. First, processing power and memory are distributed over the cells of the

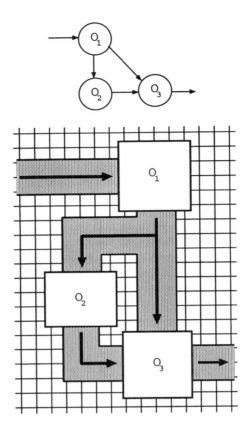

Fig. 4. Spatial mapping of a computation graph consisting of operators O_1, O_2, and O_3 on a CA.

CA, which ensures their availability anywhere in spatial computations. Secondly, CA have a homogeneous structure, which simplifies mapping of computation graphs on them, because particular hardware peculiarities can be ignored. Notwithstanding this homogeneous structure, the functionality of the cells can be adapted to any computation graph by setting the cells' states appropriately. Thirdly, the parallelism of CA allows the mappings of many computation graphs on CA simultaneously.

The efficiency by which computations can be mapped spatially on CA is the reason that CA have been used for specific applications, like image processing (see for example [21]). However, using CA to do general-purpose computations by spatial computing is less tried. Universal computability is preferred over application-specific computability because it allows the system to work autonomously without being controlled by a (von Neumann) host computer, which is likely to be a bottleneck.

Due to the nature of spatial computing, most of the time data and programs do not require to be address-accessible, because data flows along computing elements. However, in some cases there is a need to access data or program modules at a certain place somewhere in a CA, for example for copying or moving data or programs, which are operations typically done by an Operating System (OS). Though this operation is straightforward in a von Neumann computer, in a CA it is less so. Here a cell lacks sufficient memory even to store its own address, so moving data or program configurations around in a CA can only be done by exploiting the emergent behavior of collections of cells. Such behavior in the form of self-reproduction or universal constructibility was studied in [18], followed by many others, such as [3], [7], [25] (see [23] for an overview). However, self-reproduction or universal construction on CA is generally controlled by instructions on a tape of a TM that is simulated on the CA. Apart from the inefficiency of simulating a TM on CA, it necessitates encoding on a TM-tape of a program that constructs the data or program configurations to be reproduced. The resulting models are sequential, which is counter to the strengths of CA. In the context of spatial computing, reading data or programs directly in cellular space, rather than the above way of reading an encoding of them on a simulated TM tape, would be necessary to move them around. The reading operation, which has not been researched yet in literature in this form, is then to be followed by information transport- and write-operations, which are often implemented by a movable path configuration in cellular space through which information is transported, and at the end stump of which writing operations take place [18].

4 Self-Timed Hierarchical Modules

The question with regard to computers build up from a huge number of very simple self-timed identical computing cells is how to program them. We propose to address this problem by specifying a program as a hierarchy of modules that execute in a self-timed way, and that are layed out spatially on the cells of an STCA. A program in this system is thought to be built up from modules, which in their turn are built up from submodules, and so on, until the lowest level of atomic modules is reached, which constitute the cells of the STCA. The state of an STCA cell indicates the cell's function as an atomic module and the data it is processing. The hierarchical organization of modules allows programmers to view programs on different conceptual levels.

Modules are connected with each other by interfaces (see Fig. 5). Input and output to and from a module take place via the interfaces of the module, and consist of data accompanied by signals. Each interface consists of a signal kernel for transferring signals and a data kernel for transferring data. The signal kernel transfers two types of signals: a Request signal from the sender to the receiver indicates that there is data in the data kernel, and an Acknowledge signal sent back by the receiver indicates that the data has been used and may be overwritten. The ordering in which events take place is in accordance to the protocol used by the *Micropipelines* described in [26] (see also [2]).

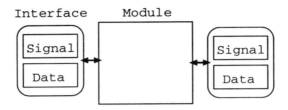

Fig. 5. A module with interfaces. The signal kernel of an interface transfers Request and Acknowledge signals, whereas the data kernel transfers data.

Modules execute their tasks upon receiving signals via their interfaces. This resembles the self-timing of the underlying STCA, and is called *signal-driven* execution. A signal may be accompanied by data on an input interface, in which case signal-driven execution is equivalent to the well-known mode of data-driven execution. Though data-driven execution is often a convenient way of processing, in some cases control will be easier if signals go their own way, separated from the flow of data. In that case the accompanying data kernels are empty. Modules sending signals have no direct control over execution in other modules and are unaware of what actions their signals trigger in other modules.

The functionality of a module is determined by the actions it takes upon receiving signals. An action's semantics depends on the level of the module: a lower level yields a simple operation, whereas a higher level yields a more complicated operation, since it is a combination of actions of lower-level modules. Examples of actions taking place in modules are the reading of input from data kernels, the writing of results to data kernels, and the sending of signals via signal kernels. A module may also ignore some signals, depending on its functionality.

A program, which can be viewed as a "super-module", receives a single Request signal to start it, and it is made to run by the combination of signal transmissions between its modules through their interfaces and the actions taken by the modules in response to those signals. Running continues until all the signals in the program die out. The semantics of a program is determined by all its atomic modules and the way they are combined into its higher level modules. Higher level modules only define the underlying structure on a conceptual level convenient to a programmer. They do not interact with data or signals. Rather, the data and signals they receive are automatically transferred to the submodules in accordance with their interconnection scheme. For example, when module M_0 in Fig. 6 receives input data via interface I_1, the input is automatically redirected to submodules M_1 and M_2, since they are connected to I_1. Module M_0 is just an empty shell encasing the submodules M_1, M_2, and M_3, and their interconnecting interfaces, so M_0 does not have any active role. Similarly, neither do the submodules of M_0 have any active role other than encasing their submodules, unless they are atomic modules.

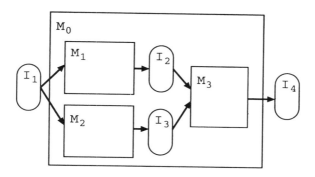

Fig. 6. Module M_0 with its submodules M_1, M_2, and M_3. Input arrives via interface I_1, and output leaves M_0 via interface I_4.

A system built up from self-timed hierarchical modules was experimented with by simulations on a von Neumann computer. As experiments two types of programs were created in hierarchical modular form: a Neural Network Simulator, and a program to read from a multitude of files in a random access way, combining the data, and displaying the results by a graphical user interface. The Neural Network Simulator could easily be expressed in a hierarchical modular way, since it processes data on-line, which is very fit to the self-timed mode employed. The other program was a bigger challenge, because it involves randomly accessing files, which is harder to deal with by self-timing.

In the process of building these programs, a few useful standard modules emerged, among which are a *Synchronizer* to impose a temporal ordering on signals and a *Repeater* to generate signals to let modules iterate through data. Fig. 7 shows a simple example in which a counter that prints the numbers from 1 to n is implemented using a Repeater. The Repeater is the equivalent of a for-loop in a conventional computer. The difference is that the Repeater can be layed out spatially on hardware elements and sends signals to other modules also layed out spatially, thus exploiting inherent parallelism of the STCA, whereas the for-loop is a control structure that causes a central processor to execute and branch sequentially. The Repeater receives a signal via its *doit* interface to start it, and as a result will send n times a signal on its *dothat* interface. The + counter receives n times a signal via its *doit* interface, and as a result will add *arg1* and *arg2*, and output *result*, which will be printed. This example shows that spatial computing does not preclude the use of control structures similar to conventional for-loops.

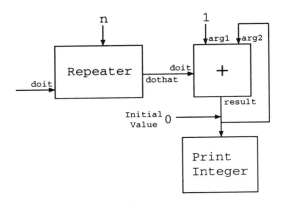

Fig. 7. Using a Repeater to print the integers from 1 to n. Upon receiving one signal via its *doit* interface, the Repeater produces n signals at its *dothat* interface, causing the counter to be incremented n times, and the integers 1 to n to be printed.

The hierarchical modular way of programming is especially suitable for programs that work asynchronously, because of the signal-driven way of processing employed. Currently, module descriptions are specified in the programming language C, though graphical specification of modules is in principle possible, and would make programming a matter of arranging graphical objects.

5 Conclusions and Discussion

There is much potential for the realization of CA-based computers with a huge computational power. Employing self-timing and spatially laying out programs on CA promises a much better efficiency than the mere mapping of TMs on CA. The challenge of programming such computers can be met by organizing programs into self-timed hierarchical modules. An important issue in our research is the design of STCA that are able to accommodate mappings of self-timed spatial programs on them, specified in a hierarchical modular way. Furthermore, copying and moving spatial configurations around in cellular space is indispensable to create a useful CA-based computer. This involves universal constructibility and self-reproduction in cellular space, as well as the design of an OS for a CA-based computer.

Statistical phenomena affect the reliability of computing on a molecular scale, necessitating built-in fault-tolerance into computer designs to suppress errors. It is still an open question as to how to achieve reliable computations on a CA of which the cells are unreliable. Partly responsible for the reduced reliability with which logical devices compute and hold their values are quantum mechanical effects. Researchers are studying the possibility of developing quantum computing techniques that would capitalize on the nonclassical behavior of devices [19]. Apart from proposals for Quantum Computers [13], [4], [10], there also have been formulations of other computation models in a quantum-mechanical framework, such as cellular automata [30], [12], [8], and neural networks [16].

Acknowledgments

This research is conducted in collaboration with Prof. Noboyuki Matsui, Teijiro Isokawa, and Noriaki Kouda of the Computer Engineering Department of Himeji Institute of Technology. I would like to thank the above three people, as well as Morio Yanagihara, Director of the Research division in SCAT, Dr. Hidefumi Sawai, head of the Human Information Section in CRL, and Susumu Adachi of the Auditory and Visual Informatics group in CRL for their proof reading of a previous version of this manuscript and for discussions. This work is financed by the Japan Ministry of Posts and Telecommunications as part of their Breakthrough 21 research project.

References

1. H. Abelson, D. Allen, D. Coore, C. Hanson, G. Homsy, T.F. Knight Jr., R. Nagpal, E. Rauch, G.J. Sussman, R. Weiss, "Amorphous Computing", Communications of the ACM, Vol. 43, No. 5, 2000, pp. 74-82.
2. Amulet group at Univ. of Manchester, "Asynchronous Logic Home Page", http://www.cs.man.ac.uk/amulet/async/index.html.
3. E.R. Banks, "Universality In Cellular Automata", Conf. Record of 1970 IEEE 11th Annual Symp. on Switching and Automata Theory, Santa Monica, California, 28-30 Oct 1970, pp.194-215.
4. P. Benioff, "Quantum mechanical Hamiltonian models of Turing machines", Journal of Statistical Physics, Vol. 29, 1982, pp. 515-546.
5. "CCA: Continuum Computer Architecture", http://www.cacr.caltech.edu/CCA/overview.html, 1997.
6. B. Chappell, "The fine art of IC design", IEEE Spectrum, Vol. 36, No. 7, 1999, pp. 30-34.
7. E.F. Codd, "Cellular Automata", New York: Academic Press, 1968.
8. W. van Dam, "A universal quantum cellular automaton", 4th Workshop on Physics and Computation, Nov. 1996, In: T. Toffoli, M. Biafore, and J. Leão (ed.), PhysComp96, New England, Complex Systems Institute, 1996, pp. 323-331.
9. A. DeHon, "Trends toward Spatial Computing Architectures", Proc. 1999 Int. Solid State Circuits Conf. (ISSCC'99), Feb'99, http://www.cs.berkeley.edu/projects/brass/documents/spatial_isscc99.html.
10. D. Deutsch, "Quantum theory, the Church-Turing principle and the universal quantum computer", Proc. of the Royal Soc. of London A, Vol. 400, 1985, pp. 97-117.
11. K.E. Drexler, "Nanosystems: Molecular Machinery, Manufacturing, and Computation", John Wiley and Sons, 1992.
12. C. Dürr, H.L. Thanh, and M. Santha, "A decision procedure for well-formed quantum linear cellular automata", Proc. of the 13th Ann. Symp. on Theoretical Aspects of Computer Science, 1996, pp. 281-292.

13. R. Feynman, "Quantum Mechanical Computers", Optic News, Vol. 11, 1985, pp. 11-29.
14. "Special Section on Configurable Computing", IEEE Transaction on Computers, vol. 48, no. 6, June 1999, pp. 553-639.
15. T. Isokawa, F. Peper, N. Kouda, N. Matsui, "Computing by Self-Timed Cellular Automata", International Workshop on Cellular Automata (CA2000), Aug. 2000, Osaka Electro-Communication Univ., H. Umeo and T. Sogabe (eds.), pp. 3-4.
16. N. Matsui, M. Takai, and H. Nishimura, "A Learning Network Based on Qubit-Like Neuron Model", Proc. 17th IASTED International Conference on Applied Informatics, Feb. 1999, Innsbruck, Austria, pp. 679-682.
17. K. Nakamura, "Asynchronous Cellular Automata and their Computational Ability", Systems, Computers, Controls, Vol. 5, No. 5, 1974, pp. 58-66 (Translated from Denshi Tsushin Gakkai Ronbunshi, Vol. 57-D, No. 10, Oct. 1974, pp. 573-580.)
18. J. von Neumann, Ed: A.W. Burks, "Theory of Self-Reproducing Automata", University of Illinois Press, 1966.
19. D.A. Patterson, "Microprocessors in 2020", Scientific American, Sept. 1995, pp. 62-67.
20. F. Peper et al., "Self-Timed Cellular Automata and their Computational Ability", In preparation, 2000.
21. A. Scarioni D., J.A. Moreno, "Border Detection in Digital Images with a Simple Cellular Automata Rule", ACRI'98 Proc. of 3rd Conf. Cellular Automata for Research and Industry, Oct. 1998, pp. 146-156.
22. J.M. Seminario, J.M. Tour, "*Ab Initio* Methods for the Study of Molecular Systems for Nanometer Technology: Toward the First-Principles of Molecular Computers", In: Annals of the New York Academy of Sciences, Vol.852, Molecular Electronics: Science and Technology, (Eds) A. Aviram, M. Ratner, 1998, pp. 68-94.
23. M. Sipper, "The Artificial Self-Replication Page", http://lslwww.epfl.ch/pages/research/papers/self_repl/home.html.
24. M. Sipper, "The Emergence of Cellular Computing", IEEE Computer, vol. 32, no. 7, 1999, pp. 18-26.
25. A.R. Smith, "Simple Nontrivial Self-Reproducing Machines", In C.G. Langton, C. Taylor, J.D. Farmer, and S. Rasmussen (Eds.), Artificial Life II, volume X of SFI Studies in the Sciences of Complexity, pages 709-725, Redwood City, CA, 1992. Addison-Wesley.
26. I.E. Sutherland, "Micropipelines", Communications of the ACM, 1989, Vol. 32, No. 6, pp. 720-738.
27. H. Terada, S. Miyata, M. Iwata, "DDMP's: Self-Timed Super-Pipelined Data-Driven Multimedia Processors", Proc. of the IEEE, Vol. 87, No. 2, 1999, pp. 282-296.
28. "National and International Technology Roadmaps for Semiconductors", http://www.sematech.org/ public/index.htm.
29. E. Waingold, M. Taylor, D. Srikrishna, V. Sarkar, W. Lee, J. Kim, M. Frank, P. Finch, R. Barua, J. Babb, S. Amarasinghe, A. Agarwal, "Baring it All to Software: Raw Machines", Computer, Sept. 1997, Vol. 30, No. 9, pp. 86-93.
30. J. Watrous, "On One-Dimensional Quantum-Cellular Automata", Proc. of the 36th Annual Symp. on Found. of Comp. Science, 1995, pp. 528-537.
31. E. Winfree, F. Liu, L.A. Wenzler, and N.C. Seeman, "Design and self-assembly of two-dimensional DNA crystals", Nature, vol. 394, 6 August 1998, pp. 539-544.

Inaccessibility in Decision Procedures

Asaki Saito[1] and Kunihiko Kaneko[2]

[1] Brain Science Institute, Saitama, Japan
[2] Department of Pure and Applied Sciences, University of Tokyo, Tokyo, Japan

Abstract. To study physical the realizability of "computational" procedures, the notion of "inaccessibility" is introduced. As specific examples, the halting set of a universal Turing machine, the Mandelbrot set, and a riddled basin, all of which are defined by decision procedures, are studied. Decision procedures of a halting set of a universal Turing machine and the Mandelbrot set are shown to be inaccessible, that is, the precision of the decision in these procedures cannot be increased asymptotically as the error is decreased to 0. On the other hand, the decision procedure of a riddled basin is shown to have different characteristics regarding (in)accessibility, from the other two instances. The physical realizability of computation models is discussed in terms of the inaccessibility.

1 Introduction

As the Turing machine is defined by analogy with physical computing machines [1–3], it is natural to study its physical aspects, especially physical realizability. For the study of unconventional models of computation, it is also important to investigate whether the model (or procedures on which the model is founded) is physically realizable or not.

In this paper, we choose to study the physical realizability of decision procedures. A decision procedure (recognizer) is one of the recursive procedures utilized for defining sets, which decides whether a given object is included in a set or not.[1] Hence, a set defined by a decision procedure is called "halting set". It includes not only a halting set of a Turing machine but also a geometric set like a basin of attraction in dynamical systems.

It is natural that the study of physical realizability is based on dynamical systems.[2] There, basins (corresponding to halting sets) and transient processes to an attractor (corresponding to decision processes) are also studied extensively and intensively. In the study of dynamical systems, not only analytic but also experimental approaches with numerical simulation are powerful, in contrast with rare use of the latter approach in classical computation theory.

---

[1] The other recursive procedure often used for defining sets is a generating procedure (generator), which generates elements of a set; however, these procedures will be not discussed in this paper.
[2] A different study of analog computation based on dynamical systems is presented in [4].

In the present paper, a halting set of a universal Turing machine from classical computation theory [1–3], the Mandelbrot set from fractal geometry [5–7], and a riddled basin from nonlinear dynamical systems study [8–10], are investigated as concrete objects, all of which are defined by decision procedures. The halting set of a universal Turing machine is a set of symbol strings, and its decision procedure is realized by a "discrete" Turing machine. However, instead of adopting the conventional treatment, we construct its geometric representation in a Euclidean space, and construct an "analog" decision procedure, by coding a symbol sequence into a real number.[3] Then its physical realizability is investigated in the same manner as in the cases of the Mandelbrot set and the riddled basin.

In this paper, we choose (in)accessibility of decision procedure as physical realizability. Focusing on the boundary of a halting set, dimension of the boundary is evaluated. When the boundary dimension is equal to the dimension of the space in which it lies, its decision procedure has so strong uncertainty that the ideal decision procedure cannot be approached by decreasing errors, no matter how small the errors are (see also [14]). Such system is qualitatively different from a system with a boundary dimension less than the space dimension, or a system with an ordinary chaotic unpredictability.

For each of the above mentioned halting sets, we study inaccessibility and its behavior against the transformation of code. To study these features, numerical approaches are borrowed from nonlinear dynamical systems.

This paper is structured as follows. In section 2, we study the code that maps a symbol sequence to a real number, to prepare for the subsequent sections. Dimensional property, as well as (in)accessibility, of (ordinary fractal) sets is shown to be variable under the transformation from one code to another. In section 3, we treat geometric representation of a halting set of a universal Turing machine, by using the code introduced in section 2. Inaccessibility and invariance of the inaccessibility against the code transformations are shown for the decision procedure of the set. In section 4, we treat the Mandelbrot set. Inaccessibility and its invariance are also shown for this case, indicating similarity with the halting set of a universal Turing machine. In section 5, we treat a riddled basin. This set, however, is shown to have a different character, e.g. against code transformations, from a halting set of a universal Turing machine and the Mandelbrot set. A discussion and conclusions are given in sections 6 and 7.

Besides the (in)accessibility, these halting sets are shown to be non-self-similar, having peculiar geometric properties that cannot simply be characterized by self-similarity.

---

[3] Coding a symbol sequence is used in [11], [12], to study a dynamical system in which a Turing machine is embedded. There, the system is shown to have a different instability other than chaotic instability (see also [13]).

2 Coding

As will be shown later, the dynamics of an automaton (e.g., universal Turing machine and pushdown automaton [2]), as well as the geometric property of its halting set, generally depends on the way of embedding a symbol sequence into a number (i.e. code). In this section, we discuss briefly the problem of coding in connection with dynamical systems, as a preparation for the subsequent sections.

2.1 The code

We use the word "code" as a mapping which transforms an infinite symbol string on a finite alphabet to a real number. Furthermore, "symbol string" indicates a finite symbol string, whereas "symbol sequence" indicates an infinite one.

Here, we restrict codes which map a symbol sequence to a real number represented by the same label of Markov partition [9] of a certain piecewise-linear map. To be concrete, a symbol sequence of n symbols is encoded into a real number on the interval [0,1] by choosing a piecewise-linear map $f : [0,1] \to [0,1]$ defined by

$$f(x) = \frac{1}{\alpha_{i+1} - \alpha_i}(x - \alpha_i) \quad \text{for} \quad x \in [\alpha_i, \alpha_{i+1}],$$

where $\alpha_0 = 0$, $\alpha_n = 1$ and $\alpha_i < \alpha_{i+1}$. Then we encode a symbol sequence to a real number which has the same label of Markov partition of this piecewise-linear map by labeling the interval between α_i and α_{i+1} as 'i'. As a result, a finite symbol string, corresponding to the set of all symbol sequences with the prefix, is encoded by this code, to an interval which has the same label of Markov partition.[4]

In particular, for a 2-symbol sequence ($n = 2$), the code has only one parameter α (i.e. α_1). When $\alpha = \frac{1}{2}$, this code corresponds to the binary encoding, using a symbol sequence as base-2 expansion of a real number.

2.2 Code transformation

Now we consider transformation of a code with 2-symbol sequences. When a 2-symbol sequence is given, one real number, denoted as x_α, is decided for a code with parameter α (code α), and another real number, x_β, is decided for another code with parameter β (code β), respectively. Then, a mapping $t_{\alpha \to \beta}$ from x_α to x_β, transforms the encoding by code α into that by code β. Thus we call such a mapping $t_{\alpha \to \beta}$ code transformation.

---

[4] See discussion about other possible choices of coding in subsection 6.2. Here we do not treat other types of codes.

When $\alpha = \frac{1}{2}$, $t_{\frac{1}{2} \to \beta}$ becomes Lebesgue's singular function L_β [15, 16]; L_β : $[0,1] \to [0,1]$ is defined by the functional equation,

$$\begin{cases} L_\beta(\frac{x}{2}) & = & \beta L_\beta(x) \\ L_\beta(\frac{1+x}{2}) & = & \beta + (1-\beta)L_\beta(x) \end{cases}$$

where $0 < \beta < 1$.[5]

Figure 1 shows $L_{\frac{1}{3}}$ (i.e. $t_{\frac{1}{2} \to \frac{1}{3}}$). When $\beta \neq \frac{1}{2}$, this "fractal" singular function L_β is shown to be continuous and strictly monotone increasing from 0 to 1, but the derivative is 0 almost everywhere [15, 16]. If $\beta = \frac{1}{2}$, then the solution of the functional equation is simply $L_{\frac{1}{2}}(x) = x$, which corresponds to the identity transformation.

Concerning the fractal (boundary) dimension[6], it should be noted that the invariance of the (boundary) dimension under the fractal transformations is not guaranteed, whereas the (boundary) dimension is preserved under the application of diffeomorphism [17]. Not only in the case of the 2-symbol code, but generally the preservation of (boundary) dimension is not guaranteed under the transformation from one code to another.

For example, consider the middle-thirds Cantor set consisting of real numbers with no '1's in their base-3 expansion. It is obtained by encoding all symbol sequences on the alphabet $\{0, 2\}$, using the Markov partition of the map f where $\alpha_0 = 0$, $\alpha_1 = \frac{1}{3}$, $\alpha_2 = \frac{2}{3}$ and $\alpha_3 = 1$.[7] The fractal dimension of this middle-thirds Cantor set is $\frac{\log 2}{\log 3}$. Since the Lebesgue measure of this set is zero, the boundary dimension of this set (also the boundary dimension of the complement of this set) is $\frac{\log 2}{\log 3}$. Then, if we transform the code using f to another 3-symbol code using the Markov partition of f' with $\alpha_0 = 0$, $\alpha_3 = 1$ and $\alpha_i < \alpha_{i+1}$, the (boundary) dimension D of the resulting Cantor set is given by the solution of equation, $\alpha_1{}^D + (1-\alpha_2)^D = 1$ [7]. Thus we can change the (boundary) dimension D by changing α_1 and α_2 (e.g., if $\alpha_1 = \frac{1}{4}$ and $\alpha_2 = \frac{3}{4}$, then $D = \frac{1}{2}$). Hence for a Cantor set, the (boundary) dimension is not preserved under the above code transformation.

---

[5] With code $\frac{1}{2}$, $\frac{x}{2}$ corresponds to the shift of a 2-symbol sequence to right with the insertion of the symbol '0' at the left most cell. The real number for this symbol sequence using code β is obtained by multiplying the real number for the original symbol sequence using code β by β. Thus $t_{\frac{1}{2} \to \beta}$ satisfies the first equation. It is easy to check $t_{\frac{1}{2} \to \beta}$ also satisfies the second equation.

[6] We treat box-counting dimension and Hausdorff dimension as fractal dimension [5–7]. The word "boundary dimension" is used as box-counting dimension or Hausdorff dimension of boundary of a set, unless otherwise mentioned.

[7] The set of symbol strings on the alphabet $\{0, 1, 2\}$, where the symbol strings end with the symbol '1' with no '1' before it, is encoded to the complement of the Cantor set with the base-3 code. This set of symbol strings is included in the class of regular sets, i.e., accepted by a finite automaton [2].

3 Halting Set of an Universal Turing Machine

In this section, by using codes introduced in the previous section, we investigate geometric and dynamical systems properties of halting set of a universal Turing machine, especially inaccessibility of its decision procedure.

3.1 The halting set

As the most important consequence of the formulation of computation by the Turing machine (TM) [1–3], it has been understood that there exists a class of an undecidable decision problems within well-defined problems. Especially, the most famous undecidable problem is the halting problem of the universal Turing machine (UTM). The halting set of UTM is known as a recursively enumerable set but not a recursive set [2], where the class of recursively enumerable sets is the most complex in the Chomsky hierarchy of the formal languages [18].

As a UTM, we choose the following Rogozhin's UTM [19]. (However, it should be noted that the study of a specific UTM is equivalent to treating all TMs including all UTMs.) Rogozhin's UTM has 24 internal states besides a halting state $\{q_1, q_2, \ldots, q_{24}\}$ and a two-way infinite tape with tape alphabet $\{0, 1\}$. q_1 is the initial state for this UTM.[8]

Since Rogozhin's UTM's tape is two-way infinite one, we map a symbol sequence of the halting set of Rogozhin's UTM to a point of a two-dimensional space.[9] [10] To be more precise, we map the halting set into two unit squares, denoted by square 0 and square 1. First, we divide a symbol sequence in the halting set into three parts, one cell (called "center cell") which is read by the head at the beginning of the computation, and the right and left sides of the symbol sequence from the center cell. Then we represent the right and left sides of the symbol sequence by a pair of real numbers given by code α, respectively. Finally, this pair is put into either square 0 or 1 depending on whether the symbol on the center cell is equal to '0' or '1', where the right side of the symbol

---

[8] See [19] for further information about Rogozhin's UTM, including its transition function.

[9] Usually, we consider a symbol string to be an input for TM. In other words, on an initial tape, we consider a special symbol called "blank symbol", written on every cell except the symbol string. This use of symbol string as an input is equivalent to the use of a symbol sequence such that blank symbols are inserted to (both) the end(s) of that symbol string. However, we should note the following points also. If a TM halts on an input, it also halts on inputs obtained by changing symbols on such cells that are not read by the tape head until the TM halts, for any symbols (not just blank symbols). Of course, the number of changeable cells is not necessary to be finite in contrast with the usual input. Thus, when we are concerned only with the halting of TM, we can consider (all) symbol sequences as inputs, instead of restricting inputs so that blank symbol is written on except a finite number of cells. Therefore, as a halting set of TM, we consider a set of symbol sequences instead of symbol strings.

[10] Another treatment of languages on infinite symbol strings is presented in e.g. [20], [21].

sequence corresponds to the horizontal axis, and the left side to the vertical axis. We name this geometric representation of the halting set of Rogozhin's UTM by using code α to be UL_α. Figure 2 shows $UL_{\frac{1}{2}}$ and $UL_{\frac{1}{3}}$.

Let us consider the structure of a halting set of a UTM. A UTM can simulate any TM. In considering a halting set, this fact indicates that a mere single halting set of UTM contains all halting sets of TM. Concerning the geometry of the undecidable halting set, from the above fact, the geometric representation of the halting set of the UTM contains different patterns, i.e., a region for each halting set of a TM, constructed by contracting the geometric representation of the halting set of the TM. (In fact, it contains various fractals which can be drawn by a digital computer.) Therefore, in strong contrast with ordinary self-similar fractals, the geometry of the halting set of the UTM has different patterns and has different fine structures on arbitrarily small scales.

3.2 Boundary dimension

Now we study the dimension of a halting set of UTM. The halting set of UTM, represented geometrically, has positive Lebesgue measure (i.e., a fat fractal [22, 8]). In fact, our UTM is numerically shown to halt for some symbol sequences on the initial tape. When the head of a UTM, moves to the right and left at most i and j sites respectively during its computation, before it halts, then the UTM with symbol sequences of the same symbols up to i and j sites and any symbols beyond them, also halts, because the UTM can only read up to i and j sites before it halts. These symbol sequences correspond to rectangular region with any of our code (in the case of $UL_{\frac{1}{2}}$, they correspond to rectangular region with sides 2^{-i} and 2^{-j}), and this region is contained in the geometric representation of the halting set of the UTM. Therefore, the halting set of the UTM has positive two-dimensional Lebesgue measure (at least $2^{-(i+j)}$ in the case of $UL_{\frac{1}{2}}$). Dots in Fig. 2 correspond to such symbol sequences.

Thus, its fractal dimension is the same as the dimension of the space (i.e. $D = 2$). Instead, we investigate the box-counting dimension of the boundary (i.e., the exterior dimension [23, 8]) of geometric representation of a halting set of UTM to study its fine structure.

The definition of the box-counting dimension of a set S in N-dimensional space is equivalent to $D_0 = N - \lim_{\varepsilon \to 0} \log V[S(\varepsilon)] / \log \varepsilon$, where $V[S(\varepsilon)]$ is the N-dimensional volume of the set $S(\varepsilon)$ created by fattening the original set S by ε [8]. In the following, based on this definition, we numerically estimate the box-counting dimension of the boundary (the boundary dimension) of geometric representation of halting set of Rogozhin's UTM. In particular, we investigate the cases with code $\frac{1}{2}$ and code $\frac{1}{3}$ (i.e. $UL_{\frac{1}{2}}$ and $UL_{\frac{1}{3}}$, respectively).

Of course, it is impossible to take an infinite time for the numerical simulations. Thus we treat the set of inputs on which UTM will halt within a given finite step n (denoted as $UL_\alpha(n)$), and investigate the boundary dimension of $UL_\alpha(n)$. Then we study the asymptotic behavior of the boundary dimension of $UL_\alpha(n)$ as n is increased (to infinity).

$UL_{\frac{1}{2}}$

The detailed procedure to study the boundary dimension D_0 of $UL_{\frac{1}{2}}$ is as follows. First we choose a point (X, Y) in either unit square 0 or 1 of a two-dimensional space at random. Then we perturb this point (X, Y) to $(X+\varepsilon, Y)$, $(X, Y+\varepsilon)$ and $(X+\varepsilon, Y+\varepsilon)$. Then we decide whether Rogozhin's UTM, starting from each of these four tapes at the center cell, will halt within n steps or not (i.e., in $UL_{\frac{1}{2}}(n)$ or not). If all four points are in $UL_{\frac{1}{2}}(n)$ or none of them are in $UL_{\frac{1}{2}}(n)$, we regard there to be no boundary in the square of length ε made from these four points. Otherwise we regard that there is a boundary in the square. We repeat this procedure for a large number of points and evaluate the fraction of squares with a boundary, denoted as $f(\varepsilon)$, which gives the estimation of $V[S(\varepsilon)]$. Varying ε, we obtain the scaling of $f(\varepsilon)$ with ε, and can evaluate $N - D_0$ (i.e. $2 - D_0$).

In Fig. 3, the log-log plot of $f(\varepsilon)$ of $UL_{\frac{1}{2}}(n)$ with ε is shown for several values of n. It can be fit as $f(\varepsilon) \sim \varepsilon^{2-D_0}$ for small ε, from which one can obtain $2 - D_0$ for each n. Fig. 4 displays the log-log plot of $2 - D_0$ versus the computation time n. It shows that $2 - D_0$ of the boundary of $UL_{\frac{1}{2}}(n)$ approaches zero roughly as $n^{-0.45}$ with the increase of the computation time n. In other words, when we increase n, the box-counting dimension of the boundary of $UL_{\frac{1}{2}}(n)$ approaches two, which is the same as the dimension of the space. Thus the box-counting dimension of the boundary of $UL_{\frac{1}{2}}$ is estimated to be two.

Here we briefly refer to the uncertainty exponent $\phi = N - D_0$ [8, 23]. Suppose there exists a set A in a certain N-dimensional space and our ability to determine the position of points has an uncertainty ε. $\bar{A}$ denotes the complement of A and S denotes the boundary separating A and $\bar{A}$. If we have to determine which set a given point lies in, the probability $f(\varepsilon)$ of making a mistake in such determination is proportional to $V[S(\varepsilon)]$. Thus, if the box-counting dimension of the boundary S is D_0, $f(\varepsilon)$ is proportional to ε^{N-D_0} ($= \varepsilon^{\phi}$). If ϕ is small, then a large decrease in ε leads to only a relatively small decrease in $f(\varepsilon)$. Thus ϕ is called uncertainty exponent.

The above result that the uncertainty exponent of the boundary of $UL_{\frac{1}{2}}(n)$ becomes smaller with the increase of computation time n indicates that it becomes more difficult to decrease mistake in determination of $UL_{\frac{1}{2}}(n)$. The null uncertainty exponent of the boundary of $UL_{\frac{1}{2}}$ (i.e. $D_0 = 2$) indicates that the probability of making a mistake (equally, the volume of ε-neighborhood of the boundary $V[S(\varepsilon)]$) does not decrease with ε.

A decision procedure is defined in an ideal condition without errors. On the other hand, robustness of a decision procedure against a noise is essentially different, depending on whether the boundary dimension of a set is equal to the space dimension or less. Indeed, if the boundary dimension is less than the space dimension with the uncertainty exponent $\phi > 0$, one can decrease the volume of ε-neighborhood of a boundary ($V[S(\varepsilon)]$) to any amount in principle by decreasing ε. In other words, one can arbitrarily approach the ideal decision procedure (i.e., ideal condition without errors) by improving accuracy in principle, even under the presence of errors. On the other hand, if the boundary dimension is equal to the space dimension with the uncertainty exponent $\phi = 0$, it is impossible to approach the ideal decision procedure (i.e. the ideal condition) as long as there exist errors no matter how small they are.

Consequently, the decision procedure of halting set of UTM by using the code $\frac{1}{2}$ has so strong uncertainty, i.e. inaccessibility, that one cannot approach the ideal decision procedure, in the presence of error.

$UL_{\frac{1}{3}}$

The estimation of the boundary dimension of $UL_{\frac{1}{3}}$ is as follows. In the same way as the case of $UL_{\frac{1}{2}}$, we choose a point (X, Y) in either unit square 0 or 1 of a two-dimensional space at random, and perturb the point (X, Y) to $(X + \varepsilon, Y)$, $(X, Y + \varepsilon)$ and $(X + \varepsilon, Y + \varepsilon)$. Then, we decide the labels of Markov partition for X, Y, $X + \varepsilon$ and $Y + \varepsilon$ (i.e., decoded symbol sequences (code $\frac{1}{3}$)$^{-1}(X)$, (code $\frac{1}{3}$)$^{-1}(Y)$, (code $\frac{1}{3}$)$^{-1}(X + \varepsilon)$ and (code $\frac{1}{3}$)$^{-1}(Y + \varepsilon)$), and compose four input symbol sequences for corresponding four points. For example, the input symbol sequence corresponding to (X, Y) in the square 0 is concatenation of (code $\frac{1}{3}$)$^{-1}(Y)$, '0', (code $\frac{1}{3}$)$^{-1}(X)$, where the symbol sequence (code $\frac{1}{3}$)$^{-1}(Y)$ is flip-flopped. The subsequent procedure is the same as for the case of $UL_{\frac{1}{2}}$.

Fig. 5 shows the log-log plot of the uncertainty exponent ϕ versus the computation time n. The exponent ϕ approaches zero with a power law (roughly as $n^{-0.70}$) as in the case of $UL_{\frac{1}{2}}$. In other words, with the increase of n, the box-counting dimension of the boundary of $UL_{\frac{1}{3}}(n)$ approaches two, i.e., the dimension of the space, in the same way as $UL_{\frac{1}{2}}$. Thus the box-counting dimension of the boundary of $UL_{\frac{1}{3}}$ is also estimated to be two.

Consequently, similarly to $UL_{\frac{1}{2}}$, the boundary dimension of $UL_{\frac{1}{3}}$ is also equal to the space dimension two, and the decision procedure of $UL_{\frac{1}{3}}$ also has inaccessibility so that one cannot approach the ideal decision procedure, in the presence of error. Thus, our result of the inaccessibility of UL_α is not based on the property of the code.[11]

Note that $UL_{\frac{1}{3}}$ is the set obtained by transforming $UL_{\frac{1}{2}}$ by the code transformation $t_{\frac{1}{2} \to \frac{1}{3}}$. As seen in the case of Cantor set in section 2, code transformation, in general, does not preserve the (boundary) dimension. However, the above results show that the property that boundary dimension is equal to space dimension, i.e., the inaccessibility, is preserved in the case of the halting set of UTM. Thus, one could say the property that a decision procedure is not robust against a noise is robust under application of even "fractal" code transformations.[12]

---

[11] The result of the boundary dimension equal to the space dimension of the halting set of UTM is independent of code α: see also subsection 5.2.

[12] In the case of a riddled basin, code transformations correspond to changing a system parameter: see section 5.

These results are also reasonable, since unlike chaotic unpredictability (i.e., sensitive dependence on initial conditions), the halting problem of UTM is undecidable even if descriptions of inputs are known exactly.

Geometric representation of halting set of other UTMs

The boundary dimension of geometric representation of halting set of other UTMs is also equal to the space dimension. Since UTMs can imitate each other, the geometric representation of halting set for any UTMs contain each other, and each has the same boundary dimension, that is equal to the space dimension.

Indeed, we have numerically studied the boundary dimension of geometric representation of halting sets of some other UTMs [14]. The boundary dimension of these sets is equal to the dimension of the space in which they lie. In these systems again, the uncertainty exponent decays to zero with some power of n.

4 Mandelbrot Set

In this section, we deal with the Mandelbrot set, to show its common features with the above halting set of a universal Turing machine.

4.1 The halting set

The Mandelbrot set [10], denoted as M, is the subset of the complex plane $\mathbf{C}$, and is defined as $M = \{c \in \mathbf{C} \mid |Q_c^n(0)| \not\to \infty\}$ for $Q_c(z) = z^2 + c$ (equivalently, $M = \{c \in \mathbf{C} \mid K_c \text{ is connected}\}$ where K_c is the filled-in Julia set of Q_c). Fig. 6 shows M.

Although M has often been studied as a self-similar structure containing small copies of M itself, M is also known as a set with extraordinarily complex structure. Indeed, M has been called one of the most intricate and beautiful objects in mathematics [10]. Likewise, M has Misiurewicz points where M and the corresponding filled-in Julia set K_c are similar. Thus, similarly with the geometric representation of halting set of UTM, M has different patterns and has a different fine structure on an arbitrarily small scale, in contrast with ordinary self-similar fractals.

There is a decision procedure for M. It is shown that if $|c| > 2$, then $|Q_c^n(0)| \to \infty$ and thus $c \notin M$. Also if $|Q_c^k(0)| > 2$ for some $k > 0$, then $|Q_c^n(0)| \to \infty$ [10]. Thus, if $c \in \mathbf{C}$ is given, the decision procedure of M (precisely, that of the complement of M) is given by deciding whether $|Q_c^k(0)| > 2$ for some $k > 0$.

Although their origins are different, M and the halting set of UTM have common intractability that there is no guarantee if their decision procedures will halt. Hence it is interesting to study if there is a relation, in some sense, between M and recursively enumerable sets [24, 25].

Here we investigate the M's boundary dimension in the same way as the halting set of UTM, to point out that (the decision procedures of) M and the halting sets of UTM have common properties.

4.2 Boundary dimension

M The Hausdorff dimension of the boundary of M, giving a lower bound for the box-counting dimension, is proven to be equal to the space dimension two [26], that is the maximal value in the two-dimensional space. To see also the asymptotic behavior of such boundary dimension, we study the box-counting dimension of the boundary of M numerically in the same way as UL_α. Here again, we cannot treat M directly by means of numerical simulation as in the case for UL_α. Thus we study the boundary dimension of $M(n)$, i.e., the set of points which are not decided not included in M within a given finite step n. The detailed procedure for this case is almost same as for UL_α's.

The asymptotic behavior of the boundary dimension of $M(n)$ with the increase of n is given in Fig. 7, where uncertainty exponent ϕ (i.e. $2 - D_0$) is plotted versus the computation time n, with a logarithmic scale. As is shown, the box-counting dimension of the boundary of $M(n)$ approaches two with the increase of the computation time n. The approach is given by $2 - cn^{-0.45}$, with a positive constant c.

$M_{\frac{1}{3}}$

Since M is already a geometric set, it intrinsically does not matter how to encode. However, it is possible to apply a fractal function which corresponds to a code transformation, to M.

Now, by using Lebesgue's singular function L_β which is a transformation from encoding with code $\frac{1}{2}$ into encoding with code β, we can define L'_β : $[-2, 2] \to [-1, 1]$

$$L'_\beta(x) = \begin{cases} L_\beta(\frac{x}{2}) & \text{for} \quad x \in [0, 2] \\ -L_\beta(-\frac{x}{2}) & \text{for} \quad x \in [-2, 0] \end{cases},$$ [13]

and also define a transformation $t_{\to \beta} : x + iy \mapsto L'_\beta(x) + iL'_\beta(y)$, where $x, y \in [-2, 2]$.

Let M_β the set obtained by applying $t_{\to \beta}$ to M (i.e. $t_{\to \beta}(M)$). In the following, we will numerically study boundary dimension of M_β in the case of $\beta = \frac{1}{3}$ (i.e. $M_{\frac{1}{3}}$: Fig. 8).

In the same way as for M, the asymptotic behavior of the boundary dimension of $M_{\frac{1}{3}}(n)$ with the increase of n is studied by considering $M_{\frac{1}{3}}(n)$ obtained by $t_{\to \frac{1}{3}}(M(n))$. Fig. 7 shows the log-log plot of uncertainty exponent ϕ versus

---

[13] In other words, the map corresponds to a transformation from a real number to the real number on Markov partition, which has the same label as the base-2 expansion of the original real number.

the computation time n as before. As is shown, the box-counting dimension of the boundary of $M_{\frac{1}{3}}(n)$ approaches two by $2 - cn^{-0.45}$ with the increase of the computation time n. Thus the box-counting dimension of the boundary of $M_{\frac{1}{3}}$ is also estimated to be two.

Consequently, similarly with the case of undecidable halting set of UTM, decision procedures of M and $M_{\frac{1}{3}}$ also indicate inaccessibility, in the sense that one cannot approach the ideal decision procedure, in the presence of error. This inaccessibility is also unchanged under application of a fractal function which corresponds to a code transformation. Now it is clear that M and halting set of UTM have common features (which were expected by naive arguments previously) from the standpoint of the inaccessibility.

5 Riddled Basin

Here we treat riddled basin structure, in the same manner. The riddled basin of a certain simple dynamical system will be shown to have different character, e.g. against "fractal" code transformations, from a halting set of a UTM and the Mandelbrot set.

5.1 The halting set

Several chaotic systems with competing attractors are known to have a riddled basin. The basin of an attractor is said to be "riddled" [27–29], when, for any point in the basin of the attractor, any of its ε-vicinity includes points with a nonzero volume in the phase space, that belong to another attractor's basin. Even qualitative behavior (i.e., which attractor is the eventual one) is unpredictable by any small error [28].

A simple model with a riddled basin is introduced by Ott et al [27]; a two-dimensional map of the region $0 \leq x \leq 1, 0 \leq y \leq 1$ by

$$x_{n+1} = \begin{cases} \frac{1}{\alpha}x_n & \text{for } x_n < \alpha \\ \frac{1}{1-\alpha}(x_n - \alpha) & \text{for } x_n > \alpha \end{cases}$$

$$y_{n+1} = \begin{cases} \gamma y_n & \text{for } x_n < \alpha \\ \delta y_n & \text{for } x_n > \alpha \end{cases}$$

where $\gamma > 1$ and $0 < \delta < 1$. The map in the region $y > 1$ is chosen so that orbits in $y > 1$ move to an attractor in $y > 1$ and thus never return to $y < 1$.

The line segment $I = \{x, y \mid y = 0, 0 \leq x \leq 1\}$ is invariant for the map. I is a chaotic attractor with a riddled basin, according to the definition of Milnor [30] if the perpendicular Lyapunov exponent $h_\perp = \alpha \ln \gamma + (1 - \alpha) \ln \delta$, is negative [27]. We focus on the basin structure of $y = 0$ attractor and $y > 1$ attractor on the horizontal line segment $\{x, y \mid y = y_0, 0 \leq x \leq 1\}$ with $0 < y_0 < 1$.

By assuming a randomly chosen initial x_0 in $(0, 1)$ with a specific initial y_0, we have a random walk in $\ln y$, starting at $\ln y_0$, where a step of size $-\ln \delta$ to the left has probability $1 - \alpha$ and a step of size $\ln \gamma$ to the right has probability α [27]. The initial point (x_0, y_0) belongs to the $y > 1$ attractor's basin if the random walk ever reaches $\ln y \geq \ln 1 (= 0)$. On the other hand, (x_0, y_0) belongs to the $y = 0$ attractor's basin if, as $n \to +\infty$, $\ln y_n \to -\infty$ without reaching $\ln y_n \geq \ln 1$ for all n.

Let 'L' denote a step to left and 'R' to right, respectively. Then consider the set of all (sample) paths of random walk starting from $\ln y_0$ until it passes $\ln 1$ for the first time. A path of such a random walk is expressed by a symbol string on the alphabet $\{L, R\}$ (e.g. RR, $LRRR$, $LRLRR$, ...), and the set of all paths are decided by γ, δ and y_0. If the ratio of step sizes between to the left and to the right is a rational number (i.e., $\frac{-\ln \delta}{\ln \gamma}$ is rational), it is easy to construct a pushdown automaton (PDA) [2] which accepts a set of random walk paths until it passes $\ln 1$ for the first time.[14] Thus, a set of these paths is a context-free language [2]. In the following, we will call the set decided by (γ, δ, y_0) as random walk context-free language (RWCFL).

Although these paths are symbol strings, we can also regard these as symbol sequences because, once a random walk passes $\ln 1$, it will move to the $y > 1$ attractor no matter what symbols follow the original symbol string (e.g. $RR * * \cdots$, $LRRR * * \cdots$, $LRLRR * * \cdots$, ...). Thus, this simple dynamical system virtually executes encoding RWCFL(γ, δ, y_0) on the horizontal line segment $\{x, y \mid y = y_0, 0 \leq x \leq 1\}$ by using the code α, as the basin of the $y > 1$ attractor.

5.2 Boundary dimension

Now we deal with dimensional property of such geometric sets. The riddled basin of the $y = 0$ attractor of the simple model is the complement of the geometric representation of RWCFL(γ, δ, y_0) with code α. Since the riddled basin is a fat fractal, we also consider the dimension of the boundary in this case, to study its fine structure. Of course, the boundary dimension of the riddled basin is equal to the boundary dimension of RWCFL(γ, δ, y_0) with code α.

The uncertainty exponent ϕ is calculated by the analysis of the simple model using the diffusion approximation to simulate the random walk in [27]. Under the condition $|\ln y_0| \gg 1$ and $|h_\perp| \ll \ln \gamma, -\ln \delta$, it is given by

$$\phi = \frac{h_\perp^2}{4D h_\parallel},$$

where $D = \frac{1}{2}\alpha(1-\alpha)(\ln \gamma - \ln \delta)^2$ is the diffusion per iterate, $h_\parallel = \alpha \ln \frac{1}{\alpha} + (1-\alpha) \ln \frac{1}{1-\alpha}$ is the Lyapunov exponent for the dynamics in I, and $h_\perp = \alpha \ln \gamma + (1-\alpha) \ln \delta$.

---

[14] Basically such a PDA is constructed as follows: First, suppose $l, r \in \mathbf{Z}$, and let the ratio of l to r equal the ratio of the step size of 'L' to that of 'R'. Then, let the PDA push l stack symbols on the stack when its input tape head reads 'L', whereas let r stack symbols be popped off when 'R' is read.

Thus, according to the above formula, for given RWCFL(γ, δ, y_0) with rational $\frac{-\ln \delta}{\ln \gamma}$, one can change the boundary dimension of the geometric representation of the RWCFL, by changing α, i.e., the code transformation. In particular, if α_c satisfies $h_\perp = 0$ (i.e., $\alpha_c \ln \gamma + (1 - \alpha_c) \ln \delta = 0$), the above formula implies that the boundary dimension of the RWCFL with code α_c (and only with that α_c) is equal to the space dimension. However, this equality is easily broken, in the present case, by code transformation (i.e. changing α). This fact is in strong contrast with the case of a halting set of a UTM and the Mandelbrot set.

If $h_\perp = 0$, however, $I = \{x, y \mid y = 0, 0 \leq x \leq 1\}$ is no longer an attractor even according to the definition of Milnor. As $h_\perp \to 0$ with $h_\perp < 0$, the measure of the riddled basin for the $y = 0$ attractor approaches zero (and the measure of the RWCFL with the corresponding code converges to one). Thus, even though the boundary dimension is equal to the space dimension with the code α_c, the decision problem of the RWCFL virtually disappears since a point in the measure zero set cannot actually be chosen. This fact is also in strong contrast with the case of the halting set of a UTM and the Mandelbrot set.

As an example, we numerically examine the boundary dimension of the RWCFL such that $\gamma \delta = 1$, $y_0 = \delta$ with code $\frac{1}{2}$ and code $\frac{1}{3}$ (i.e. $\alpha = \frac{1}{2}$ and $\alpha = \frac{1}{3}$, respectively). Here, we also treat the set of paths which are decided to be in the RWCFL within a given finite step n. In other words, the set of initial points which are decided to belong to the $y > 1$ attractor's basin within n steps is computed, with the same procedure as before.

Fig. 9 shows the asymptotic behavior of the boundary dimension for $\alpha = \frac{1}{2}$, where the uncertainty exponent ϕ (i.e. $1 - D_0$) is plotted versus n, with a logarithmic scale. As is shown, the box-counting dimension of the boundary approaches the space dimension one roughly as $1 - cn^{-1}$ with the increase of the (computation) time n,[15] similarly with the case of the halting sets of a UTM and the Mandelbrot set.

On the other hand, in Fig. 10, the boundary dimension is studied for $\alpha = \frac{1}{3}$. (This set is transformed from the set with code $\frac{1}{2}$ by the code transformation $t_{\frac{1}{2} \to \frac{1}{3}}$.) There, the log-log plot of $f(\varepsilon)$ with ε is shown for several values of n, where $f(\varepsilon)$ is the fraction of subintervals including a boundary. Unlike the case of $\alpha = \frac{1}{2}$ (binary encoding), Fig. 10 shows no asymptotic approach of the uncertainty exponent ϕ to zero with the increase of n. (As mentioned previously, ϕ is obtained by fitting $f(\varepsilon) \sim \varepsilon^\phi$.) Thus, the boundary dimension is estimated to be less than one even in the limit of $n \to \infty$. In terms of the uncertainty exponent, the probability of making a mistake (equally $V[S(\varepsilon)]$) can be

---

[15] The boundary dimension of the RWCFL($\gamma\delta = 1$, $y_0 = \delta$) with $\alpha = \frac{1}{2}$ is estimated to be

$$D_0 \approx \lim_{m \to \infty} \frac{\log_2 \frac{1}{\sqrt{\pi}}(-\arctan m^{\frac{1}{2}} + \frac{\pi}{2})}{2m + 1} + 1,$$

by adopting a random walk representation [14].

decreased to any amount, in principle, by a decrease in ε. Thus, the decision procedure of this set does not have the inaccessibility mentioned previously.

Now, it is confirmed that one can change the boundary dimension of the geometric representation of the RWCFL by changing codes. In particular, even if the boundary dimension of the RWCFL with code α_c (and only with that α_c) is equal to the space dimension, this property is easily broken by the code transformation (i.e. changing α), unlike the halting set of a UTM and the Mandelbrot set.

After all, there exist various classes in non-self-similar sets, especially in sets that have the boundary dimension equal to the space dimension. RWCFL, an example of context-free language, is ranked "between" a halting set of a UTM and a self-similar set like Cantor set which corresponds to a regular language.

So far, we have fixed a RWCFL and changed α (code). Now, let us change RWCFL by fixing α (code), i.e., a Markov partition of a piecewise-linear map characterized by α. For given α, there is a pair (γ, δ) such that $\alpha \ln \gamma + (1-\alpha) \ln \delta$ is equal to, or arbitrarily close to, zero, with rational $\frac{-\ln \delta}{\ln \gamma}$. Hence the geometric representation of the RWCFL specified by these γ and δ with appropriate y_0 using code α has the boundary dimension equal to, or arbitrarily close to, the space dimension. Thus, for any code α or any Markov partition of a piecewise-linear map, there is RWCFL such that boundary dimension is equal to, or arbitrarily close to, the space dimension.

Now, let us reconsider the halting set of a UTM. Basically, a halting set of a UTM contains all halting sets of TM. Hence it naturally contains all RWCFL. In this case, although the boundary dimension of each individual RWCFL is varied by the code transformation, within the halting set of UTM there is a RWCFL, whose boundary dimension is equal to, or arbitrarily close to the space dimension for an arbitrary code into a Markov partition of a piecewise-linear map. Thus, it is realized from this point that the boundary dimension of the halting set of a UTM is equal to, or arbitrarily close to the space dimension for an arbitrary code into a Markov partition of a piecewise-linear map.

6 Discussion

6.1 Condition for models of computation

In section 1, we mentioned the physical realizability of models of computation. Indeed several analog computation models have been studied so far (e.g., [25], [31], [32], [33]), but each model defines each computability, and relation of each other is not made so clear.

On the other hand, from a viewpoint of physical realizability, the (in)accessibility is considered to give one condition for any models of computation. Note that we have shown inaccessibility of the undecidable decision procedure in the sense of the Turing model. Here, we propose that this inaccessibility holds for any other models with undecidability and suggest the following statement: *In a model of computation, a procedure should be uncomputable (undecidable), if it has inaccessibility in the sense that the ideal procedure is not approached in the presence of error.*

This proposition is based on the following consideration: From a viewpoint of physical realizability, it is actually impossible to avoid errors in computational operation and observation. Thus, it is irrelevant to count a procedure with the inaccessibility as computable, since one cannot even approach the ideal procedure on which the model is founded, in the presence of error. Such procedure having the inaccessibility should be included in uncomputable procedures.

Concerning the Mandelbrot set M, Penrose suggests that M (and also the complement of M) is not "recursive" (i.e. "undecidable") [24]. By our condition of physical realizability, M should also be undecidable. A remark should be made here: Blum, Shub and Smale have already pointed out and proved the undecidability of M over $\mathbf{R}$ in their well-known formulation of analog computation [25]. However, all fractals (like a typical Julia set) are also undecidable according to their formulation. Taking into account of the viewpoint of the precision and introducing the inaccessibility, we can properly distinguish in our criterion between ordinary fractals and more complex sets that are hard to handle.

6.2 Class of appropriate codes

About the coding, we have introduced the mapping from a symbol sequence to a real number represented by the same label of Markov partition of a certain piecewise-linear map, so that the effect of farther cells gets smaller in the real number. Otherwise, it is impossible to discuss the distance in the symbol sequence necessary to consider geometric and dynamical systems properties. A code satisfying this condition is considered to be appropriate, but this condition may be too strong. The question about the appropriate condition for coding remains unanswered yet, and a wider class of codes seems to exist.

Clarifying the appropriate class of coding will be directly connected with the explicit statement of the "physical" Church-Turing thesis [34], since it will reveal how much we can utilize analog states, in order to accomplish a higher computability than the discrete Turing machine.

7 Conclusions

Our main results in this paper are as follows:

- A boundary dimension equal to space dimension indicates that decision procedure of a set has so strong uncertainty that one cannot approach the ideal decision procedure, in the presence of error. The decision procedure of the sgeometric representation of halting set of a universal Turing machine has this inaccessibility, and this property is preserved under code

transformations. This result is also reasonable from the viewpoint of undecidability of the halting problem of a universal Turing machine.
- The Mandelbrot set, known as a set with extraordinarily complex structure, can be connected with undecidable sets by both the inaccessibility and its invariance against "fractal" function which corresponds to a code transformation.
- The riddled basin of a certain simple dynamical system, having different unpredictability from chaotic unpredictability, represents geometrically a certain context-free language. The riddled basin has a different character, e.g. against code transformations, and is ranked between an ordinary fractal and a halting set of a universal Turing machine or the Mandelbrot set. Therefore, there exist various classes in non-self-similar sets that are not simply characterized by self-similarity, especially in sets that have boundary dimension equal to space dimension.

Acknowledgments

The authors would like to thank M. Taiji, T. Ikegami, S. Sasa, O. Watanabe, M. Shishikura and I. Shimada for useful discussions. This work is partially supported by a Grant-in-Aid for Scientific Research from the Ministry of Education, Science, and Culture of Japan.

References

1. A. M. Turing, "On computable numbers with an application to the Entscheidungsproblem," Proc. London Math. Soc. **42** (1936), 230.
2. J. E. Hopcroft and J. D. Ullman, *Introduction to automata theory, languages and computation* (Addison-Wesley, Reading, Mass., 1979).
3. M. Davis, *Computability & unsolvability* (Dover, New York, 1982).
4. H. T. Siegelmann and S. Fishman, "Analog computation with dynamical systems," Physica D **120** (1998), 214.
5. B. B. Mandelbrot, *The fractal geometry of nature* (W. H. Freeman, New York, 1983).
6. M. F. Barnsley, *Fractals everywhere* (Academic Press, Boston, 1988).
7. K. J. Falconer, *The geometry of fractal sets* (Cambridge University Press, Cambridge, 1985).
8. E. Ott, *Chaos in dynamical systems* (Cambridge University Press, Cambridge, 1993).
9. J. Guckenheimer and P. Holmes, *Nonlinear oscillations, dynamical systems, and bifurcations of vector fields* (Springer-Verlag, New York, 1983).
10. R. L. Devaney, *A first course in chaotic dynamical systems: theory and experiment* (Addison-Wesley, Reading, Mass., 1992); *An introduction to chaotic dynamical systems* (Addison-Wesley, Redwood City, Calif., 1989).
11. C. Moore, "Generalized shifts: unpredictability and undecidability in dynamical systems," Nonlinearity **4** (1991), 199; "Unpredictability and undecidability in dynamical systems," Phys. Rev. Lett. **64** (1990), 2354.
12. I. Shimada, Talk at international symposium on information physics 1992, Kyushu institute of technology.

13. S. Wolfram, "Undecidability and intractability in theoretical physics," Phys. Rev. Lett. **54** (1985), 735.
14. A. Saito and K. Kaneko, "Geometry of undecidable systems," Prog. Theor. Phys. **99** (1998), 885; "Inaccessibility and Undecidability in Computation, Geometry and Dynamical Systems," submitted to Physica D.
15. T. Kamae and S. Takahashi, *Ergodic theory and fractals* (Springer, Tokyo, 1993).
16. G. de Rham, "Sur quelques courbes définies par des équations fonctionnelles," Rend. Sem. Mat. Torino **16** (1957), 101.
17. P. Grassberger, "Generalized dimensions of strange attractors," Phys. Lett. **97A** (1983), 227.
18. N. Chomsky, "Three models for the description of language," IRE Trans. on Information Theory **2** (1956), 113; "On certain formal properties of grammars," Information and Control **2** (1959), 137.
19. Y. Rogozhin, "Small universal Turing machines," Theor. Comput. Sci. **168** (1996), 215.
20. L. Staiger, "ω-Languages," in *Handbook of formal languages 3*, edited by G. Rozenberg and A. Salomaa (Springer, Berlin, 1997).
21. W. Thomas, "Automata on infinite objects," in *Handbook of theoretical computer science B*, edited by J. van Leeuwen (Elsevier, Amsterdam, 1990).
22. J. D. Farmer, "Sensitive dependence on parameters in nonlinear dynamics," Phys. Rev. Lett. **55** (1985), 351.
23. C. Grebogi et al., "Exterior dimension of fat fractals," Phys. Lett. A **110** (1985), 1.
24. R. Penrose, *The emperor's new mind* (Oxford University Press, Oxford, 1989).
25. L. Blum, et al., *Complexity and real computation* (Springer, New York, 1998).
26. M. Shishikura, "The boundary of the Mandelbrot set has Hausdorff dimension two," Astérisque **222** (1994), 389.
27. E. Ott, et al., "The transition to chaotic attractors with riddled basins," Physica D **76** (1994), 384.
28. J. C. Sommerer and E. Ott, "A physical system with qualitatively uncertain dynamics," Nature **365** (1993), 138.
29. E. Ott et al., "Scaling behavior of chaotic systems with riddled basins," Phys. Rev. Lett. **71** (1993), 4134.
30. J. Milnor, "On the concept of attractor," Commun. Math. Phys. **99** (1985), 177.
31. M. B. Pour-El and J. I. Richards, *Computability in analysis and physics* (Springer Verlag, Berlin, 1989).
32. H. T. Siegelmann and E. D. Sontag, "Analog computation via neural networks," Theor. Comput. Sci. **131** (1994), 331.
33. C. Moore, "Recursion theory on the reals and continuous-time computation," Theor. Comput. Sci. **162** (1996), 23.
34. R. Penrose, *Shadows of the mind* (Oxford University Press, Oxford, 1994).

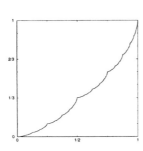

Fig. 1. Lebesgue's singular function $L_{\frac{1}{3}}$ (code transformation $t_{\frac{1}{2} \to \frac{1}{3}}$).

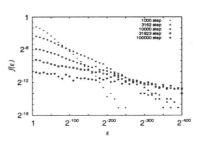

Fig. 3. Log-log plot of $f(\varepsilon)$ of $UL_{\frac{1}{2}}(n)$ with ε for computation time $n = 1000$, 3162, 10000, 31623 and 100000. The slope of $f(\varepsilon)$ becomes smaller with the increase of n.

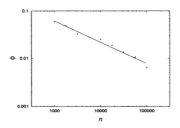

Fig. 4. Log-log plot of $2 - D_0$ (i.e., uncertainty exponent ϕ) of $UL_{\frac{1}{2}}(n)$ versus the computation time n. The data are least-square fit to a straight line, $1.4n^{-0.45}$.

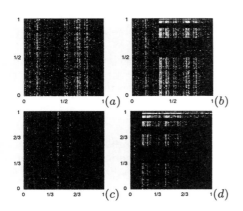

Fig. 2. The halting set of Rogozhin's UTM, represented by code $\frac{1}{2}$ and code $\frac{1}{3}$. Dots correspond to inputs halted within 500 steps ($UL_{\frac{1}{2}}(500)$ and $UL_{\frac{1}{3}}(500)$ in the text). (a) Square 0 of $UL_{\frac{1}{2}}(500)$. (b) Square 1 of $UL_{\frac{1}{2}}(500)$. (c) Square 0 of $UL_{\frac{1}{3}}(500)$. (d) Square 1 of $UL_{\frac{1}{3}}(500)$. Inputs are chosen at random in each grid cell (1024 × 1024 grid and 729 × 729 grid, respectively).

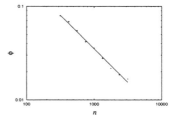

Fig. 5. Log-log plot of uncertainty exponent ϕ (i.e., $2 - D_0$) of $UL_{\frac{1}{3}}(n)$ versus the computation time n.

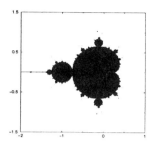

Fig. 6. The Mandelbrot set M (precisely, $M(500)$).

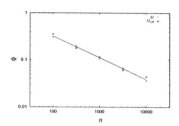

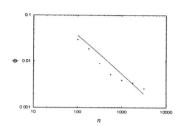

Fig. 9. Log-log plot of uncertainty exponent ϕ (i.e., $1 - D_0$) of the RWCFL with code $\frac{1}{2}$ versus the computation time n. The dashed line shows an analytic estimate [14], given by $\phi = -\frac{\log_2 \pi^{-\frac{1}{2}}(-\arctan(\frac{n-1}{2})^{\frac{1}{2}} + \frac{\pi}{2})}{n}$.

Fig. 7. Log-log plot of uncertainty exponent ϕ (i.e., $2 - D_0$) of $M(n)$ and $M_{\frac{1}{3}}(n)$ versus the computation time n. The exponents ϕ for both cases approach zero roughly as $n^{-0.45}$ with the increase of the computation time n. This power-law decay of the exponent ϕs is in common with the case of halting set of UTM.

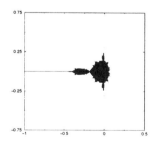

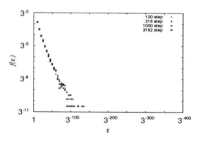

Fig. 10. Log-log plot of $f(\varepsilon)$ of the RWCFL with code $\frac{1}{3}$ with ε for computation time $n = 100, 316, 1000$ and 3162. The boundary dimension remains less than one even with the increase of n. Note the difference from the case in Fig. 3.

Fig. 8. $M_{\frac{1}{3}}$ obtained by $t_{\to \frac{1}{3}}(M)$ (precisely, $M_{\frac{1}{3}}(500)$).

On the Power of Nonlinear Mappings in Switching Map Systems

Yuzuru Sato[1] *, Makoto Taiji[2] and Takashi Ikegami[1]

[1] Institute of Physics, Graduate school of Arts and Sciences, University of Tokyo, Tokyo, Japan
[2] The Institute of Statistical Mathematics, Tokyo, Japan

Abstract. A dynamical systems based model of computation is studied, and we demonstrate the computational ability of nonlinear mappings. There exists a switching map system with two types of baker's map to emulate any Turing machine. The baker's maps correspond to the elementary operations of computing such as left/right shift and reading/writing of symbols in Turing machines. Taking other mappings with second-order nonlinearity (e.g. the Hénon map) as elementary operations, the switching map system becomes an effective analog computer executing parallel computation similarly to MRAM. Our results show that, with bitwise Boolean AND, it has PSPACE computational power. Without bitwise Boolean AND, it is expected that its computational power is between classes RP and PSPACE. These dynamical systems execute more unstable computation than that of classical Turing machines, and this suggests a trade-off principle between stability of computation and computational power.

1 Introduction

Dynamical systems have recently been actively studied in terms of computing processes in physical systems (see e.g., [6] [9] [16]). In [9] it is shown that any Turing machine is equivalent to a class of 2 dimensional piecewise-linear maps (Generalized Shifts, GSs), which can be embedded in smooth flows in $\mathbf{R}^3$.

Here we introduce another class of dynamical systems called *switching map systems* and demonstrate their computational capability. If defined on the space $\mathbf{R}$, they can be a model of real number computation [10] [4], while if the space is restricted to the set of the dyadic rationals with finite binary expansions, they are regarded as a model of classical computation.

There exists a switching map system with two types of baker's map to emulate any Turing machine [12]. The baker's maps correspond to the elementary operations of Turing machines such as left/right head-moving and reading/writing symbols. An advantage of this system is that it found to be a 'programmable' GS, by means of which we can easily construct various concrete

---

* Corresponding author. E-mail: ysato@sacral.c.u-tokyo.ac.jp.

examples. Since this system is also a class of input-output mappings defined in the BSS model [4], a connection between GSs and input-output maps is also shown.

We suggest that other types of nonlinear mappings can also be the basic components of 'natural' effective procedures from the dynamical systems point of view. Taking mappings with second-order nonlinearity (e.g. the Hénon map) as elementary operations, it is expected that the switching map system will show a new model of computation with the nonlinearity as an oracle. Here we show that the computational power of the oracle is PSPACE under the condition that unrestricted usage of mod-operations or bitwise Boolean operations would be allowed. It is also expected that the computational power of switching map systems with polynomial mappings with rational coefficients is between classes RP and PSPACE. It is remarkable that recursive application of simple nonlinear operations with mod-operations or bitwise Boolean operations results in far stronger computational power than classical Turing machines. On the other hand, these dynamical systems execute more unstable computation than that of classical Turing machines, which implies the existence of a trade-off principle between stability of computation and computational power.

2 The Switching Map Systems and Turing Machines

Here we consider the correspondence between a class of dynamical systems called switching map systems and Turing machines. We introduce a symbolic dynamics to interpret dynamical systems as Turing machines.

2.1 The switching map systems

Let $f_1, f_2, \ldots, f_M$ be M mappings on $X \subset \mathbf{R}^n$. A set of internal states $S = \{0, \ldots, N-1\}$ and N branching functions labeled by the states $g_0, g_1, \ldots, g_{N-1}$ are given in advance. $g_n : X \to S$ are mappings from a value of x to the label of the next state. We denote a switching map system with $f_1, f_2, \ldots, f_M$ as $F(f_1, f_2, \ldots, f_M)$:

$$F : S \times X \to S \times X : (n, x) \mapsto (n', x') = (g_n(x), h_n(x)) \ (n \in S),$$

where $h_n \in \{f_1, f_2, \ldots, f_M\}$. F maps each state/space pair (n, x) deterministically to the unique next state/space pair (n', x'). Here in practice, mappings are switched and applied successively to an initial configuration $(0, x_0) \in S \times X$ as follows:

$$\begin{aligned}
(0, x_0) &\mapsto (n_1, x_1) = (g_0(x_0), h_0(x_0)) \\
&\mapsto (n_2, x_2) = (g_{n_1}(x_1), h_{n_1}(x_1)) \\
&\vdots \\
&\mapsto (n_{t+1}, x_{t+1}) = (g_{n_t}(x_t), h_{n_t}(x_t))
\end{aligned}$$

Here g_{n_t} is applied to x_t and $g_{n_t}(x_t)$ is used as function label n_{t+1}, in the same way, h_{n_t} is applied to x_t, and $h_{n_t}(x_t)$ is substituted as initial value x_{t+1} for $h_{n_{t+1}}$, where $n_1, n_2, \ldots$ is the state of the system at time $t = 1, 2, \ldots$ respectively. This process is applied iteratively to determine the successive functional form.

We can regard a switching map system $F(f_1, f_2, \ldots, f_M)$ with suitable branching function g_n as a program, and its trajectories as a process of computation. The structures of its attractors correspond to the output results. Since we can use periodic orbits or attractors as output results, the halting states are not always required. With this framework, we can define a process of computation as a dynamics and interpret properties of dynamical systems such as measures, dimensions and topological structures from a computation theoretic point of view. Additionally, a switching map system can be represented by a skew product of dynamical systems [5] and it is also introduced in terms of information processing in higher-dimensional dynamical systems [17].

2.2 The horseshoe map, symbolic dynamics and Turing machines

A dynamical system is called chaotic if the set of the orbit in the phase space embeds Smale's topological horseshoe (Fig. 1). The horseshoe map is described in terms of an invertible planar map which can be thought of as a Poincaré map arising from a 3 dimensional autonomous differential equation or a forced oscillator such as the Duffing equation. The set of non-wandering points of the horseshoe map Λ forms a square Cantor set, in which each point corresponds to a sequence in $\{0, 1\}^{\mathbb{N}}$. The half (left or right) of the square to which the point belongs gives a rule of the gray code. The action of the map induces a shift on a bi-infinite sequence of two symbols where $\overline{a_i}$ denotes a bit-flipping:

$$H_0 : \Lambda \to \Lambda : \ldots a_{-2}a_{-1}.a_0a_1a_2\ldots \mapsto \ldots a_{-2}.a_{-1}a_0a_1a_2\ldots$$

$$H_1 : \Lambda \to \Lambda : \ldots a_{-2}a_{-1}.a_0a_1a_2\ldots \mapsto \ldots a_{-2}.\overline{a_{-1}}a_0a_1a_2\ldots$$

$$H_0^{-1} : \Lambda \to \Lambda : \ldots a_{-2}a_{-1}.a_0a_1a_2\ldots \mapsto \ldots a_{-2}a_{-1}a_0.a_1a_2\ldots$$

Regarding the decimal point as the position of the head of a Turing machine, these actions correspond to the elementary operations of Turing machines, such as (1) left-shifting ($H_0 = H_1 H_0^{-1} H_1$); (2) right-shifting (H_0^{-1}); (3) bit-flipping and left-shifting (H_1); and (4) bit-flipping and right-shifting ($H_0^{-1} H_1 H_0^{-1}$). Thus, we can use suitable combinations of H_0^{-1} and H_1 to emulate arbitrary computation processes. When a halting state is required, we can use an identity map I as the halting operation.

In 2 dimensional discrete dynamical systems on a unit space $[0, 1] \times [0, 1]$, we can substitute the baker's map for the horseshoe map. It is a measure-preserving map that conjugates to the shift on two symbols. In this case, the

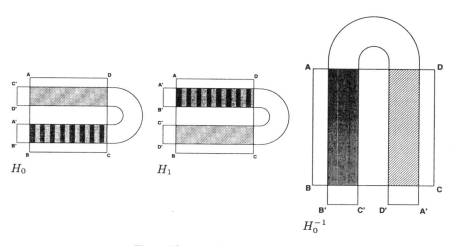

Fig. 1. The Smale's Horseshoe Map.

left and right halves of a point's sequence are simply the binary expansions of its x and y coordinates, respectively. Here we use a standard metric on the unit space. In practice, Turing machines with N internal states can be emulated by the following set of baker's mappings:

$$B_0(x,y) = \begin{cases} (2x, \frac{y}{2}) & x \in [0, \frac{1}{2}) \\ (2x-1, \frac{y+1}{2}) & x \in [\frac{1}{2}, 1) \end{cases} \quad (1)$$

$$B_1(x,y) = \begin{cases} (2x, \frac{y+1}{2}) & x \in [0, \frac{1}{2}) \\ (2x-1, \frac{y}{2}) & x \in [\frac{1}{2}, 1) \end{cases} \quad (2)$$

$$B_0^{-1}(x,y) = \begin{cases} (\frac{x}{2}, 2y) & y \in [0, \frac{1}{2}) \\ (\frac{x+1}{2}, 2y-1) & y \in [\frac{1}{2}, 1) \end{cases} \quad (3)$$

Theorem 1. *For any Turing machine M, there exists a switching map systems $F(B_0^{-1}, B_1, I)$ conjugate to M via a map Φ.*

Proof. Note that $B_0 = B_1 B_0^{-1} B_1$. For further details, see [12]. □

3 Nonlinear Computation

Previously we have adopted baker's maps for the elementary operations of Turing machines. However, we can also take other types of nonlinear mappings as 'natural' elementary operations from the dynamical systems point of view. Taking second-order nonlinear mappings (e.g. Hénon maps) instead of baker's maps, it is expected that switching map systems can show a new model of computation with the nonlinearity as an oracle.

3.1 Hénon map as nonlinear operations

The Hénon map T_0 is given by $T_0(x, y) = (a - x^2 + by, x)$ where $|b| \leq 1$ and a are given as control parameters. It conjugates to Smale's horseshoe and its invariant set is hyperbolic if it has strong nonlinearity. In practice, the range $a > (5 + 2\sqrt{5})(1 + |b|)^2/4$ corresponds to the hyperbolic case.

$$T_0(x, y) = (a - x^2 + by, x) \tag{4}$$
$$T_1(x, y) = (a - x^2 + by, -x) \tag{5}$$
$$T_0^{-1}(x, y) = (y, (x - a + y^2 + by)/b) \tag{6}$$

We introduce here the adjoint Hénon map T_1 corresponding to H_1 in Figure 1, and let the invariant sets of T_0 and T_1 be Λ_0 and Λ_1, respectively. Turing machines can be emulated as the symbolic dynamics on the domain including $\Lambda_0 \cap \Lambda_1$ using the Hénon mappings with a large value of a.

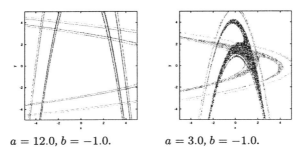

$a = 12.0, b = -1.0.$ $a = 3.0, b = -1.0.$

Fig. 2. The Hénon Map and its Inverse Map.

Lowering the control parameter a to the point where the first homoclinic tangency occurs, we observe that the map transits from a hyperbolic map to a non-hyperbolic one (Fig. 2). This transition can easily break the conjugate relationship between the switching map systems and Turing machines.[1] However, as we show in simple theorems below, it opens a new model of computation. That is, it is thought that this property shows a potential transition from

---

[1] The invariant set associated with a Hénon map and that with an adjoint Hénon map can certainly have overlaps. However, we have not examined whether the overlaps can embed any sequences, so that we do not have to constrain the computational ability of these switching map systems with Hénon maps. This problem will be discussed elsewhere. Most probably, we have to bound the computation time in accordance with the value of nonlinearity a. The lower the value, the more the time that certificates correct computation is bounded. Apparently, the case of an infinitely large value of a coincides with the switching map systems with Smale's horseshoe. On the other hand, when we lower the parameter to have homoclinic tangency, the computation process corresponding to a Turing machine instantly breaks up.

an ordinary Turing machine to a 'nonlinear' one. We refer to nonlinear mappings like Hénon maps as nonlinear operations, algorithms based on them as nonlinear algorithms (or nonlinear Turing machines), and computation processes based on them as nonlinear computation.

3.2 The computational power of nonlinear mappings

We consider here the computational power of nonlinear Turing machines.

First we show the equivalence between classical computation and nonlinear computation in terms of computability. In order to compare the computational power of nonlinear computation with that of classical computation, rational functions with dyadic rational coefficients $f : \mathbf{Q} \to \mathbf{Q}$ should be used as nonlinear operations. Otherwise we could obtain a real number, that is, an infinite binary sequence, within one step of computation. We assume that the nonlinear operations execute within one step of computation.

Theorem 2. *The computational power of classical Turing machines and nonlinear Turing machines are equivalent in terms of computability.*

Proof. The theorem holds directly from the assumptions that the operations are all rational functions (computable functions), and that input sequences are all rational numbers (computable numbers), and Theorem 1. □

Next we consider the computational power of the nonlinear Turing machines in terms of computational complexity. We show that the power of nonlinear computation is stronger than that of Turing machines, by embedding a random access machine with multiplication (MRAM) into a switching map system with (3 dimensional) nonlinear operations. MRAM is one of the second class machines described by the models of weak parallel computation and its computational power is stronger than that of Turing machines (see [8],[3]). We denote the MRAMs with extended instruction Q as M$(Q)^2$ and the class of the languages accepted by the polynomial-time bounded machines M is denoted PTIME(M). We also denote PTIME(F) for polynomial-time bounded switching map systems.

Here we consider the results in [8], [3] and [15];

$$PTIME(M(\times, \div)) = PTIME(M(\times, \wedge)) = PSPACE,$$

where $\div$ and $\wedge$ mean integer division and bitwise Boolean AND, respectively. We have to be careful to deal with the mod-operations (corresponding to integer division) and the bitwise Boolean AND in the dynamical systems point of view. Unrestricted usage of these operations implies the application of the exponentially complex mappings in Fig. 3, 4, under our binary expansion

---

[2] The standard instructions of random access machines (RAMs) are assignment, indirect addressing, sum, proper subtraction, Boolean operations, conditional jump, and accept/reject.

coding. We refer to the mod-operation V_1: $z = x/y$ (mod 1) and the bitwise Boolean AND V_2: $z = x \wedge y$ as the 'masking maps.' We can show that the computational power of the switching map systems achieves PSPACE with these masking maps V_1, V_2.

Theorem 3. *There exists a switching map system with 3 dimensional nonlinear operations and a masking map V_1 or V_2 that can recognize the PSPACE-complete languages in polynomial time. In practice, for PSPACE computational power, we need a switching map systems with the following nonlinear mappings in addition to B_0^{-1}, B_1, I;*

$$T_{xy}(x,y,z) = (y,x,z), \quad T_{zx}(x,y,z) = (z,y,x),$$
$$L_1(x,y,z) = (0,y,z), \quad L_1'(x,y,z) = (\frac{1}{2},y,z),$$
$$L_2(x,y,z) = (\frac{x}{2},y,z), \quad L_2'(x,y,z) = (2x,y,z) \ (mod\,1),$$
$$L_3(x,y,z) = (x,y+x,z), \quad L_3'(x,y,z) = (x,y-x,z),$$
$$L_4(x,y,z) = (y,y,z), \quad V_1(x,y,z) = (x,y,\frac{z}{x}) \ (mod\,1)$$
$$N_1(x,y,z) = (x^2,y,z), \quad N_2(x,y,z) = (x,y,zx).$$

$$PTIME(F(B_0^{-1}, B_1, I; T_{xy}, T_{zx}, L_1, L_1', L_2, L_2', L_3, L_3', L_4, N_1, N_2, V_1))$$
$$= PTIME(F(B_0^{-1}, B_1, I; T_{xy}, T_{zx}, L_1, L_1', L_2, L_2', L_3, L_3', L_4, N_1, N_2, V_2))$$
$$= PSPACE.$$

Proof. This is proved by showing that we can solve QSAT using a switching map system with 3 dimensional nonlinear operations. The multiplication and integer division for MRAM M($\times, \div$) are executed by the nonlinear operations in one step of computation, by assumption. There exist a switching map system with 3 dimensional nonlinear operations and a masking map V_1 that can emulate MRAM M($\times, \div$) (see [13]). We can solve QSAT with it in polynomial time. V_2 is emulated by V_1 in polynomial time. For further details, see [8][3]. □

Corollary 1. *We assume that O(1) times application of the masking map V_1 or V_2 is allowed. Then there exists a switching map system with 3 dimensional nonlinear operations that can recognize the NP-complete languages.*

Proof. We can solve kSAT with switching map system with 3 dimensional nonlinear operations allowing O(1) times application of the masking map V_1 or V_2 in polynomial time. For further details, see [15]. □

Nonlinear Mappings in Switching Map Systems   241

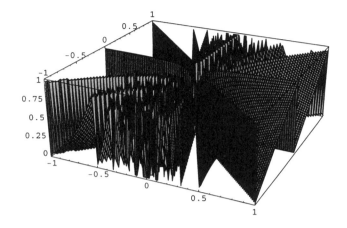

Fig. 3. V_1: $z = x/y \pmod{1}$.

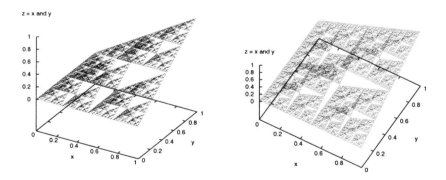

Fig. 4. V_2: $z = x \wedge y$.

These results are not trivial because we can give examples of dynamical systems with PSPACE computational power applying a complex mapping in polynomial time. It is thought that the emulation of the masking maps with the baker's map in polynomial time means P=PSPACE in terms of computational theory. The masking maps V_1, V_2 work to pick up the information of a single bit in an exponential number of bits, masking the needless bits. Theorem 3 tells us that the computational power of nonlinear mappings with a masking map is stronger than classical Turing machine computations. Note that

$$PTIME(F(B_0^{-1}, B_1, I, V_2)) = P$$

depending on the following relationships:[3]

$$PTIME(M(\wedge)) = P.$$

The multiplication of the two variables of the system xy is the potential source of MRAM's weak parallelism and it directly indicates 'nonlinearity' of the systems. Thus we can say that the strong bit-mixing property of the second-order nonlinearity is a cause of the PSPACE computational power.

We next consider the computational power of switching map systems without the masking maps. Here we restrict the elementary operations to polynomial mappings.

For the integers k, l and m which are bounded to polynomial order for input size n, we give f_i ($i = 1, 2, \ldots, l$) as the k-dimensional mappings, which are described by the mth order polynomials with rational coefficients. Here all coefficients of f_is are dyadic rationals with finite binary expansions. We are now interested in the class $PTIME(F(B_0^{-1}, B_1, I; f_i))$. Using the results in [15]:

$$PTIME(M(=, \times)) \subseteq RP,{}^4$$

and Theorem 3, it is expected that

$$RP \subseteq PTIME(F(B_0^{-1}, B_1, I; f_i)) \subseteq PSPACE \ (i = 1, 2, \ldots, l).$$

It is defined that the probabilistic Turing machines with RP computational power should reject languages with perfect decision (in our model, a perfect comparison for $x \in [0, 1]$ with 0.5 in the phase space), but that seems to be incompatible with the unpredictability of chaotic systems. Thus, it is expected that $F(B_0^{-1}, B_1, I; f_i)$ cannot be emulated by any probabilistic Turing machines with RP computational power in polynomial time, and $RP \subsetneq PTIME(F(B_0^{-1}, B_1, I; f_i))$.

Suppose that an automaton tries to emulate a given dynamical system's behavior. If the automaton fails to emulate the behavior, it implies that the

---

[3] For 'rational number' division with V_1, it is unclear whether the same results would hold or not.

[4] Note that inequalities are not allowed, while we do allow them as branching functions.

dynamical system has more than P computational power. As has been argued by [14], for example, the pseudo-orbit tracing property does not hold for non-hyperbolic dynamical systems. That is, an infinite number of bits or long time correlation are needed to correctly emulate such systems. We thus argue that non-hyperbolic dynamical systems are another candidate to have more than RP computational power.

It is known that, given random oracles, P=RP=BPP. If we could use chaos with an initial condition with finite precision, as a 'perfect' random oracle, there exists a (deterministic) switching map system $F(B_0^{-1}, B_1, I;\ f_i)$ that can emulate probabilistic Turing machines with BPP computational power in polynomial time. However, non-hyperbolic mappings have intrinsic statistical properties such as the non-exponential decay of correlations. Thus, if we simply assume chaos as a source of random numbers, it does not work as a random oracle, and its computational power would be lower than BPP.

The precise estimation of the lower bound of the computational power of $F(B_0^{-1}, B_1, I;\ f_i)$ would be worth studying in terms of both dynamical systems theory and computational theory. We can possibly define a new class of computational complexity based on dynamical systems and analyse the chaos from a computation theoretic point of view.

While masking maps can be allowed in logical systems, they would not be allowed in physical systems, since natural dynamical systems are smooth and continuous. This implies that 'physical' analog computation is weaker than logical computation. Here we just analyse the simple case, for the polynomial mappings. Studies on general nonlinear mapping cases are left as problems for future work.

3.3 Stability of computation

We have the results that there exists a nonlinear dynamical system with stronger computational power than Turing machines. However, we cannot execute the algorithms because the microscopic noise of order $\sim 2^{-2^n}$ can completely destroy the computation process (see Fig. 5, 6). The second-order nonlinearity also causes the expansion of the effects of microscopic noise. Even if the noise flips one or more bits below the threshold of measurement, its effect can reach bits high enough to affect accuracy of measurement within a number of steps of polynomial order, and macroscopic behavior will completely change. Shift maps do not have this property, so neither do Turing machines; if the noise flips one or more bits below the threshold of measurement, it takes a number of steps of exponential order for its effects to appear at a macroscopic scale.[5] Therefore, the nonlinear computations are more unstable than classical ones. We can say that there would be a trade-off principle between stability

---

[5] In arithmetic operations, it takes $O(\log n)$ steps for the bit blowing-up effect to appear in higher bits. Thus, the time scale of noise effect to macroscopic scale for noise of order $\sim 2^{-2^n}$ is $Exp(n)$ in the shift map, and is $Poly(n)$ in second order nonlinear mappings.

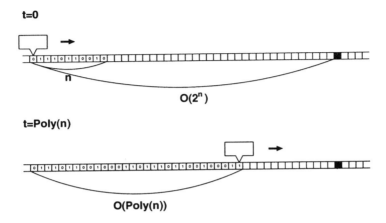

Fig. 5. The effects of noise on Turing machines.

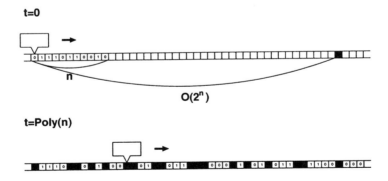

Fig. 6. The effects of noise on nonlinear Turing machines.

of computation and computational power. This property corresponds to the time-space trade-off in computational complexity for our binary expansion coding.

If we secure fixed N bits accuracy by using some error-correcting method, the computational power reduces to class P, which is that of ordinary (polynomial-time bounded) Turing machines. How can we use a digital computer to simulate such analog computation processes? A huge memory capacity does no help, because we would need a memory of 10 billion Gigabytes for the computation of just a 64-bit problem. Consequently, it is impossible to execute exact nonlinear computation.

4 Conclusion and Future Works

We have investigated dynamical systems called switching map systems as a new model of computation. If defined on the space **R**, they can be a model of real number computation, while if the space is restricted to the set of the dyadic rationals with finite binary expansions, they are regarded as a model of classical computation. A connection between generalized shifts [9] and input-output mappings in the BSS model [4] has been shown with our model of computation. Our model is a class of input-output mappings, but at the same time it is a class of GSs when it is constructed using a single 2-dimensional piecewise-linear mapping. Note that the BSS model allows for a linearly growing number of variables, while the switching map systems have a fixed number of variables.

We have suggested that the generic nonlinear mappings can also be basic components of 'natural' effective procedures, and have defined a new notion of computation, nonlinear computation, with the switching map systems. As a result, we have shown that switching map systems with 3-dimensional second-order nonlinear mappings have PSPACE computational power, if we assume the unrestricted usage of the mod-operations or bitwise Boolean operations. It is also expected that the power of switching map systems based on polynomial mappings with rational coefficients is between class RP and PSPACE. These dynamical systems execute more unstable computation than that of classical Turing machines, which implies that there would be a trade-off principle between stability of computation and computational power. This property corresponds to the time-space trade-off in computational complexity for our binary expansion coding. Consequently, it is impossible to execute exact nonlinear computation.

Similar results on the power of analog computation in neural networks are known [16]. For the weights of the network w, if $w \in \mathbf{Z}$ the deterministic neural networks can recognize regular languages, in case $w \in \mathbf{Q}$, the power of the system is in class P, and in case $w \in \mathbf{R}$, it is in class P/poly. It is natural that nonlinear Turing machines have computational power greater than P, because our model contains Siegelmann's neural networks with piecewise-linear mappings as threshold functions. Here we have pointed out that the nonlinearity (multiplication of the valuables) can possibly result in computational power greater than RP in deterministic computations on $\mathbf{Q}$. Though there has been criticism of the zero-measured oracles that cause P/poly computational power for $w \in \mathbf{R}$, our model (without the masking maps) is an effective one that consists of rational functions and contains no oracles.

We list below some of the directions of future study.

(1) It is thought that non-hyperbolicity of dynamical systems can possibly increase their computational power. The notion of computational complexity class may be useful for studies of non-hyperbolic dynamical systems [7]. The pseudo-orbit tracing property can also be analysed with computation theoretic approaches. We can possibly define a new class of computational complexity based on dynamical systems and analyse chaos in a computation theoretic point of view.

(2) The problem of deterministic chaotic maps has a complete solution at the statistical level, where the intrinsic probabilistic aspect of the dynamics is fully understood (see e.g. [1]). Thus, we can investigate probabilistic Turing machines with our deterministic switching map systems based on chaotic maps.

(3) A GS is topologically embedded into a smooth flow in $\mathbf{R}^3$ with suspension on the extended phase space [9]. In the same way, switching map systems can be embedded into networks of ordinary differential equations (see Fig. 7). Our model can possibly give a design principle of dynamical systems, and we can discuss their constructibility. Studies on the construction of hyperchaos [11] and hybrid systems [2] would be related to these schemes.

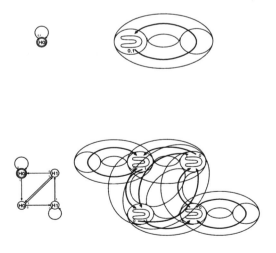

Fig. 7. The schematic view of construction of higher-dimensional Flow.

Acknowledgements

One of the authors (Y.S.) would like to thank Prof. Ichiro Tsuda (Hokkaido Univ.), and Prof. Cristian S. Calude (Univ. of Auckland). This work was partially supported by the research fellowships for young scientists of the Japan society for the promotion of science (Grant No. 09352).

References

1. Antoniou, I., Tasaki, S.: Spectral decompositions of the Renyi map. Journal of Physics **A26** (1993) 73–94
2. Asartin, E., Maler, O., and Pnueli, A.: Reachability analysis of dynamical systems with piecewise-constant derivatives. Theoretical Computer Science **138** (1995) 35–66
3. Bertoni, A., Mauri, G., and Sabadini, N.: Simulations among classes of random access machines and equivalence among numbers succinctly represented. Ann. Discrete Mathematics **25** (1985) 65–90
4. Blum, L., Cucker, F., Shub, M. and Smale, S.: Complexity and Real Computation. Springer-Verlag / New York (1998)
5. Cornfeld, I. P., Fomin, S. V., and Sinai, Y. G.: Ergodic theory. Springer Verlag / New York (1982)
6. Crutchfield, J. P., Young, K.: Computation at the onset of chaos. Complexity, Entropy and the Physics of information (1990) 223–269
7. Cvitanovic, P., Gnaratne, G. H., and Procaccia, I.: Topological and metric properties of Hénon type strange attractors. Phys. Rev. **38A** 3 (1988) 1503–1520
8. Hartmanis, J., Simon, J.: On the power of multiplication in Random Access Machines. Proc. of 15th IEEE Symp. on Switching and Automata Theory (1974) 13–23
9. Moore, C.: Generalized shifts: Unpredictability and Undecidability in Dynamical Systems. Nonlinearity **4** (1991) 199–230
10. Pour-El, M. B., Richards, J. I.: Computability in Analysis and Physics. Springer-Verlag / Berlin (1989).
11. Rössler, O. E.: Chaotic oscillations: An example of hyperchaos in Nonlinear Oscillations in Biology. Lectures in Applied Mathematics **17** (1979) 141–156
12. Sato, Y., Ikegami, T.: Nonlinear Computation with Switching Map Systems. to appear in Journal of Universal Computer Science (2000)
13. Sato, Y.: Doctoral dissertation at University of Tokyo (2000)
14. Sauer, T., Grebogi, C., and Yorke, J. A.: How Long Do Numerical Chaotic Solutions Remain Valid? Phys. Rev. Lett. **79** (1997) 59–63
15. Shönage, A.: On the power of Random Access Machines. Proc. 6th ICALP, Lect. Not. in Comp. Sci. **71** (1979) 520–529
16. Siegelmann, H. T.: Neural Networks and Analog Computation. Birkhauser / Boston (1999).
17. Tsuda, I.: Chaotic itinerancy as a dynamical basis of hermeneutics in brain and mind. World Futures **32** (1991) 167–173

Quantum Information: The New Frontier

Karl Svozil

Institut für Theoretische Physik, University of Technology Vienna, Vienna, Austria,

Abstract. Quantum information and computation is the new hype in physics. It is promising, mindboggling and even already applicable in cryptography, with good prospects ahead. A brief, rather subjective outline is presented.

1 Is Nature Telling Us Something?

Friends in experimental physics tell me that the essence of their observations are clicks in some counter. There is a click or there is none. This is experimental physics in a nutshell. There may be some magic in properly designing experiments and some magic in interpreting those clicks, but that is all there is.

A single click represents some elementary physical proposition. It is also an answer to a question which might not even have been posed consciously. It is tempting to state that "Nature wants to tell us something" with these clicks about some formal structures, symmetries, music or numbers beyond the phenomena. Maybe that is the case, and most physicists tend to believe so; but maybe we are just observing crap, erratic emanations devoid of any meaning [1].

Anyway, we have to deal with those clicks, and one way to deal with them is to interpret them as information. For example, in an experimental input-output scheme, information is received, transformed and communicated by the system. One might think of a physical system as a black box with an input and an output interface [2]. The experimenter inputs some information and the black box responds with some information as output.

If we are dealing with mechanical systems, all the conceivable gadgets inside the black box can be isomorphically translated into a sheet or a tape of paper on which finite computations are performed and vice versa. This was Turing's insight.

But if the black box is essentially quantum driven, then the paper metaphor becomes questionable. The quantum is illusive and highly nonintuitive. In the words of John Archibald Wheeler, one is capturing a "smoky [[quantum]] dragon" [3] inside the black box. Or, in Danny Greenberger's dictum, "quantum mechanics is magic" [4]. In addition, quantized systems such as the quantized electromagnetic field have "more" degrees of freedom as compared to their classical correspondents. Therefore, any isomorphic translation into classical mechanistic devices remains very expensive in terms of paper consumption, at best. To make things worse, under certain reasonable side assumption, it can be proven that a complete "mechanical" paper set of all quantum answers is inconsistent.

Because of these novel non-classical features it is so exiting to pursue the quantum information concept. But even if we look aside and do not want to be bothered with the quantum, the quantum catches up on us: due to the progressing miniaturization of circuits forming logical gates, we shall soon be confronted with quantum phenomena there. In the following, some of the recent developments are reviewed below; and some speculations and prospects are mentioned.

2 Formalization of Quantum Information

In order to be applicable, any formalization of information has to be based on its proper realization in physical terms; i.e., as states of a physical system. In this view, information theory is part of physics; or conversely, physics is part of information theory. And just as the classical bit represents the distinction between two classical physical states, the quantum bit, henceforth often abbreviated by the term 'qubit,' represents the conceivable states of the most elementary quantized system. As we shall see, qubits feature quantum mechanics 'in a nutshell.' Quantum bits are more general structures than classical bits. That is, classical bits can be represented as the limit of qubits, but not vice versa.

Classical information theory is based on the classical bit as fundamental atom. This classical bit, henceforth called 'cbit,' is in one of two classical states t (often interpreted as "true") and f (often interpreted as "false"). It is customary to code the classical logical states by $\#(t) = 1$ and $\#(f) = 0$ ($\#(s)$ stands for the code of s). The states can, for instance, be realized by some condenser which is discharged ($\equiv$ cbit state 0) or charged ($\equiv$ cbit state 1).

In quantum information theory (see Appendix A for a brief outline of quantum mechanics) qubits can be physically represented by a 'coherent superposition' of the two orthonormal[1] states t and f. The qubit states

$$x_\alpha = \alpha t + \beta f \qquad (1)$$

form a continuum, with $|\alpha|^2 + |\beta|^2 = 1$, $\alpha, \beta \in \mathbf{C}$.

What is a coherent superposition? Formally it is just a sum of two elements (representing quantum states) in Hilbert space, which results in an element (a quantum state) again per definition. So, formally we are on the safe side. Informally speaking, a coherent superposition of two different and classically distinct states contains them both. Now classically this sounds like outright nonsense! A classical bit cannot be true and false at the same time. This would be inconsistent, and inconsistencies in physics sound as absurd as in mathematics [5].

---
[1] $(t,t) = (f,f) = 1$ and $(t,f) = 0$.

Yet, quantum mechanics (so far consistently) achieves the implementation of classically inconsistent information into a single quantum bit. Why is that possible? Maybe we get a better feeling for this when we take up Erwin Schrödinger's interpretation of the quantum wave function (in our terms: of the qubit states) as a sort of "catalogue of expectation values" [6]. That is, the qubit appears to be a *representation of the state of our knowledge* about a physical system rather than what may be called "its true state." (Indeed we have to be extremely careful here with what we say. The straightforward classical pretension that quantum systems must have "a true state", albeit hidden to us, yields to outright contradictions!)

Why have I mentioned quantum superpositions here? Because they lie at the heart of quantum parallelism. And quantum parallelism lies at the heart of the quantum speedups which caused so much hype recently.

The coding of qubits is discussed in Appendix C.

The classical and the quantum mechanical concept of information differ from each other in several aspects. Intuitively and classically, a unit of information is context-free. That is, it is independent of what other information is or might be present. A classical bit remains unchanged, no matter by what methods it is inferred. It obeys classical logic. It can be copied. No doubts can be left.

By contrast, to mention just a few nonclassical properties of qubits:

- Qubits are contextual [7]. A quantum bit may appear different, depending on the method by which it is inferred.
- Qubits cannot be copied or "cloned" [8–13]. This due to the fact that the quantum evolution is reversible, i.e., one-to-one.
- Qubits do not necessarily satisfy classical tautologies such as the distributive law [14, 15].
- Qubits obey quantum logic [16] which is different from classical logic.
- Qubits are coherent superpositions of classically distinct, contradicting information.
- Qubits are subject to complementarity.

3 Complementarity and Quantum Cryptography

Before we proceed to quantum computing, which makes heavy use of the possibility to superpose classically distinct information, we shall mention an area of quantum information theory which has already matured to the point where the applications have almost become commercially available: quantum cryptography. At the moment, this might be seen as *the* "killer app" of quantum information theory.

Quantum cryptography (for a detailed review see [17]) is based on the quantum mechanical feature of *complementarity*. A formalization of quantum complementarity has been attempted by Edward Moore [18] who started finite automata theory with this. (Recent results are contained in Ref. [19] and [20, chapter 10]; see also Appendix B.)

Informally speaking, quantum complementarity stands for the principal impossibility to measure two observables at the same time with arbitrary position. If you decide to precisely measure the first observable, you "loose control" over the second one and vice versa. By measuring one observable, the state of the system undergoes a "state reduction" or, expressed differently, "the wave function collapses" and becomes different from the original one. This randomizes a subsequent measurement of the second, complementary observable: in performing the subsequent measurement, one obtains some measurement results (i.e., clicks, you remember?), but they do not tell us much about the original qubit, they are unusable crap. There is no other way of recovering the original state than by completely "undoing" the first measurement in such a way that no trace is left of the previous measurement result; not even a copy of the "classical measurement"[2] result!

So how can this quantum property of complementarity can be put to use in cryptography? The answer is straightforward (if one knows it already): By taking advantage of complementarity, the sender "Alice" of a secret and the receiver "Bob" are able to monitor the secure quantum communication channel and to know when an eavesdropper is present.

This can be done as follows. Assume that Alice sends Bob a qubit and an eavesdropper is present. This eavesdropper is in an inescapable dilemma: neither can the qubit be copied, nor can it be measured. The former case is forbidden in quantum information theory and the letter case would result in a state reduction which modifies Alice's qubit to the point where it is nonsense for Bob. Bob and Alice can realize this by comparing some of their results over a classical (insecure) channel.[3] The exact protocol can for instance be found in [17]. Another scheme [21] operates with entangled pairs of qubits. Here entanglement means that whatever measurement of a particular type is performed on one qubit, if you perform the same measurement on the other qubit of the pair, the result is the same.

Actually, in the real world, the communication over the insecure classical channel has to go back and forth, and they have to constantly compare a certain amount of their measured qubits in order to be able to assure a guaranteed amount of certainty that no eavesdropper is present. That is by no means trivial [22]. But besides this necessary overhead, the quantum channel can be certified to be secure, at least up to some desired amount of certainty and up to the point where someone comes up with a theory which is "better than quantum mechanics" and which circumvents complementarity somehow. Of course, the contemporaries always believe and assure the authorities that there will never be such a theory!

---

[2] I put a quote here because if one is able to "undo a measurement", then this process cannot be classical: per definition, classicality means irreversibility, many-to-oneness.

[3] Actually, if the eavesdropper has total control over the classical channel, this might be used for a reasonable attack strategy.

Quantum cryptographic schemes of the above type have already been demonstrated to work for distances of 1000m (and longer) and net key sizes (after error correction) of 59000 Bits at sustained (105 s) production rates of 850 Bits/s [23]. Yet there is no commercially available solution so far.

4 Quantum Computing

Quantum computers operate with qubits. We have dealt with qubits already. Now what about the operation of quantum computers on qubits? We have to find something similar than Turing's "paper-and-pencil-operations" on paper or tape. The most natural candidate for a formalization is the unitary time evolution of the quantum states. This is all there is (maybe besides measurement [24]), because there is nothing beyond the unitary time evolution. Unitary operators stand for generalized rotations in complex Hilbert spaces. Therefore, a universal quantum computer can just be represented by the most general unitary operator!

That is a straightforward concept: given a finite dimensional Hilbert space of, say, dimension n, then the most general unitary operator $U(n)$ can for instance be parameterized by composition of unitary operations in two (sub)dimensions $U(2)$ [25]. Now we all know how $U(2)$ looks like (cf. Appendix D), so we know how $U(n)$ looks like. Hence we all know how to properly formalize a universal quantum computer!

This looks simple enough, but where is the advantage? Of course one immediate answer is that it is perfectly all right to simulate a quantized system with a quantum computer — we all know that every system is a perfect copy of itself!

But that is not the whole story. What is really challenging here is that we may be able to use quantum parallelism for speedups. And, as mentioned already, at the heart of quantum parallelism is the superposition principle and quantum entanglement. Superposition enables the quantum programmer to "squeeze" 2^N classical bits into N qubits. In processing 1 qubit state $\alpha t + \beta f$, the computer processes 2 classical bit states t and f at once. In processing N qubit states, the computer may be able to processes 2^N classical bit states at once. Many researchers in quantum computing interpret this (in the so-called "Everett interpretation of quantum mechanics") as an indication that 2^N separate computer run in 2^N separate worlds (one computer in each world); thereby running through each one of the computational passes in parallel. That might certainly be a big advantage as compared to a classical routine which might only be able to process the cases consecutively, one after the other.

There are indeed indications that speedups are possible. The most prominent examples are Shor's quantum algorithm for prime factoring [26, 27] and Grover's search algorithm [28] for a single item satisfying a given condition in an unsorted database. A detailed review of the suggested quantum algorithms exceeds the scope of this brief discussion and can for instance be found in Gruska's book [29].

One fundamental feature of the unitary evolution is its bijectivity, its one-to-oneness. This is the reason why copying is not allowed, but this is also the reason why there is no big waste basked where information vanishes into oblivion or nirvana forever. In a quantum computer, one and the same "message" is constantly permutated. It always remains the same but expresses itself through different forms. Information is neither created nor discarded but remains constant at all times.[4]

Is there a price to be paid for parallelism? Let me just mention one important problem here: the problem of the readout of the result. This is no issue in classical computation. But in quantum computation, to use the Everett metaphor, it is by no means trivial how the many parallel entangled universes communicate with each other in such a way that the classical result can be properly communicated. In many cases one has to make sure that, through positive interference, the proper probability amplitudes indicating this result build up. One may even speculate that there is no sufficient buildup of the states if the problem allows for many nonunique solutions [30, 31].

Another problem is the physical overhead which has to be invested in order for the system to remain "quantum" and not turn classical [32]. One may speculate that the necessary equipment to process more qubits grows exponentially with the number of qubits. If that would be the case, then the advantages of quantum parallelism would be essentially nullified.

5 Summary and Outlook

Let me close with a few observations. So far, quantum information theory has applied the quantum features of complementarity, entanglement and quantum parallelism to more or less real-world applications. Certain other quantum features such as contextuality have not been put to use so far.

There are good prospects for quantum computing; if not for other reasons but because our computer parts will finally reach the quantum domain. We may be just at the very beginning, having conceived the quantum analogies of classical tubes (e.g., quantum optical devices). Maybe in the near future someone comes up with a revolutionary design such as a "quantum transistor" which will radically change the technology of information processing.

This is a very exciting and challenging new field of physics and computer sciences.

---

[4] This implicit time symmetry spoils the very notion of "progress" or "achievement," since what is a valuable output is purely determined by the subjective meaning the observer associates with it and is devoid of any syntactic relevance.

Appendix A: All (and Probably More That) You Ever Wanted to Know about Quantum Mechanics

"Quantization" has been introduced by Max Planck around 1900 [33–35]. In a courageous, bold step Planck assumed a *discretization* of the total energy U_N of N linear oscillators ("Resonatoren"),

$$U_N = P\epsilon \in \{0, \epsilon, 2\epsilon, 3\epsilon, 4\epsilon, \ldots\},$$

where $P \in \mathbf{N}_0$ is zero or a positive integer and ϵ stands for the *smallest quantum of energy*. ϵ is a linear function of frequency ω and proportional to Planck's fundamental constant $\hbar \approx 10^{-34}$ Js; i.e.,

$$\epsilon = \hbar\omega.$$

That was a bold step in a time of the predominant continuum models of classical mechanics.

In extension of Planck's discretized resonator energy model, Einstein [36] proposed a quantization of the electromagnetic field. According to the light quantum hypothesis, energy in an electric field mode characterized by the frequency ω can be produced, absorbed and exchanged only in a discrete number n of "lumps" or "quanta" or "photons"

$$E_n = n\hbar\omega, \quad n = 0, 1, 2, 3, \ldots.$$

The following is a very brief introduction to the principles of quantum mechanics for logicians and computer scientists, as well as a reminder for physicists.[5] To avoid a shock from a too early exposure to "exotic" nomenclature prevalent in physics — the Dirac bra-ket notation — the notation of Dunford-Schwartz [50] is adopted.[6]

Quantum mechanics, just as classical mechanics, can be formalized in terms of a linear space structure, in particular by Hilbert spaces [46]. That is, all objects of quantum physics, in particular the ones used by quantum logic, ought to be expressed in terms of objects based on concepts of Hilbert space theory—scalar products, linear summations, subspaces, operators, measures and so on.

---

[5] Introductions to quantum mechanics can be found in Feynman, Leighton & M. Sands [37], Harris [38], Lipkin [39], Ballentine [40], Messiah [41], Davydov [42], Dirac [43], Peres [44], Mackey [45], von Neumann [46], and Bell [47], among many other expositions. The history of quantum mechanics is reviewed by Jammer [48]. Wheeler & Zurek [49] published a helpful resource book.

[6] The bra-ket notation introduced by Dirac is widely used in physics. To translate expressions into the bra-ket notation, the following identifications work for most practical purposes: for the scalar product, "$($ ≡ $($", "$)$ ≡ $)$", "$,$ ≡ $|$". States are written as $|\psi\rangle \equiv \psi$, operators as $\langle i | A | j \rangle \equiv A_{ij}$.

Unless stated differently, only finite-dimensional Hilbert spaces are considered.[7]

A quantum mechanical *Hilbert space* is a linear vector space $\mathcal{H}$ over the field **C** of complex numbers (with vector addition and scalar multiplication), together with a complex function $(\cdot, \cdot)$, the *scalar* or *inner product*, defined on $\mathcal{H} \times \mathcal{H}$ such that (i) $(x,x) = 0$ if and only if $x = 0$; (ii) $(x,x) \geq 0$ for all $x \in \mathcal{H}$; (iii) $(x+y, z) = (x, z) + (y, z)$ for all $x, y, z \in \mathcal{H}$; (iv) $(\alpha x, y) = \alpha(x, y)$ for all $x, y \in \mathcal{H}, \alpha \in \mathbf{C}$; (v) $(x, y) = (y, x)^*$ for all $x, y \in \mathcal{H}$ (α^* stands for the complex conjugate of α); (vi) If $x_n \in \mathcal{H}, n = 1, 2, \ldots$, and if $\lim_{n,m \to \infty}(x_n - x_m, x_n - x_m) = 0$, then there exists an $x \in \mathcal{H}$ with $\lim_{n \to \infty}(x_n - x, x_n - x) = 0$.

We shall make the following identifications between physical and theoretical objects (a *caveat:* this is an incomplete list).

(0) The dimension of the Hilbert space corresponds to the number of degrees of freedom.

(I) A *pure physical state* x is represented either by the one-dimensional linear subspace (closed linear manifold) $^{(x)} = \{y \mid y = \alpha x, \alpha \in \mathbf{C}, x \in \mathcal{H}\}$ spanned by a (normalized) vector x of the Hilbert space $\mathcal{H}$ or by the orthogonal projection operator E_x onto $^{(x)}$. Thus, a vector $x \in \mathcal{H}$ represents a pure physical state.

Every one-dimensional projection E_x onto a one-dimensional linear subspace $^{(x)}$ spanned by $x \in \mathcal{H}$ can be represented by the dyadic product $E_x = |x)(x|$.

If two nonparallel vectors $x, y \in \mathcal{H}$ represent pure physical states, their vector sum $z = x + y \in \mathcal{H}$ is again a vector representing a pure physical state. This state z is called the *superposition* of state x and y.[8]

Elements $b_i, b_j \in \mathcal{H}$ of the set of orthonormal base vectors satisfy $(b_i, b_j) = \delta_{ij}$, where $\delta_{ij} = \begin{cases} 1 \text{ if } i = j \\ 0 \text{ if } i \neq j \end{cases}$ is the Kronecker delta function. Any pure state x can be written as a linear combination of the set of orthonormal base vectors $\{b_1, b_2, \cdots\}$, i.e., $x = \sum_{i=1}^{n} \beta_i b_i$, where n is the dimension of $\mathcal{H}$ and $\beta_i = (b_i, x) \in \mathbf{C}$. In the Dirac bra-ket notation, unity is given by $1 = \sum_{i=1}^{n} |b_i)(b_i|$.

In the nonpure state case, the system is characterized by the density operator ρ, which is nonnegative and of trace class.[9] If the system is in a

---

[7] Infinite dimensional cases and continuous spectra are nontrivial extensions of the finite dimensional Hilbert space treatment. As a heuristic rule, which is not always correct, it might be stated that the sums become integrals, and the Kronecker delta function δ_{ij} becomes the Dirac delta function $\delta(i - j)$, which is a generalized function in the continuous variables i, j. In the Dirac bra-ket notation, unity is given by $1 = \int_{-\infty}^{+\infty} |i)(i| \, di$. For a careful treatment, see, for instance, the books by Reed and Simon [51, 52].

[8] $x + y$ is sometimes referred to as "coherent" superposition to indicate the difference to "incoherent" mixtures of state vectors, in which the absolute squares $|x|^2 + |y|^2$ are summed up.

[9] Nonnegativity means $(\rho x, x) = (x, \rho x) \geq 0$ for all $x \in \mathcal{H}$, and trace class means $\text{trace}(\rho) = 1$.

nonpure state, then the preparation procedure does not specify the decomposition into projection operators (depending on the choice of basis) precisely. ρ can be brought into its spectral form $\rho = \sum_{i=1}^{n} P_i E_i$, where E_i are projection operators and the P_i's are the associated probabilities (nondegenerate case[10]).

(II) *Observables A* are represented by hermitian operators A on the Hilbert space $\mathcal{H}$ such that $(Ax, y) = (x, Ay)$ for all $x, y \in \mathcal{H}$. (Observables and their corresponding operators are identified.) In matrix notation, the adjoint matrix $A^\dagger$ is the complex conjugate of the transposed matrix of A; i.e., $(A^\dagger)_{ij} = (A^*)_{ji}$. Hermiticity means that $(A^\dagger)_{ij} = A_{ij}$.
Any hermitian operator has a spectral representation $A = \sum_{i=1}^{n} \alpha_i E_i$, where the E_i's are orthogonal projection operators onto the orthonormal eigenvectors a_i of A (nondegenerate case).

Note that the projection operators, as well as their corresponding vectors and subspaces, have a double rôle as pure state and elementary proposition (that the system is in that pure state).

Observables are said to be *compatible* or *comeasurable* if they can be defined simultaneously with arbitrary accuracy. Compatible observables are polynomials (Borel measurable functions in the infinite dimensional case) of a single "Ur"-observable.

A criterion for compatibility is the *commutator*. Two observables A, B are compatible if their *commutator* vanishes; i.e., if $[A, B] = AB - BA = 0$. In this case, the hermitian matrices A and B can be simultaneously diagonalized, symbolizing that the observables corresponding to A and B are simultaneously measurable.[11]

It has recently been demonstrated that (by an analog embodiment using particle beams) every hermitian operator in a finite dimensional Hilbert space can be experimentally realized [53].

Actually, one can also measure normal operators N which can be decomposed into the sum of two commuting operators A, B according to $N = A + iB$, with $[A, B] = 0$.

(III) The result of any single measurement of the observable A on an arbitrary state $x \in \mathcal{H}$ can only be one of the real eigenvalues of the corresponding hermitian operator A. (Actually, one can also measure normal operators which can be decomposed into the sum of two commuting If $x = \beta_1 a_1 + \cdots + \beta_i a_i + \cdots + \beta_n a_n$ is in a superposition of eigenstates $\{a_1, \ldots, a_n\}$ of A, the particular outcome of any such single measurement

---

[10] If the same eigenvalue of an operator occurs more than once, it is called *degenerate*.

[11] Let us first diagonalize A; i.e., $A_{ij} = \text{diag}(A_{11}, A_{22}, \ldots, A_{nn})_{ij} = \begin{cases} A_{ii} & \text{if } i = j \\ 0 & \text{if } i \neq j \end{cases}$.
Then, if A commutes with B, the commutator $[A, B]_{ij} = (AB - BA)_{ij} = A_{ik}B_{ki} - B_{ik}A_{kj} = (A_{ii} - A_{jj})B_{ij} = 0$ vanishes. If A is nondegenerate, then $A_{ii} \neq A_{jj}$ and thus $B_{ij} = 0$ for $i \neq j$. In the degenerate case, B can only be block diagonal. That is, each one of the blocks of B corresponds to a set of equal eigenvalues of A such that the corresponding subblockmatrix of A is proportional to the unit matrix. Thus, each block of B can be diagonalized separately without affecting A [44, p. 71].

is indeterministic; i.e., it cannot be predicted with certainty. As a result of the measurement, the system is in the state a_i which corresponds to the associated real-valued eigenvalue α_i which is the measurement outcome; i.e.,

$$x \to a_i.$$

The arrow symbol "$\to$" denotes an irreversible measurement; usually interpreted as a "transition" or "reduction" of the state due to an irreversible interaction of the microphysical quantum system with a classical, macroscopic measurement apparatus. This "reduction" has given rise to speculations concerning the "collapse of the wave function (state)."
As has been argued recently (e.g., by Greenberger and YaSin [54], and by Herzog, Kwiat, Weinfurter and Zeilinger [55]), it is possible to reconstruct the state of the physical system before the measurement; i.e., to "reverse the collapse of the wave function," if the process of measurement is reversible. After this reconstruction, no information about the measurement is left, not even in principle.
How did Schrödinger, the creator of wave mechanics, perceive the quantum physical state, or, more specifically, the ψ-function? In his 1935 paper "Die gegenwärtige Situation in der Quantenmechanik" ("The present situation in quantum mechanics" [6, p. 823]), Schrödinger states,[12]

> *The ψ-function as expectation-catalog:* ... In it [[the ψ-function]] is embodied the momentarily-attained sum of theoretically based future expectation, somewhat as laid down in a *catalog.* ... For each measurement one is required to ascribe to the ψ-function (=the prediction catalog) a characteristic, quite sudden change, which *depends on the measurement result obtained,* and so *cannot be foreseen;* from which alone it is already quite clear that this second kind of change of the ψ-function has nothing whatever in common with its orderly development *between* two measurements. The abrupt change [[of the ψ-function (=the prediction

---

[12] *Die ψ-Funktion als Katalog der Erwartung:* ... Sie [[die ψ-Funktion]] ist jetzt das Instrument zur Voraussage der Wahrscheinlichkeit von Maßzahlen. In ihr ist die jeweils erreichte Summe theoretisch begründeter Zukunftserwartung verkörpert, gleichsam wie in einem *Katalog* niedergelegt. ... Bei jeder Messung ist man genötigt, der ψ-Funktion (=dem Voraussagenkatalog) eine eigenartige, etwas plötzliche Veränderung zuzuschreiben, die von der *gefundenen Maßzahl* abhängt und sich *nicht vorhersehen läßt;* woraus allein schon deutlich ist, daß diese zweite Art von Veränderung der ψ-Funktion mit ihrem regelmäßigen Abrollen *zwischen* zwei Messungen nicht das mindeste zu tun hat. Die abrupte Veränderung durch die Messung ... ist der interessanteste Punkt der ganzen Theorie. Es ist genau *der* Punkt, der den Bruch mit dem naiven Realismus verlangt. Aus *diesem* Grund kann man die ψ-Funktion *nicht* direkt an die Stelle des Modells oder des Realdings setzen. Und zwar nicht etwa weil man einem Realding oder einem Modell nicht abrupte unvorhergesehene Änderungen zumuten dürfte, sondern weil vom realistischen Standpunkt die Beobachtung ein Naturvorgang ist wie jeder andere und nicht per se eine Unterbrechung des regelmäßigen Naturlaufs hervorrufen darf.

catalog)]] by measurement ... is the most interesting point of the entire theory. It is precisely *the* point that demands the break with naive realism. For *this* reason one cannot put the ψ-function directly in place of the model or of the physical thing. And indeed not because one might never dare impute abrupt unforeseen changes to a physical thing or to a model, but because in the realism point of view observation is a natural process like any other and cannot *per se* bring about an interruption of the orderly flow of natural events.

It therefore seems not unreasonable to state that, epistemologically, quantum mechanics appears more as a theory of knowledge of an (intrinsic) observer rather than the Platonic physics "God knows." The wave function, i.e., the state of the physical system in a particular representation (base), is a representation of the observer's knowledge; it is a representation or name or code or index of the information or knowledge the observer has access to.

(IV) The probability $P_x(y)$ to find a system represented by a normalized pure state x in some normalized pure state y is given by

$$P_x(y) = |(x,y)|^2, \quad |x|^2 = |y|^2 = 1.$$

In the nonpure state case, The probability $P(y)$ to find a system characterized by ρ in a pure state associated with a projection operator E_y is

$$P_\rho(y) = \text{trace}(\rho E_y).$$

(V) The *average value* or *expectation value* of an observable A represented by a hermitian operator A in the normalized pure state x is given by

$$\langle A \rangle_x = \sum_{i=1}^{n} \alpha_i |(x, a_i)|^2, \quad |x|^2 = |a_i|^2 = 1.$$

The *average value* or *expectation value* of an observable A represented by a hermitian operator A in the nonpure state ρ is given by

$$\langle A \rangle = \text{trace}(\rho A) = \sum_{i=1}^{n} \alpha_i \text{trace}(\rho E_i).$$

(VI) The dynamical law or equation of motion between subsequent, irreversible, measurements can be written in the form $x(t) = Ux(t_0)$, where $U^\dagger = U^{-1}$ ("$\dagger$ stands for transposition and complex conjugation) is a linear *unitary evolution operator*.[13] Per definition, this evolution is reversible;

---

[13] Any unitary operator $U(n)$ in finite-dimensional Hilbert space can be represented by the product — the serial composition — of unitary operators $U(2)$ acting in twodimensional subspaces [25, 53].

i.e., bijective, one-to-one. So, in quantum mechanics we have to distinguish between unitary, reversible evolution of the system inbetween measurements, and the "collapse of the wave function" at an irreversible measurement.

The *Schrödinger equation* $i\hbar \frac{\partial}{\partial t}\psi(t) = H\psi(t)$ for some state ψ is obtained by identifying U with $U = e^{-iHt/\hbar}$, where H is a hermitian Hamiltonian ("energy") operator, by partially differentiating the equation of motion with respect to the time variable t; i.e., $\frac{\partial}{\partial t}\psi(t) = -\frac{iH}{\hbar}e^{-iHt/\hbar}\psi(t_0) = -\frac{iH}{\hbar}\psi(t)$. In terms of the set of orthonormal base vectors $\{b_1, b_2, \ldots\}$, the Schrödinger equation can be written as $i\hbar\frac{\partial}{\partial t}(b_i, \psi(t)) = \sum_j H_{ij}(b_j, \psi(t))$. For stationary states $\psi_n(t) = e^{-(i/\hbar)E_n t}\psi_n$, the Schrödinger equation can be brought into its time-independent form $H\psi_n = E_m\psi_m$ (nondegenerate case). Here, $i\hbar\frac{\partial}{\partial t}\psi_m(t) = E_m\psi_m(t)$ has been used; E_m and ψ_m stand for the m'th eigenvalue and eigenstate of H, respectively.

Usually, a physical problem is defined by the Hamiltonian H and the Hilbert space in question. The problem of finding the physically relevant states reduces to finding a complete set of eigenvalues and eigenstates of H.

Appendix B: Complementarity and Automaton Logic

A systematic, formal investigation of the black box system or any finite input/output system can be given by finite automata. Indeed, the study of finite automata was motivated from the very beginning by their analogy to quantum systems [18]. Finite automata are universal with respect to the class of computable functions. That is, universal networks of automata can compute any effectively (Turing-) computable function. Conversely, any feature emerging from finite automata is reflected by any other universal computational device. In this sense, they are "robust". All rationally conceivable finite games can be modeled by finite automata.

Computational complementarity, as it is sometimes called [56], can be introduced as a game between Alice and Bob. The rules of the game are as follows. Before the actual game, Alice gives Bob all he needs to know about the intrinsic workings of the automaton. For example, Alice tells Bob, *"if the automaton is in state 1 and you input the symbol 2, then the automaton will make a transition into state 2 and output the symbol 0,"* and so on. Then Alice presents Bob a black box which contains a realization of the automaton. Attached to the black box are two interfaces: a keyboard for the input of symbols, and an output display, on which the output symbols appear. Again, no other interfaces are allowed. In particular, Bob is not allowed to "screw the box open."

Suppose now that Alice chooses some initial state of the automaton. She may either throw a dice, or she may make clever choices using some formalized system. In any case, Alice does not tell Bob about her choice. All Bob has at his disposal are the input-output interfaces.

Bob's goal is to find out which state Alice has chosen. Alice's goal is to fool Bob.

Bob may simply guess or rely on his luck by throwing a dice. But Bob can also perform clever input-output experiments and analyze his data in order to find out. Bob wins if he gives the correct answer. Alice wins if Bob's guess is incorrect. (So, Alice has to be really mean and select worst-case scenarios).

Suppose that Bob tries very hard. Is cleverness sufficient? Will Bob always be able to uniquely determine the initial automaton state?

The answer to that question is "no." The reason is that there may be situations when Bob's input causes an irreversible transition into a black box state which does not allow any further queries about the initial state.

What has been introduced here as a game between Alice and Bob is what the mathematicians have called the *state identification problem* [18, 57–59]: given a finite deterministic automaton, the task is to locate an unknown initial state. Thereby it is assumed that only *a single* automaton copy is available for inspection. That is, no second, identical, example of the automaton can be used for further examination. Alternatively, one may think of it as choosing at random a single automaton from a collection of automata in an ensemble differing only by their initial state. The task then is to find out which was the initial state of the chosen automaton.

The logico-algebraic structure of the state identification problem has been introduced in [60], and subsequently studied in [60–66, 19]. We shall deal with it next.

Step 1: Computation of the experimental equivalence classes. In the propositional structure of sequential machines, state partitions play an important rôle. Indeed, the set of states is partitioned into equivalence classes with respect to a particular input-output experiment.

Suppose again that the only unknown feature of an automaton is its initial state; all else is known. The automaton is presented in a black box, with input and output interfaces. The task in this *complementary game* is to find (partial) information about the initial state of the automaton [18].

To illustrate this, consider the Mealy automaton M_s discussed above. Input/output experiments can be performed by the input of just one symbol i (in this example, more inputs yield no finer partitions). Suppose again that Bob does not know the automaton's initial state. So, Bob has to choose between the input of symbols 1,2, or 3. If Bob inputs, say, symbol 1, then he obtains a definite answer whether the automaton was in state 1 — corresponding to output 1; or whether the automaton was not in state 1 — corresponding to output 0. The latter proposition "not 1" can be identified with the proposition that the automaton was either in state 2 or in state 3.

Likewise, if Bob inputs symbol 2, he obtains a definite answer whether the automaton was in state 2 — corresponding to output 1; or whether the automaton was not in state 2 — corresponding to output 0. The latter proposition "not 2" can be identified with the proposition that the automaton was either in state 1 or in state 3. Finally, if Bob inputs symbol 3, he obtains a definite answer whether the automaton was in state 3 — corresponding to output 1; or whether the automaton was not in state 3 — corresponding to output 0. The latter proposition "not 3" can be identified with the proposition that the automaton was either in state 1 or in state 2.

Recall that Bob can actually perform only one of these input-output experiments. This experiment will irreversibly destroy the initial automaton state (with the exception of a "hit"; i.e., of output 1). Let us thus describe the three possible types of experiment as follows.

- Bob inputs the symbol 1.
- Bob inputs the symbol 2.
- Bob inputs the symbol 3.

The corresponding observable propositions are:

$p_{\{1\}} \equiv \{1\}$: On input 1, Bob receives the output symbol 1.
$p_{\{2,3\}} \equiv \{2,3\}$: On input 1, Bob receives the output symbol 0.
$p_{\{2\}} \equiv \{2\}$: On input 2, Bob receives the output symbol 1.
$p_{\{1,3\}} \equiv \{1,3\}$: On input 2, Bob receives the output symbol 0.
$p_{\{3\}} \equiv \{3\}$: On input 3, Bob receives the output symbol 1.
$p_{\{1,2\}} \equiv \{1,2\}$: On input 3, Bob receives the output symbol 0.

Note that, in particular, $p_{\{1\}}, p_{\{2\}}, p_{\{3\}}$ are not comeasurable. Note also that, for $\epsilon_{ijk} \neq 0$, $p'_{\{i\}} = p_{\{j,k\}}$ and $p_{\{j,k\}} = p'_{\{i\}}$; or equivalently $\{i\}' = \{j,k\}$ and $\{j,k\} = \{i\}'$.

In that way, we naturally arrive at the notion of a *partitioning* of automaton states according to the information obtained from input/output experiments. Every element of the partition stands for the proposition that the automaton is in (one of) the state(s) contained in that partition. Every partition corresponds to a quasi-classical Boolean block. Let us denote by $v(x)$ the block corresponding to input (sequence) x. Then we obtain

no input:
$$v(\emptyset) = \{\{1,2,3\}\},$$

one input symbol:

input	output 1	output 0
$v(1) =$	$\{\{1\}$	$,\{2,3\}\}$
$v(2) =$	$\{\{2\}$	$,\{1,3\}\}$
$v(3) =$	$\{\{3\}$	$,\{1,2\}\}.$

Conventionally, only the finest partitions are included into the set of state partitions.

Step 2: Pasting of the partitions. Just as in quantum logic, the *automaton propositional calculus* and the associated *partition logic* is the *pasting* of all the blocks of partitions $v(i)$ on the atomic level. That is, elements of two blocks are identified if and only if the corresponding atoms are identical.

The automaton partition logic based on *atomic* pastings differs from previous approaches [60–66, 19]. Atomic pasting guarantees that there is no mixing of elements belonging to two different order levels. Such confusions can give rise to the nontransitivity of the order relation [60] in cases where both $p \to q$ and $q \to r$ are operational but incompatible, i.e., complementary, and hence $p \to r$ is not operational.

For the Mealy automaton M_s discussed above, the pasting renders just the horizontal sum — only the least and greatest elements $0, 1$ of each 2^2 is identified—and one obtains a "Chinese lantern" lattice MO_3. The Hasse diagram of the propositional calculus is drawn in Figure 1.

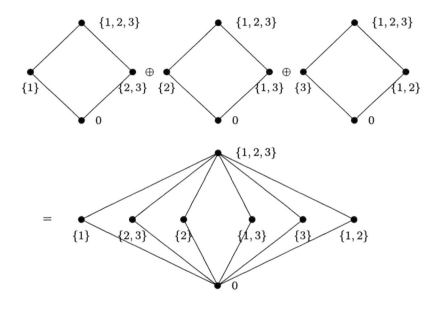

Fig. 1. Hasse diagram of the propositional calculus of the Mealy automaton.

Let us give a formal definition for the procedures sketched so far. Assume a set S and a family of partitions $\mathcal{B}$ of S. Every partition $E \in \mathcal{B}$ can be identified with a Boolean algebra B_E in a natural way by identifying the elements of the partition with the atoms of the Boolean algebra. The pasting of the Boolean algebras $B_E, E \in \mathcal{B}$ on the atomic level is called a partition logic, denoted by $(S, \mathcal{B})$.

The logical structure of the complementarity game (initial-state identification problem) can be defined as follows. Let us call a proposition concerning the initial state of the machine *experimentally decidable* if there is an experiment E which determines the truth value of that proposition. This can be done by performing E, i.e., by the input of a sequence of input symbols $i_1, i_2, i_3, \ldots, i_n$ associated with E, and by observing the output sequence

$$\lambda_E(s) = \lambda(s, i_1), \lambda(\delta(s, i_1), i_2), \ldots, \lambda(\underbrace{\delta(\cdots \delta(s, i_1) \cdots, i_{n-1})}_{n-1 \text{ times}}, i_n).$$

The most general form of a prediction concerning the initial state s of the machine is that the initial state s is contained in a subset P of the state set S. Therefore, we may identify propositions concerning the initial state with subsets of S. A subset P of S is then identified with the proposition that the initial state is contained in P.

Let E be an experiment (a preset or adaptive one), and let $\lambda_E(s)$ denote the obtained output of an initial state s. λ_E defines a mapping of S to the set of output sequences O^*. We define an equivalence relation on the state set S by

$$s \stackrel{E}{\equiv} t \text{ if and only if } \lambda_E(s) = \lambda_E(t)$$

for any $s, t \in S$. We denote the partition of S corresponding to $\stackrel{E}{\equiv}$ by $S/\stackrel{E}{\equiv}$. Obviously, the propositions decidable by the experiment E are the elements of the Boolean algebra generated by $S/\stackrel{E}{\equiv}$, denoted by B_E.

There is also another way to construct the experimentally decidable propositions of an experiment E. Let $\lambda_E(P) = \bigcup_{s \in P} \lambda_E(s)$ be the direct image of P under λ_E for any $P \subseteq S$. We denote the direct image of S by O_E; i.e., $O_E = \lambda_E(S)$.

It follows that the most general form of a prediction concerning the outcome W of the experiment E is that W lies in a subset of O_E. Therefore, the experimentally decidable propositions consist of all inverse images $\lambda_E^{-1}(Q)$ of subsets Q of O_E, a procedure which can be constructively formulated (e.g., as an effectively computable algorithm), and which also leads to the Boolean algebra B_E.

Let $\mathcal{B}$ be the set of all Boolean algebras B_E. We call the partition logic $R = (S, \mathcal{B})$ an *automaton propositional calculus*.

Appendix C: Quantum Coding

In the usual Hilbert space formalization, qubits can then be written as

$$\#(x_\alpha) = e^{i\varphi}(\sin \omega, e^{i\delta} \cos \omega) \in \mathbf{C}^2, \qquad (2)$$

with $\alpha = \alpha(\omega, \varphi, \delta), \omega, \varphi, \delta \in \mathbf{R}$ Qubits can be identified with cbits as follows

$$\#(x_{\alpha(\pi/2, \varphi, \delta)}) = (a, 0) \equiv 1 \text{ and } \#(x_{\alpha(0, \varphi, \delta)}) = (0, b) \equiv 0 \quad , \quad |a|, |b| = 1 \quad , \tag{3}$$

where the complex numbers a and b are of modulus one. The quantum mechanical states associated with the classical states 0 and 1 are mutually orthogonal.

Notice that, provided that $\alpha, \beta \neq 0$, a qubit is not in a pure classical state. Therefore, any practical determination of the qubit x_α amounts to a measurement of the state amplitude of t or f. According to the quantum postulates, any such *single* measurement will be indeterministic (provided again that $\alpha, \beta \neq 0$). That is, the outcome of a single measurement occurs unpredictably. The probabilities that the qubit x_α is measured in states t and f are $P_t(x_\alpha) = |(x_\alpha, t)|^2$ and $P_f(x_\alpha) = |(x_\alpha, f)|^2 = 1 - P_t(\alpha, \beta)$, respectively.

Appendix D: Universal Manipulation of a Single Qubit: The $U(2)$-Gate

It is well known that any n-dimensional unitary matrix U can be composed from elementary unitary transformations in two-dimensional subspaces of $\mathbf{C}^n$. This is usually shown in the context of parameterization of the n-dimensional unitary groups (cf. [25, chapter 2] and [53, 67]). Thereby, a transformation in n-dimensional spaces is decomposed into transformations in 2-dimensional subspaces. This amounts to a successive array of $U(2)$ elements, which in their entirety forms an arbitrary time evolution $U(n)$ in n-dimensional Hilbert space.

Hence, all quantum processes and computation tasks which can possibly be executed must be representable by unitary transformations. Indeed, unitary transformations of qubits are a necessary and sufficient condition for quantum computing. *The group of unitary transformations in arbitrary- but finite-dimensional Hilbert space is a model of universal quantum computer.*

It remains to be shown that the universal $U(2)$-gate is physically operationalizable. This can be done in the framework of Mach-Zehnder interferometry. Note that the number of elementary $U(2)$-transformations is polynomially bounded and does not exceed $\binom{n}{2} = n(n-1)/2 = O(n^2)$.

In what follows, a lossless *Mach-Zehnder* interferometer drawn in Fig. 2 is discussed. The computation proceeds by successive substitution (transition) of states; i.e.,

$$S_1 : a \to (b + ic)/\sqrt{2} \ , \tag{4}$$
$$P : b \to be^{i\varphi} \ , \tag{5}$$
$$S_2 : b \to (e + id)/\sqrt{2} \ , \tag{6}$$
$$S_2 : c \to (d + ie)/\sqrt{2} \ . \tag{7}$$

The resulting transition is

$$a \to \psi = i\left(\frac{e^{i\varphi}+1}{2}\right) d + \left(\frac{e^{i\varphi}-1}{2}\right) e \ . \tag{8}$$

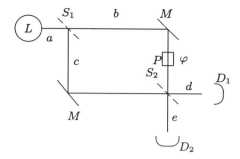

Fig. 2. Mach-Zehnder interferometer. A single quantum (photon, neutron, electron *etc*) is emitted in L and meets a lossless beam splitter (half-silvered mirror) S_1, after which its wave function is in a coherent superposition of b and c. In beam path b a phase shifter shifts the phase of state b by φ. The two beams are then recombined at a second lossless beam splitter (half-silvered mirror) S_2. The quant is detected at either D_1 or D_2, corresponding to the states d and e, respectively.

Assume that $\varphi = 0$, i.e., there is no phase shift at all. Then, equation (8) reduces to $a \to id$, and the emitted quant is detected only by D_1. Assume that $\varphi = \pi$. Then, equation (8) reduces to $a \to -e$, and the emitted quant is detected only by D_2. If one varies the phase shift φ, one obtains the following detection probabilities:

$$P_{D_1}(\varphi) = |(d,\psi)|^2 = \cos^2(\frac{\varphi}{2}) \quad , \quad P_{D_2}(\varphi) = |(e,\psi)|^2 = \sin^2(\frac{\varphi}{2}) \quad . \tag{9}$$

For some "mindboggling" features of Mach-Zehnder interferometry, see [68].

The elementary quantum interference device $\mathbf{T}^{bs}_{21}$ depicted in Fig. (3.a) is just a beam splitter followed by a phase shifter in one of the output ports.

Alternatively, the action of a lossless beam splitter may be described by the matrix $\begin{pmatrix} T(\omega) & iR(\omega) \\ iR(\omega) & T(\omega) \end{pmatrix} = \begin{pmatrix} \cos\omega & i\sin\omega \\ i\sin\omega & \cos\omega \end{pmatrix}$. A phase shifter in a two-dimensional Hilbert space is represented by either $\begin{pmatrix} e^{i\varphi} & 0 \\ 0 & 1 \end{pmatrix}$ or $\begin{pmatrix} 1 & 0 \\ 0 & e^{i\varphi} \end{pmatrix}$. The action of the entire device consisting of such elements is calculated by multiplying the matrices in reverse order in which the quanta pass these elements [69, 70].

$$P_1 : \mathbf{0} \to \mathbf{0} e^{i\alpha+\beta} \quad , \tag{10}$$
$$P_2 : \mathbf{1} \to \mathbf{1} e^{i\beta} \quad , \tag{11}$$
$$S : \mathbf{0} \to T\mathbf{1'} + iR\mathbf{0'} \quad , \tag{12}$$
$$S : \mathbf{1} \to T\mathbf{0'} + iR\mathbf{1'} \quad , \tag{13}$$
$$P_3 : \mathbf{0'} \to \mathbf{0'} e^{i\varphi} \quad . \tag{14}$$

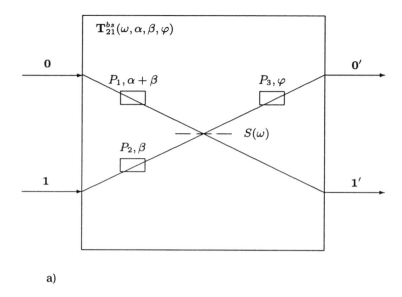

Fig. 3. Elementary quantum interference device. An elementary quantum interference device can be realized by a 4-port interferometer with two input ports $0, 1$ and two output ports $0', 1'$. Any two-dimensional unitary transformation can be realized by the devices. a) shows a realization by a single beam splitter $S(T)$ with variable transmission t and three phase shifters P_1, P_2, P_3; b) shows a realization with 50:50 beam splitters $S_1(\frac{1}{2})$ and $S_2(\frac{1}{2})$ and four phase shifters P_1, P_2, P_3, P_4.

If $\mathbf{0} \equiv \mathbf{0'} \equiv \begin{pmatrix} 1 \\ 0 \end{pmatrix}$ and $\mathbf{1} \equiv \mathbf{1'} \equiv \begin{pmatrix} 0 \\ 1 \end{pmatrix}$ and $R(\omega) = \sin\omega$, $T(\omega) = \cos\omega$, then the corresponding unitary evolution matrix which transforms any coherent superposition of $\mathbf{0}$ and $\mathbf{1}$ into a superposition of $\mathbf{0'}$ and $\mathbf{1'}$ is given by

$$\begin{aligned}\mathbf{T}_{21}^{bs}(\omega,\alpha,\beta,\varphi) &= \left[e^{i\beta}\begin{pmatrix} i\,e^{i(\alpha+\varphi)}\sin\omega & e^{i\alpha}\cos\omega \\ e^{i\varphi}\cos\omega & i\sin\omega \end{pmatrix}\right]^{-1} \\ &= e^{-i\beta}\begin{pmatrix} -i\,e^{-i(\alpha+\varphi)}\sin\omega & e^{-i\varphi}\cos\omega \\ e^{-i\alpha}\cos\omega & -i\sin\omega \end{pmatrix} .\end{aligned} \quad (15)$$

The elementary quantum interference device $\mathbf{T}_{21}^{MZ}$ depicted in Fig. (3.b) is a (rotated) Mach-Zehnder interferometer with *two* input and output ports and three phase shifters. According to the "toolbox" rules, the process can be quantum mechanically described by

$$P_1 : \mathbf{0} \to \mathbf{0}e^{i\alpha+\beta} , \quad (16)$$
$$P_2 : \mathbf{1} \to \mathbf{1}e^{i\beta} , \quad (17)$$
$$S_1 : \mathbf{1} \to (b + i\,c)/\sqrt{2} , \quad (18)$$
$$S_1 : \mathbf{0} \to (c + i\,b)/\sqrt{2} , \quad (19)$$
$$P_3 : c \to c e^{i\omega} , \quad (20)$$
$$S_2 : b \to (\mathbf{1'} + i\,\mathbf{0'})/\sqrt{2} , \quad (21)$$
$$S_2 : c \to (\mathbf{0'} + i\,\mathbf{1'})/\sqrt{2} , \quad (22)$$
$$P_4 : \mathbf{0'} \to \mathbf{0'}e^{i\varphi} . \quad (23)$$

When again $\mathbf{0} \equiv \mathbf{0'} \equiv \begin{pmatrix} 1 \\ 0 \end{pmatrix}$ and $\mathbf{1} \equiv \mathbf{1'} \equiv \begin{pmatrix} 0 \\ 1 \end{pmatrix}$, then the corresponding unitary evolution matrix which transforms any coherent superposition of $\mathbf{0}$ and $\mathbf{1}$ into a superposition of $\mathbf{0'}$ and $\mathbf{1'}$ is given by

$$\mathbf{T}_{21}^{MZ}(\alpha,\beta,\omega,\varphi) = -i\,e^{-i(\beta+\frac{\omega}{2})}\begin{pmatrix} -e^{-i(\alpha+\varphi)}\sin\frac{\omega}{2} & e^{-i\varphi}\cos\frac{\omega}{2} \\ e^{-i\alpha}\cos\frac{\omega}{2} & \sin\frac{\omega}{2} \end{pmatrix} . \quad (24)$$

The correspondence between $\mathbf{T}_{21}^{bs}(T(\omega),\alpha,\beta,\varphi)$ with $\mathbf{T}_{21}^{MZ}(\alpha',\beta',\omega',\varphi')$ in equations (15) (24) can be verified by comparing the elements of these matrices. The resulting four equations can be used to eliminate the four unknown parameters $\omega' = 2\omega$, $\beta' = \beta - \omega$, $\alpha' = \alpha - \pi/2$, $\beta' = \beta - \omega$ and $\varphi' = \varphi - \pi/2$; i.e.,

$$\mathbf{T}_{21}^{bs}(\omega,\alpha,\beta,\varphi) = \mathbf{T}_{21}^{MZ}(\alpha - \frac{\pi}{2}, \beta - \omega, 2\omega, \varphi - \frac{\pi}{2}) . \quad (25)$$

Both elementary quantum interference devices are *universal* in the sense that *every* unitary quantum evolution operator in two-dimensional Hilbert space can be brought into a one-to-one correspondence to $\mathbf{T}_{21}^{bs}$ and $\mathbf{T}_{21}^{MZ}$; with corresponding values of $T, \alpha, \beta, \varphi$ or $\alpha, \omega, \beta, \varphi$. This can be easily seen by a similar calculation as before; i.e., by comparing equations (15) (24) with the "canonical" form of a unitary matrix, which is the product of a $U(1) = e^{-i\beta}$ and of the unimodular unitary matrix $SU(2)$ [25]

$$\mathbf{T}(\omega, \alpha, \varphi) = \begin{pmatrix} e^{i\alpha} \cos\omega & -e^{-i\varphi} \sin\omega \\ e^{i\varphi} \sin\omega & e^{-i\alpha} \cos\omega \end{pmatrix} , \qquad (26)$$

where $-\pi \leq \beta, \omega \leq \pi$, $-\frac{\pi}{2} \leq \alpha, \varphi \leq \frac{\pi}{2}$. Let

$$\mathbf{T}(\omega, \alpha, \beta, \varphi) = e^{-i\beta} \mathbf{T}(\omega, \alpha, \varphi) \quad . \qquad (27)$$

A proper identification of the parameters $\alpha, \beta, \omega, \varphi$ yields

$$\mathbf{T}(\omega, \alpha, \beta, \varphi) = T_{21}^{bs}(\omega - \frac{\pi}{2}, -\alpha - \varphi - \frac{\pi}{2}, \beta + \alpha + \frac{\pi}{2}, \varphi - \alpha + \frac{\pi}{2}) \quad . \qquad (28)$$

Let us examine the realization of a few primitive logical "gates" corresponding to (unitary) unary operations on qubits. The "identity" element **I** is defined by $0 \to 0, 1 \to 1$ and can be realized by

$$\mathbf{I} = T_{21}^{bs}(-\frac{\pi}{2}, -\frac{\pi}{2}, \frac{\pi}{2}, \frac{\pi}{2}) = T_{21}^{MZ}(-\pi, \pi, -\pi, 0) = \begin{pmatrix} 1 & 0 \\ 0 & 1 \end{pmatrix} \quad . \qquad (29)$$

The "not" element is defined by $0 \to 1, 1 \to 0$ and can be realized by

$$\mathbf{not} = T_{21}^{bs}(0, 0, 0, 0) = T_{21}^{MZ}(-\frac{\pi}{2}, 0, 0, -\frac{\pi}{2}) = \begin{pmatrix} 0 & 1 \\ 1 & 0 \end{pmatrix} \quad . \qquad (30)$$

The next element, "$\sqrt{\mathrm{not}}$" is a truly quantum mechanical; i.e., nonclassical, one, since it converts a classical bit into a coherent superposition of 0 and 1. $\sqrt{\mathrm{not}}$ is defined by $0 \to 0 + 1, 1 \to -0 + 1$ and can be realized by

$$\sqrt{\mathrm{not}} = T_{21}^{bs}(-\frac{\pi}{4}, -\frac{\pi}{2}, \frac{\pi}{2}, \frac{\pi}{2}) = T_{21}^{MZ}(-\pi, \frac{3\pi}{4}, -\frac{\pi}{2}, 0) = \frac{1}{\sqrt{2}} \begin{pmatrix} 1 & -1 \\ 1 & 1 \end{pmatrix} \quad . \qquad (31)$$

Note that $\sqrt{\mathrm{not}} \cdot \sqrt{\mathrm{not}} = \mathrm{not} \cdot \mathrm{diag}(1, -1) = \mathrm{not} \pmod 1$. The relative phases in the output ports showing up in $\mathrm{diag}(1, -1)$ can be avoided by defining

$$\sqrt{\mathrm{not}}' = T_{21}^{bs}(-\frac{\pi}{4}, 0, \frac{\pi}{4}, 0) = T_{21}^{MZ}(-\frac{\pi}{2}, \frac{\pi}{2}, -\frac{\pi}{2}, -\frac{\pi}{2}) = \frac{1}{2} \begin{pmatrix} 1+i & 1-i \\ 1-i & 1+i \end{pmatrix} \quad . \qquad (32)$$

With this definition, $\sqrt{\mathrm{not}}' \sqrt{\mathrm{not}}' = \mathrm{not}$.

It is very important that the elementary quantum interference device realizes an arbitrary quantum time evolution of a two-dimensional system. The performance of the quantum interference device is determined by four parameters, corresponding to the phases $\alpha, \beta, \varphi, \omega$.

References

1. Cristian Calude and F. Walter Meyerstein. Is the universe lawful? *Chaos, Solitons & Fractals*, 10(6):1075–1084, 1999.
2. Karl Svozil. Quantum interfaces. e-print arXiv:quant-ph/0001064 available http://arxiv.org/abs/quant-ph/0001064, 2000.
3. John A. Wheeler. Law without law. In John A. Wheeler and W. H. Zurek, editors, *Quantum Theory and Measurement*, pages 182–213. Princeton University Press, Princeton, 1983. [49].
4. Daniel M. Greenberger. Private communication.
5. David Hilbert. Über das Unendliche. *Mathematische Annalen*, 95:161–190, 1926.
6. Erwin Schrödinger. Die gegenwärtige Situation in der Quantenmechanik. *Naturwissenschaften*, 23:807–812, 823–828, 844–849, 1935. English translation in [71] and [49, pp. 152-167].
7. Simon Kochen and Ernst P. Specker. The problem of hidden variables in quantum mechanics. *Journal of Mathematics and Mechanics*, 17(1):59–87, 1967. Reprinted in [72, pp. 235–263].
8. W. K. Wooters and W. H. Zurek. A single quantum cannot be cloned. *Nature*, 299:802–803, 1982.
9. D. Dieks. Communication by EPR devices. *Physics Letters*, 92A(6):271–272, 1982.
10. L. Mandel. Is a photon amplifier always polarization dependent? *Nature*, 304:188, 1983.
11. Peter W. Milonni and M. L. Hardies. Photons cannot always be replicated. *Physics Letters*, 92A(7):321–322, 1982.
12. R. J. Glauber. Amplifiers, attenuators and the quantum theory of measurement. In E. R. Pikes and S. Sarkar, editors, *Frontiers in Quantum Optics*. Adam Hilger, Bristol, 1986.
13. C. M. Caves. Quantum limits on noise in linear amplifiers. *Physical Review*, D26:1817–1839, 1982.
14. Simon Kochen and Ernst P. Specker. Logical structures arising in quantum theory. In *Symposium on the Theory of Models, Proceedings of the 1963 International Symposium at Berkeley*, pages 177–189, Amsterdam, 1965. North Holland. Reprinted in [72, pp. 209–221].
15. Simon Kochen and Ernst P. Specker. The calculus of partial propositional functions. In *Proceedings of the 1964 International Congress for Logic, Methodology and Philosophy of Science, Jerusalem*, pages 45–57, Amsterdam, 1965. North Holland. Reprinted in [72, pp. 222–234].
16. Garrett Birkhoff and John von Neumann. The logic of quantum mechanics. *Annals of Mathematics*, 37(4):823–843, 1936.
17. Charles H. Bennett, F. Bessette, G. Brassard, L. Salvail, and J. Smolin. Experimental quantum cryptography. *Journal of Cryptology*, 5:3–28, 1992.
18. Edward F. Moore. Gedanken-experiments on sequential machines. In C. E. Shannon and J. McCarthy, editors, *Automata Studies*. Princeton University Press, Princeton, 1956.
19. Cristian Calude, Elena Calude, Karl Svozil, and Sheng Yu. Physical versus computational complementarity I. *International Journal of Theoretical Physics*, 36(7):1495–1523, 1997.
20. K. Svozil. *Quantum Logic*. Springer, Singapore, 1998.
21. Artur Ekert. Quantum cryptography based on Bell's theorem. *Physical Review Letters*, 67:661–663, 1991.

22. M. Hamrick G. Gilbert. Practical quantum cryptography: A comprehensive analysis (part one). MITRE report MTR 00W0000052 and e-print arXiv:quant-ph/0009027 available at http://arxiv.org/abs/quant-ph/0009027, 2000.
23. T. Jenewein, G. Weihs, C. Simon, H. Weinfurter, and A. Zeilinger. Poster, 1998.
24. Karl Svozil. The information interpretation of quantum mechanics. e-print arXiv:quant-ph/0006033 available http://arxiv.org/abs/quant-ph/0006033, 2000.
25. F. D. Murnaghan. *The Unitary and Rotation Groups*. Spartan Books, Washington, 1962.
26. Peter W. Shor. Algorithms for quantum computation: discrete logarithms and factoring. In *Proceedings of the 35th Annual Symposium of on Foundations of Computer Science, Santa Fe, NM, Nov. 20-22, 1994*. IEEE Computer Society Press, November 1994. arXiv:quant-ph/9508027.
27. Artur Ekert and Richard Jozsa. Quantum computation and Shor's factoring algorithm. *Reviews of Modern Physics*, 68(3):733–753, 1996.
28. L. K. Grover. A fast quantum mechanical algorithm for database search. In *Proceedings of the Twenty-Eighth Annual ACM Symposium on the Theory of Computing*, pages 212–219. 1996.
29. Josef Gruska. *uantum Computing*. McGraw-Hill, London, 1999.
30. Georg Gottlob. Private communication.
31. Cristian S. Calude, Michael J. Dinneen, and Karl Svozil. Reflections on quantum computing. *Complexity*, 2000. in print; e-print http://www.cs.auckland.ac.nz/CDMTCS//researchreports/130cris.pdf.
32. Günther Mahler. private communication.
33. Max Planck. Ueber eine Verbesserung der Wien'schen Spectralgleichung. *Verhandlungen der deutschen physikalischen Gesellschaft*, 2:202, 1900. See also [34].
34. Max Planck. Ueber das Gesetz der Energieverteilung im Normalspectrum. *Annalen der Physik*, 4:553–566, 1901.
35. Max Planck. Zur Theorie des Gesetzes der Energieverteilung im Normalspectrum. *Verhandlungen der deutschen physikalischen Gesellschaft*, 2:237, 1900. See also [34].
36. Albert Einstein. Über einen die Erzeugung und Verwandlung des Lichtes betreffenden heuristischen Gesichtspunkt. *Annalen der Physik*, 17:132–148, 1905.
37. Richard P. Feynman, Robert B. Leighton, and Matthew Sands. *The Feynman Lectures on Physics. Quantum Mechanics*, volume III. Addison-Wesley, Reading, MA, 1965.
38. E. G. Harris. *A Pedestrian Approach to Quantum Field Theory*. Wiley-Interscience, New York, 1971.
39. H. J. Lipkin. *Quantum Mechanics, New Approaches to Selected Topics*. North-Holland, Amsterdam, 1973.
40. L. E. Ballentine. *Quantum Mechanics*. Prentice Hall, Englewood Cliffs, NJ, 1989.
41. A. Messiah. *Quantum Mechanics*, volume I. North-Holland, Amsterdam, 1961.
42. A. S. Davydov. *Quantum Mechanics*. Addison-Wesley, Reading, MA, 1965.
43. P. A. M. Dirac. *The Principles of Quantum Mechanics*. Oxford University Press, Oxford, 1947.
44. Asher Peres. *Quantum Theory: Concepts and Methods*. Kluwer Academic Publishers, Dordrecht, 1993.
45. George W. Mackey. *The Mathematical Foundations of Quantum Mechanics*. W. A. Benjamin, Reading, MA, 1963.

46. John von Neumann. *Mathematische Grundlagen der Quantenmechanik.* Springer, Berlin, 1932. English translation: *Mathematical Foundations of Quantum Mechanics*, Princeton University Press, Princeton, 1955.
47. John S. Bell. *Speakable and Unspeakable in Quantum Mechanics.* Cambridge University Press, Cambridge, 1987.
48. Max Jammer. *The Philosophy of Quantum Mechanics.* John Wiley & Sons, New York, 1974.
49. John Archibald Wheeler and Wojciech Hubert Zurek. *Quantum Theory and Measurement.* Princeton University Press, Princeton, 1983.
50. N. Dunford and J. T. Schwartz. *Linear Operators I.* Interscience Publishers, New York, 1958.
51. Michael Reed and Barry Simon. *Methods of Mathematical Physics I: Functional Analysis.* Academic Press, New York, 1972.
52. Michael Reed and Barry Simon. *Methods of Mathematical Physics II: Fourier Analysis, Self-Adjointness.* Academic Press, New York, 1975.
53. M. Reck, Anton Zeilinger, H. J. Bernstein, and P. Bertani. Experimental realization of any discrete unitary operator. *Physical Review Letters*, 73:58–61, 1994. See also [25].
54. Daniel B. Greenberger and A. YaSin. "Haunted" measurements in quantum theory. *Foundation of Physics*, 19(6):679–704, 1989.
55. Thomas J. Herzog, Paul G. Kwiat, Harald Weinfurter, and Anton Zeilinger. Complementarity and the quantum eraser. *Physical Review Letters*, 75(17):3034–3037, 1995.
56. David Finkelstein and Shlomit R. Finkelstein. Computational complementarity. *International Journal of Theoretical Physics*, 22(8):753–779, 1983.
57. Gregory J. Chaitin. An improvement on a theorem by E. F. Moore. *IEEE Transactions on Electronic Computers*, EC-14:466–467, 1965.
58. J. H. Conway. *Regular Algebra and Finite Machines.* Chapman and Hall Ltd., London, 1971.
59. W. Brauer. *Automatentheorie.* Teubner, Stuttgart, 1984.
60. Karl Svozil. *Randomness & Undecidability in Physics.* World Scientific, Singapore, 1993.
61. Martin Schaller and Karl Svozil. Partition logics of automata. *Il Nuovo Cimento*, 109B:167–176, 1994.
62. Martin Schaller and Karl Svozil. Automaton partition logic versus quantum logic. *International Journal of Theoretical Physics*, 34(8):1741–1750, August 1995.
63. Martin Schaller and Karl Svozil. Automaton logic. *International Journal of Theoretical Physics*, 35(5):911–940, May 1996.
64. Anatolij Dvurečenskij, Sylvia Pulmannová, and Karl Svozil. Partition logics, orthoalgebras and automata. *Helvetica Physica Acta*, 68:407–428, 1995.
65. Karl Svozil and Roman R. Zapatrin. Empirical logic of finite automata: microstatements versus macrostatements. *International Journal of Theoretical Physics*, 35(7):1541–1548, 1996.
66. Karl Svozil and Josef Tkadlec. Greechie diagrams, nonexistence of measures in quantum logics and Kochen–Specker type constructions. *Journal of Mathematical Physics*, 37(11):5380–5401, November 1996.
67. M. Reck and Anton Zeilinger. Quantum phase tracing of correlated photons in optical multiports. In F. De Martini, G. Denardo, and Anton Zeilinger, editors, *Quantum Interferometry*, Singapore, 1994. World Scientific.
68. Charles H. Bennett. Night thoughts, dark sight. *Nature*, 371:479–480, 1994.

69. B. Yurke, S. L. McCall, and J. R. Klauder. SU(2) and SU(1,1) interferometers. *Physical Review*, A33:4033–4054, 1986.
70. R. A. Campos, B. E. A. Saleh, and M. C. Teich. Fourth-order interference of joint single-photon wave packets in lossless optical systems. *Physical Review*, A42:4127, 1990.
71. J. D. Trimmer. The present situation in quantum mechanics: a translation of Schrödinger's "cat paradox". *Proc. Am. Phil. Soc.*, 124:323–338, 1980. Reprinted in [49, pp. 152-167].
72. Ernst Specker. *Selecta*. Birkhäuser Verlag, Basel, 1990.

Quantum Computation Relative to Oracles

Christino Tamon[1] and Tomoyuki Yamakami[2]

[1] Department of Mathematics and Computer Science, Clarkson University, Potsdam, New York, U.S.A.
[2] School of Information Technology and Engineering, University of Ottawa, Ottawa, Ontario, Canada

Abstract. The study of the power and limitations of quantum computation remains a major challenge in complexity theory. Key questions revolve around the quantum complexity classes **EQP**, **BQP**, **NQP** and their derivatives. This paper presents new relativized worlds in which (i) co-**RP** $\not\subseteq$ **NQE**, (ii) **P** = **BQP** and **UP** = **EXP**, (iii) **P** = **EQP** and **RP** = **EXP**, and (iv) **EQP** $\not\subseteq \Sigma_2^P \cup \Pi_2^P$. We also show a partial answer to the question of whether Almost-**BQP** = **BQP**.

1 Introduction

A major question in quantum complexity theory is the power and limitations of a quantum computer for solving intractable problems. Since its inception by Benioff [6] and Feynman [14], the idea of a quantum mechanical computer was investigated in various early works by Deutsch, Jozsa, and others [13, 7, 8, 30, 1], culminating in the seminal discovery by Shor [31] on an efficient quantum factoring algorithm. Much of the research has also focused on understanding the power of quantum complexity classes, such as **EQP** and **BQP**, in comparison with its classical counterparts. In fact, the early works by Berthiaume and Brassard [7], Simon [30], and Bennett et al. [10] have been done using *black-box* models with query mechanism. These early works gave fundamental relativization results for both **EQP** and **BQP**.

The notion of *relativization* was first introduced to computational complexity theory by Baker, Gill, and Solovay [2] to discuss a variety of relationships among central complexity classes, such as **P** and **NP**. Although its implication to the unrelativized (i.e., real) world is debatable, it surely supplies, in light of structural differences of complexity classes, a useful insight into the behavior of query computation.

In this paper, we will focus on oracle quantum computation with query mechanism. We use recent techniques from both structural complexity theory and quantum complexity theory.

First, we consider a simple application of the *polynomial method* [27, 3] to show that nondeterministic quantum linear-exponential time does not contain the complement of the random complexity class **RP**, i.e., co-**RP** $\not\subseteq$ **NQE**, relative to a certain oracle. Here, **NQE** is the collection of sets computable in nondeterministic linear-exponential time by well-formed quantum Turing machines. As an immediate consequence, we have a single oracle relative to which $\mathbf{P} \neq \mathbf{RP} \neq \text{co-}\mathbf{RP} \neq \mathbf{BPP}$, $\mathbf{P} \neq \mathbf{NP} \neq \text{co-}\mathbf{NP} \neq \Delta_2^P$, and $\mathbf{EQP} \neq \mathbf{RQP} \neq \text{co-}\mathbf{RQP} \neq \mathbf{BQP}$.

Next, we prove that $\mathbf{BQP}$ is exponentially easier than $\mathbf{UP}$, i.e., $\mathbf{P} = \mathbf{BQP}$ and $\mathbf{UP} = \mathbf{EXP}$, relative to a recursive oracle. To obtain this result, we adapt a recent technique developed by Beigel, Buhrman, and Fortnow [5], who proved the reverse direction of ours; namely, $\mathbf{P} = \mathbf{UP}$ and $\mathbf{BQP} = \mathbf{EXP}$ relative to a certain oracle. Our result also extends the result of Fortnow and Rogers [16] on $\mathbf{P}^A = \mathbf{BQP}^A \neq \mathbf{UP}^A \cap \text{co-}\mathbf{UP}^A$ for a certain *non-recursive* oracle A.

Moreover, by combining the above two oracle construction methods, we can show in §5 that $\mathbf{P} = \mathbf{EQP}$ and $\mathbf{RP} = \mathbf{EXP}$ relative to a certain oracle.

As our next results, we investigate the power of $\mathbf{EQP}$ and $\mathbf{BQP}$ in comparison with the first two levels of the polynomial-time hierarchy. We present an oracle relative to which $\mathbf{EQP} \not\subseteq \Sigma_2^P \cup \Pi_2^P$. Our result greatly strengthens an early result of Berthiaume and Brassard [7], who constructed an oracle relative to which $\mathbf{EQP} \not\subseteq \mathbf{NP} \cup \text{co-}\mathbf{NP}$. We obtain our result by using an argument based on circuit complexity and a recent improvement [11] on Simon's test language [30]. In this paper, we introduce two variants of Simon's test language, which we call L_1^A and L_2^A, and show that, while both of these languages belong to $\mathbf{EQP}^A$ for all *good* oracles A, $L_1^A \not\in \text{co-}\mathbf{NP}^A$ and $L_2^A \not\in \Pi_2^P(A)$ for a single *good* oracle A. We contrast our result to an earlier one by Fortnow and Rogers [16] that $\mathbf{PH}^A$ is infinite but $\mathbf{P}^A = \mathbf{BQP}^A$ for a certain oracle A. Our result also implies that $\mathbf{BPP} \subsetneq \mathbf{BQP}$ relative to a certain oracle.

Finally, we study the role of randomization in quantum complexity via random oracles. The notion of random oracles was introduced in a classic work of Bennett and Gill [9]. For a relativizable complexity class $\mathcal{C}$, Almost-$\mathcal{C}$ denotes the collection of sets A such that A is in $\mathcal{C}^X$ relative to a random oracle X with probability 1. Bennett and Gill [9] and Kurtz [25] proved that Almost-$\mathbf{P}$ = Almost-$\mathbf{BPP}$ = $\mathbf{BPP}$. A decade later, using the idea of pseudorandom generators, Nisan and Wigderson [28] showed that Almost-$\mathbf{NP}$ = $\mathbf{AM}$ and Almost-$\mathbf{PH}$ = $\mathbf{PH}$. In this paper, we explore Nisan and Wigderson's technique to facilitate the quantum setting, and raise the conjecture that Almost-$\mathbf{BQP}$ = $\mathbf{BQP}$. We provide a partial result towards resolving this conjecture.

2 Preliminaries

Let $\mathbb{N}, \mathbb{Z}, \mathbb{Q}, \mathbb{R}$, and $\mathbb{C}$ denote the set of naturals, integers, rationals, reals, and complex numbers, respectively. Let $\mathbb{R}^+ = \{r \in \mathbb{R} \mid r \geq 0\}$. We use $\log n$ (or $\log(n)$) to denote $\log_2 n$ and also follow the convention $\log 0 = 0$. For simplicity, we write $\mathrm{ilog}(n)$ to denote $\lceil \log n \rceil$.

A real number r is called *polynomial-time approximable*[1] if there exists a deterministic polynomial-time Turing machine M, which outputs dyadic in-

---

[1] Ko and Friedman [22] first introduced this notion under the name "polynomial-time computable." To avoid the reader's confusion, we prefer to use the term "polynomial-time approximable" instead.

tegers, such that $|M(0^n) - r| \leq 2^{-n}$ for every number $n \in \mathbb{N}$. Let $\tilde{\mathbb{C}}$ be the set of complex numbers whose real and imaginary parts are both polynomial-time approximable.

The binary alphabet $\Sigma = \{0, 1\}$ is used throughout the paper. Denote λ as the empty string. Let 2^{Σ^*} be the power set of our alphabet Σ^*. For $n \in \mathbb{N}$ and $A \subseteq \Sigma^*$, let $A^{\leq n}, A^{=n}$ denote the set of all strings in A of length at most n and exactly n, respectively. For $i > 0$ and $s \in \{0, 1\}^*$, let $s_{(i)}$ denote the ith bit of s. For simplicity, we write s_i^n to denote the $(i + 1)$th string in $\{0, 1\}^{i\log(n)}$. In particular, $s_0^n = 0^{i\log(n)}$. For a string $s \in \{0, 1\}^*$, we denote $\#_1(s)$ to be the number of 1s in s. We identify a set S with its characteristic function, that is, $S(x) = 1$ if $x \in S$ and $S(x) = 0$ otherwise.

Classical Computation. We assume the reader's familiarity with Boolean circuits and classical Turing machines. We use standard complexity classes, such as **P, UP, NP, RP, BPP, $\oplus$P**, and the polynomial-time hierarchy $\{\Delta_k^P, \Sigma_k^P, \Pi_k^P \mid k > 0\}$ with $\mathbf{PH} = \bigcup_{k>0} \Sigma_k^P$. For more complete descriptions of these classes, refer to, e.g., [29]. We also consider the **UP**-hierarchy, where $\mathbf{U}\Delta_0^P = \mathbf{P}$ and $\mathbf{U}\Delta_{k+1}^P = \mathbf{UP}^{\Sigma_k^P}$, for each $k \geq 0$ (see [18]).

Following the convention, we identify a set $A \in 2^{\Sigma^*}$ with a binary real number in $[0, 1)$. Let **m** denote *Lebesgue measure* on the unit real interval $[0, 1]$. An oracle dependent property $\mathcal{P}(A)$ *holds relative to a random oracle A with probability* 1 if $\mathbf{m}(\{A \mid \mathcal{P}(A) \text{ holds }\}) = 1$. See, e.g., [9] for more information.

Lemma 1. [9] *Let $\mathcal{A}$ be any nonempty subset of 2^{Σ^*}. Let L^A be an oracle-dependent set and let $\mathcal{C}^A = \{L(M_i, A) \mid i \in \mathbb{N}\}$ be a collection of oracle-dependent sets such that a certain enumeration of oracle Turing machines $\{M_i\}_{i\in\mathbb{N}}$ satisfying $\mathcal{C}^A = \{L(M_i, A) \mid i \in \mathbb{N}\}$ fulfills Conditions 1–4 in [9, p.98]. If there exists a positive constant $\epsilon > 0$ such that $\mathbf{m}_\mathcal{A}(\{A \mid L^A \neq L(M_i, A)\}) > \epsilon$ holds for each $i \in \mathbb{N}$, then $\mathbf{m}_\mathcal{A}(\{A \mid L^A \in \mathcal{C}^A\}) = 0$.*

In this paper, we consider only circuits in a tree form with gates of unbounded fanin. The *top gate* of a circuit is its root and a *bottom gate* is one of its gates attached to the leaves.

For $k, n > 0$, a $\Sigma_k(n)$-*circuit* C is a depth-$(k + 1)$ circuit such that (i) C has alternating levels of AND and OR gates with a top OR gate, (ii) the number of gates at each level from level 1 to level k is at most 2^n, and (iii) the fanin of each bottom gate is at most n. Similarly, we define a $\Pi_k(n)$-circuit by interchanging the roles of AND and OR.

We write ρ_A to denote the restriction ρ such that $\rho(v_z) = A(z)$ for all $z \in \{0, 1\}^*$, where v_z is the variable corresponding to a string z. For more details, see, e.g., [17, 20, 24].

Quantum Computation. A k-tape *quantum Turing machine*[2] (referred to as QTM) M is a quintuple $(Q, q_0, q_f, \Sigma_1 \times \Sigma_2 \times \cdots \times \Sigma_k, \delta)$, where each Σ_i is

---

[2] The notion of multi-tape QTMs has been discussed elsewhere and is known, from [32, 26], to be "equivalent" to the model proposed in [8].

a finite alphabet with a distinguished blank symbol #, Q is a finite set of internal states including an initial state q_0 and a final state q_f, and δ is a multi-valued, *quantum transition function* from $Q \times \Sigma_1 \times \Sigma_2 \times \cdots \times \Sigma_k$ to $\mathbb{C}^{Q \times \Sigma_1 \times \Sigma_2 \times \cdots \times \Sigma_k \times \{R,N,L\}^k}$. A QTM has two-way infinite tapes of cells indexed by $\mathbb{Z}$ and read/write tape heads that move along the tapes. Directions R and L mean that a head steps right and left, respectively, and direction N mean that a head makes no movement. A QTM has *K-amplitudes* if the entries of its time-evolution matrix are all drawn from set K. A QTM is in *normal form* if there exists a fixed direction $d \in \{L, N, R\}^k$ such that $\delta(q_f, \sigma) = |q_0\rangle|\sigma\rangle|d\rangle$ for any symbols σ. The *running time* of M on x is defined to be the minimal number T such that, at time T, all computation paths of M on x reach final configurations. Let $\text{Time}_M(x)$ denote the running time of M on x if one exists; otherwise, it is undefined. We say that M on input x *halts in time* T if $\text{Time}_M(x)$ exists and $\text{Time}_M(x) = T$. A QTM is *well-formed* if its time-evolution operator preserves the ℓ_2-norm [8].

For a well-formed QTM M and a string x, the notation $M(|x\rangle)$ denotes the final superposition of M on input x. The notation $M(x)$, however, represents the majority outcome (either 0 or 1) of $M(|x\rangle)$; that is, we can observe $M(x)$ in the start cell of M's output tape with probability more than $1/2$. For an oracle TM or QTM M, let $L(M, A)$ be the set accepted by M with oracle A.

Assume $K \subseteq \mathbb{C}$. Let $\#\mathbf{QP}_K$ denote the set of functions from Σ^* to the unit interval $[0, 1]$, each of which, on input x, outputs the acceptance probability of a polynomial-time QTM with K-amplitudes on input x [33]. A set S is in $\mathbf{EQP}_K$ if there is a $\#\mathbf{QP}_K$-function f such that $f(x) = S(x)$ for all x [8]. A set S is in $\mathbf{NQP}_K$ if there exists a function f in $\#\mathbf{QP}_K$ such that, for every x, $f(x) > 0$ iff $x \in S$ [1]. The class $\mathbf{BQP}_K$ is the collection of all sets S such that, for an appropriate function $f \in \#\mathbf{QP}_K$, $|f(x) - S(x)| < 1/3$ holds for all x [8]. Note that we can choose $2^{-p(n)}$, where p is a polynomial, instead of $1/3$. A set S is in $\mathbf{RQP}_K$ if there exists a function $f \in \#\mathbf{QP}_K$ such that, for every x, $f(x) > 1/2$ if $x \in S$, and $f(x) = 0$ otherwise. We also take $1/p(n)$, where p is a polynomial, instead of $1/2$. Similarly, we define $\mathbf{EQE}_K$ and $\mathbf{NQE}_K$ by replacing the phrase "polynomial time" with "linear exponential time" (that is, 2^{cn} for some constant $c > 0$). For brevity, we drop script K when $K = \tilde{\mathbb{C}}$.

The following lemma is an extension of the result by Ko [23], who showed that $\mathbf{NP} \subseteq \mathbf{BPP}$ iff $\mathbf{RP} = \mathbf{NP}$. We leave the proof to the interested reader.

Lemma 2. *Assume* $\mathbb{Q} \subseteq K \subseteq \mathbb{C}$. $\mathbf{NP} \subseteq \mathbf{BQP}_K$ *if and only if* $\mathbf{NP} \subseteq \mathbf{RQP}_K$.

For an oracle A and a subset S of strings, $A^{(S)}$ denotes the oracle satisfying that, for every y, $A(y) = A^{(S)}(y)$ if and only if $y \notin S$. For a well-formed oracle Turing machine M, an oracle A from an oracle collection $\mathcal{A}$, and $\epsilon > 0$, let the ϵ-*block sensitivity*, $bs_\epsilon^{\mathcal{A}}(M, A, x)$, of M with oracle A on input x be the maximal integer ℓ such that there are ℓ nonempty, disjoint sets $\{S_i\}_{i=1}^{\ell}$ such that, for each $i \in [\ell]$, (i) $A^{(S_i)} \in \mathcal{A}$ and (ii) $|\rho_M^{A^{(S_i)}}(x) - \rho_M^A(x)| \geq \epsilon$, where $\rho_M^A(x)$ is the acceptance probability of M with oracle A on input x.

Lemma 3. [10, 3] $bs_\epsilon^A(M, A, x) \leq 8T(x)^2/\epsilon^2$, where $T(x)$ is the running time of M with A on x.

Let $X = x_0 x_1 \cdots x_{n-1}$ be a *string variable* over $\{0,1\}^n$. A *quantum network* U *of input size* m *and query size* n *with* t *queries* (or simply, an (m, n, t)-*quantum network*) is a series of unitary operators of the following form: $U^X = U_{t+1} O^X U_t O^X \cdots U_1 O^X U_0$, where each U_i is a unitary operator independent of X and O^X is an operator that maps $|s_j^n\rangle|b\rangle|z\rangle$ to $|s_j^n\rangle|b \oplus x_j\rangle|z\rangle$, where $0 \leq j < n$, $b \in \{0,1\}$ and $z \in \{0,1\}^{m-1-\mathrm{ilog}(n)}$. Note that the first $\mathrm{ilog}(n)$ qubits fed into operator O^X are used to locate the query bit. Assume that the initial quantum state of the network is $|0^m\rangle$. At the end of the computation, we observe the first qubit of the final quantum state. We say that the quantum network U *outputs bit* b *(with bounded error)* if $|b\rangle$ is observed with probability at least $2/3$; that is, $\|\langle b|U^X|0^m\rangle\|^2 \geq 2/3$. See [3] for more details.

The next lemma states that an oracle QTM can be simulated by a quantum network.

Lemma 4. *For every well-formed oracle QTM M running in time $T(|x|)$ on input x, there exists a family $\{U_n\}_{n\in\mathbb{N}}$ of $(T(n), 2^{T(n)}, T(n))$-quantum networks such that $\rho_M^A(x) = \|\langle 1|U_{|x|}^A|0^{T(|x|)}\rangle\|^2$ for every x, where $\rho_M^A(x)$ denotes the acceptance probability of M on input x with oracle A.*

3 CoRP Is Harder Than NQE

Beigel showed in [4] that $\mathbf{NP}^A \not\subseteq \mathbf{C_=P}^A$ relative to a certain oracle A. From recent results of Yamakami and Yao [34] and Fenner et al. [15], who proved that $\mathbf{NQP}^A = \mathbf{NQP}^A = \text{co-}\mathbf{C_=P}^A$ for all oracles A, Beigel's result implies that co-$\mathbf{NP}^A \not\subseteq \mathbf{NQP}^A$. Later, Green [19] improved Beigel's result by showing that co-$\mathbf{NP}^A \not\subseteq \mathbf{NQP}^{\oplus \mathbf{P}^A}$.

In this section, we show that co-$\mathbf{RP}^A \not\subseteq \mathbf{NQE}^A \cup \mathbf{NQP}^{\oplus \mathbf{P}^A}$ relative to a certain oracle A. Our proof is a simple application of the polynomial method of Beals et al. [3], who showed that a $\mathbb{C}$-valued multilinear polynomial over $\{0,1\}$ can characterize a computation of a quantum network. A characterization of oracle quantum computation follows in a similar fashion.

Lemma 5. *Let M be a well-formed, oracle QTM running in time $T(x)$ on input x independent of the choice of oracles. Let A be a subset of Σ^*. For each x, there exists a $\mathbb{R}$-valued, $2^{T(x)^2}$-variate, multilinear polynomial $p^{(x)}(v_{x_1}, v_{x_2}, \ldots, v_{x_m})$, where $m = 2^{T(x)}$, of degree at most $T(x)^2$ such that $p^{(x)}(A(x_1), A(x_2), \ldots, A(x_m))$ is the probability that M accepts x with oracle A, where v_y is a Boolean variable indexed y in $\{0,1\}^*$.*

Now, we state our main result in this section.

Proposition 1. *There exists a set A such that* co-$\mathbf{RP}^A \not\subseteq \mathbf{NQE}^A \cup \mathbf{NQP}^{\oplus \mathbf{P}^A}$.

Proof sketch: In this proof, we show only co-$\mathbf{RP}^A \not\subseteq \mathbf{NQE}^A$. Let $\mathcal{A} = \{A \mid \forall n(|A^{=n}| = 0 \text{ or } |A^{=n}| \geq 2^{n-1})\}$. For each A, set $L^A = \{0^n \mid |A^{=n^2}| = 0\}$. It is obvious that L^A belongs to co-$\mathbf{RP}^A$ for every oracle A in $\mathcal{A}$. It suffices to show that L^A does not belong to $\mathbf{NQE}^A$ for a certain oracle A in $\mathcal{A}$.

Let $\{M_i\}_{i \in \mathbb{N}}$ be an enumeration of all oracle $\mathbf{NQE}$-machines with $\mathbb{Q}$-amplitudes such that each M_i halts in time $2^{c_i n}$ on inputs of length n for any choice of oracles.

We recursively construct the desired oracle A. Set $A = \emptyset$ and $n_0 = 0$. Assume that n_i is already defined. We define n_{i+1} to be the minimal n such that $2^{2c_{i+1}n} < \frac{1}{2} 2^{n^2}$ and $n_i < n$. For simplicity, we write n for n_{i+1} and set $m = 2^{n^2}$. Assume that, for every extension B of A, M_{i+1} with oracle B on input 0^n correctly computes $L^A(0^n)$. Let $p(x_1, x_2, \ldots, x_m)$ be a multilinear polynomial defined by Lemma 5. Note that the degree of p is at most $(2^{c_{i+1}n})^2$.

Let p^{sym} be the *symmetrization* of p defined as in [3]. Consider the univariate polynomial q_p obtained from the polynomial p^{sym} such that $q_p(\#_1(\boldsymbol{x})) = p^{sym}(\boldsymbol{x})$ for all $\boldsymbol{x} = (x_1, x_2, \ldots, x_m)$ [3]. Note that $deg(q_p) = deg(p^{sym})$.

Since $q_p(\#_1(\boldsymbol{x})) = 0$ for all $\boldsymbol{x}$ with $\frac{1}{2}m \leq \#_1(\boldsymbol{x}) \leq m$, the degree of q_p must be at least $\frac{1}{2}m - m + 1$, which is larger than $2^{c_{i+1}n}$, a contradiction. Hence, there exists an oracle B, $B \subseteq \Sigma^{<m}$, such that M_{i+1} with B on input 0^n cannot compute $L^A(0^n)$. Take such a B and set it as the new A. □

As an immediate corollary, we can show the following separation result.

Corollary 1. *There exists a set A such that (i) $\mathbf{P} \neq \mathbf{RP} \neq$ co-$\mathbf{RP} \neq \mathbf{BPP}$, (ii) $\mathbf{P} \neq \mathbf{NP} \neq$ co-$\mathbf{NP} \neq \Delta_2^P$, and (iii) $\mathbf{EQP}^A \neq \mathbf{RQP}^A \neq$ co-$\mathbf{RQP}^A \neq \mathbf{BQP}^A$.*

4 BQP Is Exponentially Easier Than UP

In this section, we exhibit an oracle A that makes $\mathbf{BQP}$ exponentially easier than $\mathbf{UP}$. We achieve this by showing an oracle A relative to which $\mathbf{P} = \mathbf{BQP}$ and $\mathbf{UP} = \mathbf{EXP}$. This result complements the recent result of Fortnow and Rogers [16] showing that $\mathbf{P} = \mathbf{BQP} \neq \mathbf{UP} \cap$ co-$\mathbf{UP}$ via a non-recursive oracle. Here, we provide an alternative, recursive construction of such an oracle.

Our proof method is adapted from a technique of Beigel, Buhrman, and Fortnow [5], who constructed a recursive oracle A satisfying both $\mathbf{P}^A = \oplus \mathbf{P}^A$ and $\mathbf{RP}^A = \mathbf{EXP}^A$. Note that, since $\mathbf{UP}^A \subseteq \oplus \mathbf{P}^A$ and $\mathbf{RP}^A \subseteq \mathbf{BQP}^A$, the same oracle shows that $\mathbf{P}^A = \mathbf{UP}^A$ and $\mathbf{BQP}^A = \mathbf{EXP}^A$, a reverse of our result.

Theorem 1. *There exists an oracle A such that $\mathbf{P}^A = \mathbf{BQP}^A$ and $\mathbf{UP}^A = \mathbf{EXP}^A$.*

Proof sketch: In this proof, we fix any polynomial-time computable pairing function $\langle \cdot, \cdot, \cdot \rangle$ with three arguments such that $|\langle x, y, z \rangle| > \max\{|x|, |y|, |z|\}$ for all x, y, z. To show $\mathbf{P}^A = \mathbf{BQP}^A$, it suffices to consider only well-formed, oracle QTMs, each of which on input $(0^i, x)$, (i) runs in polynomial time on any input and with any oracle, (i) makes at most $|x|$ queries of length smaller than $|x|$, and (ii) has error probability smaller than $2^{-i-|x|}$. We fix an effective enumeration of such polynomial-time, well-formed, oracle QTMs $\{M_k\}_{k\in\mathbb{N}}$.

Let L^A be any $\mathbf{EXP}^A$-complete set (under $\mathbf{P}^A$-m-reduction) and let N be a deterministic oracle TM that recognizes L^A in exponential time. Without loss of generality, we can assume that N runs in time $2^{|x|}$ on input x. For convenience, we also view N as an $\mathbf{EQE}$-machine.

To show the claim, we construct A so that, for any sufficiently large n and all strings x of length n, and all k with $0 \leq k < n$,

1. if M_k accepts $(0^n, x)$ with probability at least $1 - 2^{-2n}$, then $\langle s_k^n, x, 1^{n^2}\rangle \in A$;
2. if M_k rejects $(0^n, x)$ with probability at least $1 - 2^{-2n}$, then $\langle s_k^n, x, 1^{n^2}\rangle \notin A$;
3. $x \in L^A$ implies $|\{y \in \Sigma^{n^2} \mid \langle 1^{i\log(n)}, x, y\rangle \in A\}| = 1$; and
4. $x \notin L^A$ implies $|\{y \in \Sigma^{n^2} \mid \langle 1^{i\log(n)}, x, y\rangle \in A\}| = 0$.

Conditions 1 and 2 ensure that $\mathbf{P}^A = \mathbf{BQP}^A$ and conditions 3 and 4 imply that $\mathbf{UP}^A = \mathbf{EXP}^A$. For simplicity, we say that w is a k-string (ω-string, resp.) if w is of the form $\langle s_k^{|u|}, u, 1^{|u|^2}\rangle$ with $0 \leq k < |u|$ ($\langle 1^{i\log(|u|)}, u, v\rangle$ with $|v| = |u|^2$, resp.). Note that $|w| > |u|^2$.

In what follows, we will construct such an oracle A by stages. The oracle is best described as a partial oracle σ_A at each stage. Initially (at Stage λ), we set $\sigma_A(x) = 0$ for all strings x that are neither k-strings nor ω-strings. Let x be any nonempty string.

Stage x: we will expand σ_A that has been defined by the previous stage. Assume by induction hypothesis that all strings $\langle \cdot, z, w\rangle$ for $z < x$ are already determined (that is, $\text{dom}(\sigma_A)$ includes all such strings). Let $n = |x|$. For each string w, let v_w denote the "quantum state variable" that is indexed by w. We denote by z the sequence of all string variables indexed with ω-strings of the form $\langle 1^{i\log(|u|)}, u, v\rangle$, with $|u| = |x|$, not in $\text{dom}(\sigma_A)$. We want to define a quantum function $q_x(z)$ that computes $N^A(|x\rangle)$. To do so, we first assign to each string w, in the range between $\langle s_0^0, \lambda, \lambda\rangle$ to $\langle s_{2^n-1}^{2^n}, 1^{2^n}, 1^{2^{2n}}\rangle$, a quantum function p_w recursively.

> Round w: (o) case where w is neither k-string nor ω-string: in this case, define $p_w(z) = |0\rangle$.
> (i) case where w is an ω-string of the form $\langle 1^{i\log(|u|)}, u, v\rangle$: note that if $u < x$ then $w \in \text{dom}(\sigma_A)$. We define $p_w(z)$ as follows. If $w \notin \text{dom}(\sigma_A)$, then define $p_w(z)$ to be the "quantum state variable" v_w, if $|u| = |x|$, or the constant $|0\rangle$, if $|u| > |x|$, and we mark v_w. If $w \in \text{dom}(\sigma_A)$, then define $p_w(z) = |\sigma_A(w)\rangle$.
> (ii) case where w is a k-string, $0 \leq k < m$, of the form $\langle s_k^m, u, 1^{m^2}\rangle$ with $m = |u|$: if $w \in \text{dom}(\sigma_A)$, then set $p_w(z) = |\sigma_A(w)\rangle$. Assume otherwise that $w \notin \text{dom}(\sigma_A)$. Consider the set $Q(M_k, A, (0^m, u))$ of all query words made by M_k on input $(0^m, u)$ with oracle A. Note that $|Q(M_k, A, (0^m, u))| \leq 2^{|u|}$. Let $Q(M_k, A, (0^m, u)) = \{y_1, y_2, \ldots, y_\ell\}$, where $0 \leq \ell \leq 2^m$, and let $q(v_{y_1}, v_{y_2}, \ldots, v_{y_\ell})$ be the outcome

$M_k^A(|0^m, u\rangle)$. Note that $|y_i| < |u| < \sqrt{|w|}$. Define $p_w(z) = q(p_{y_1}(z), \ldots, p_{y_\ell}(z))$. Note that p_w can be viewed as another well-formed oracle QTM.

Remember that we consider only the case where z takes a value either zero $0 = 0^{2^{n^2}}$ or a unit vector $d_e = 0^e 1 0^{2^{n^2} - e - 1}$ ($0 \leq e < 2^{n^2}$). Finally, we define $q_x(z)$ to be the quantum state $N(|x\rangle)$ with the condition that, whenever N makes a query of the form w, we compute $p_w(z)$ instead.

By our definition, $q_x(z)$ can be viewed as being constructed inductively "level by level" by subcomputations p_w's. Note that the number of levels needed to build q_x is at most $\lfloor \log n \rfloor - 1$. Thus, the number of times q_x makes queries along each computation path is at most $\prod_{k=0}^{\lfloor \log n \rfloor} (2^n)^{1/2^k}$, which is less than 2^{2n-1}. It is important to note that the success probability so far to compute $q_x(z)$ is more than $(1 - 2^{-2n})^{2^{2n-1}} > \sqrt{1/e} > 2/3$ (without counting on any wrong answers given by some p_w's).

Note that each $p_w(z)$ used to construct $q_x(z)$ is either *improper* (i.e., having a value in the interval $(2^{-2n}, 1 - 2^{-2n})$) or *proper*. We define a *replacement* for $q_x(d)$ as an assignment of either $|0\rangle$ or $|1\rangle$ to all improper $p_w(d)$'s that appear in the definition $q_x(d)$ by induction on the construction of q_x. We write $q_x^\tau(d)$ to denote the quantum state obtained by applying replacement τ to $q_x(d)$.

It is easy to show that either we can set z to zero 0 or we find a unit vector d_e that will complete the construction of the oracle A at stage x. The most crucial case is that $\|\langle 1 | q_x^\tau(0)\rangle\|^2 \geq 2/3$ and $\|\langle 1 | q_x^\tau(d_e)\rangle\|^2 \leq 1/3$ for every replacement τ and every unit vector d_e, where $\langle 1 | q_x(z)\rangle$ denotes the quantum state resulting from observing 1 on the start cell of an output tape of $q_x(z)$. This, however, contradicts Lemma 3. □

5 RP Is Exponentially Harder Than EQP

Combining two oracle constructions demonstrated in §3 and §4, we can show the existence of an oracle relative to which $\mathbf{P} = \mathbf{EQP}$ and $\mathbf{RP} = \mathbf{EXP}$.

We first prepare an enumeration $\{M_k\}_{k \in \mathbb{N}}$ of polynomial-time, well-formed, oracle QTMs and take the same $\mathbf{EXP}^A$-complete set L^A used in the proof of Theorem 1. Here, we use the following four requirements:

1. if M_k accepts x with probability 1 then $\langle s_k^n, x, 1^{n^2}\rangle \in A$;
2. if M_k rejects x with probability 1 then $\langle s_k^n, x, 1^{n^2}\rangle \in A$;
3. if $x \in L^A$ then $|\{y \in \Sigma^{n^2} \mid \langle 1, x, y\rangle \in A\}| \geq 2^{n^2}/2$; and
4. if $x \notin L^A$ then $|\{y \in \Sigma^{n^2} \mid \langle 1, x, y\rangle \in A\}| = 0$.

Note that the first two requirements ensure $\mathbf{P}^A = \mathbf{EQP}^A$ and the rest imply $\mathbf{RP}^A = \mathbf{EXP}^A$. Following a similar construction in the proof of Theorem 1, we define $q_x(z)$. A key case is that, for every replacement τ and every nonzero vector $d \in W_n$, it holds that $\|\langle 1 | q_x^\tau(0)\rangle\|^2 = 0$ and $\|\langle 1 | q_x^\tau(d)\rangle\|^2 = 1$, where $W_n =$

$\{s \in \{0,1\}^{2^{n^2}} \mid \#_1(s) \geq 2^{n^2}/2 \text{ or } \#_1(s) = 0\}$. By the polynomial method as in the proof of Proposition 1, however, this case is shown to be impossible.

Therefore, we obtain our main theorem.

Theorem 2. *There exists an oracle A relative to which* $\mathbf{P}^A = \mathbf{EQP}^A$ *and* $\mathbf{RP}^A = \mathbf{EXP}^A$.

6 EQP Is Harder Than $\Sigma_2^P \cup \Pi_2^P$

In this section, we investigate the power of **EQP** in contrast to $\Sigma_2^P \cup \Pi_2^P$, the second level of the polynomial-time hierarchy. We show a recursive oracle relative to which $\mathbf{EQP} \not\subseteq \Sigma_2^P \cup \Pi_2^P$.

To show our result, we will introduce an oracle-dependent language, which lies in the gap $\mathbf{EQP}^A - \Pi_2^P(A)$ for a certain oracle A. Our proof relies on adapting Simon's test language in stages.

Simon [30] was the first who presented an oracle-dependent test language which is in $\mathbf{BQP}^A$ but not in $\mathrm{BPTIME}[2^{\epsilon n}]^A$ for any nonnegative constant $\epsilon < 1$. Subsequently, Brassard and Høyer [11] showed that Simon's language is actually in $\mathbf{EQP}^A$. A recent result of Hemaspaandra et al. [21] achieves almost-everywhere superiority of a slight modification of Simon's language against the class $\mathrm{BPTIME}[2^{\epsilon n}]^A$.

For a string $w \in \Sigma^*$, let $Parity(w) = \#_1(w) \bmod 2$. Define the basic function $\eta_{1,k}^A$, where $k \in \{1, 2, 3\}$, as follows: letting $\boldsymbol{x} = (x_1, x_2, \ldots, x_n)$,

$$\eta_{1,k}^A(\boldsymbol{x}) = \begin{cases} \tilde{A}(0^k 1 x_1 x_2 \cdots x_k 01)\tilde{A}(0^k 1 x_1 x_2 \cdots x_k 01^2) \cdots \tilde{A}(0^k 1 x_1 x_2 \cdots x_k 01^{nk}) \\ \quad \text{if } n = |x_1| = |x_2| = \cdots = |x_k|, \\ \lambda \quad \text{otherwise,} \end{cases}$$

where $\tilde{A}(0^k 1 x_1 \cdots x_k 01^i) = Parity(x_1) \oplus A(0^k 1 x_1 \cdots x_k 01^i)$. Define $\eta_{2,1}^A(x) = \tilde{S}_{1,2}^A(x, \bar{1})\tilde{S}_{1,2}^A(x, \bar{2}) \cdots \tilde{S}_{1,2}^A(x, \bar{n})$, where $\bar{j} = 0^{n-j} 1 0^{j-1}$ and $\tilde{S}_{1,2}^A(x, w) = Parity(x) \oplus S_{1,2}^A(x, w)$ and $S_{1,2}^A = \{(x_1, x_2) \mid \exists y, z \in \Sigma^{|x_1|}[y \neq z \wedge \eta_{1,3}^A(x_1, x_2, y) = \eta_{1,3}^A(x_1, x_2, z)]\}$. Finally, we define our test languages L_i^A as follows: for $i \in \{1, 2\}$,

$$L_i^A = \{0^n \mid \exists a \in \Sigma^n \setminus \{0^n\}[\eta_{i,1}^A(0^n) = \eta_{i,1}^A(a) \& Parity(a) = 1]\}.$$

To guarantee that L_i^A is in $\mathbf{EQP}^A$, we need to define $\mathcal{A}$, the set of "good" oracles:

$$\mathcal{A} = \{A \mid \forall n \forall (i,k) \in W \exists a \in \Sigma^n \forall \boldsymbol{x} \in (\Sigma^n)^{k-1} \text{ s.t.}$$
$$\forall u, v \in \Sigma^n [\eta_{i,k}^A(\boldsymbol{x}, u) = \eta_{i,k}^A(\boldsymbol{x}, v) \implies u = v \vee u = v \oplus a]\},$$

where $W = \{(1,1), (1,2), (1,3), (2,1)\}$. The unique non-zero element a (if it exists) is called the *period* of $\eta_{i,k}^A$ on $\{0,1\}^n$. We say a period *even* (*odd*, resp.) if $Parity(a) = 1$ ($Parity(a) = 0$, resp.). Note that, for every $A \in \mathcal{A}$, every $k, n >$

0, and every $x \in (\Sigma^n)^{k-1}$, $\lambda z.\eta_{i,k}^A(x,z)$ is either a one-to-one or two-to-one function from $\{0,1\}^n$ to $\{0,1\}^n$.

Clearly, $L_i^A \in \mathbf{U}\Delta_i^P(A)$ holds for any $i \in \{1,2\}$ and for any oracle $A \in \mathcal{A}$. Note that L_1^A is basically Simon's test language.

The following lemma follows from the facts that, for every $A \in \mathcal{A}$, $L_1^A \in \mathbf{EQP}^A$ [11], $L_2^A \in \mathbf{EQP}^{L_1^A}$, and $\mathbf{EQP}^{\mathbf{EQP}^A} = \mathbf{EQP}^A$.

Lemma 6. *For any $k \in \{1,2\}$, L_k^A belongs to $\mathbf{EQP}^A$ for all oracles A in $\mathcal{A}$.*

Now, we state the key proposition of this section.

Proposition 2.
1. *There exists a set $A \in \mathcal{A}$ such that $L_1^A \notin$ co-NTIME$[o(2^{n/2})]^A$. Thus, $\mathbf{EQP}^A \cap \mathbf{UP}^A \not\subseteq$ co-$\mathbf{NP}^A$.*
2. *There exists a set $B \in \mathcal{A}$ such that $L_2^B \notin$ co-NTIME$[o(2^{n/3})]^{\text{NTIME}[n]^A}$. Thus, $\mathbf{EQP}^B \cap \mathbf{U}\Delta_2^P(B) \not\subseteq \Pi_2^P(B)$.*

Proposition 2(1) implies $\mathbf{EQP}^A \not\subseteq \mathbf{NP}^A \cup$ co-$\mathbf{NP}^A$ and $\mathbf{NP}^A \neq$ co-$\mathbf{NP}^A$. Note that Berthiaume and Brassard [7] already proved that $\mathbf{EQP}^A \not\subseteq \mathbf{NP}^A \cup$ co-$\mathbf{NP}^A$ for a certain oracle A. Proposition 2(2), however, yields the following improvement.

Theorem 3. $\mathbf{EQP}^A \not\subseteq \Sigma_2^P(A) \cup \Pi_2^P(A)$ and $\Sigma_2^P(A) \neq \Pi_2^P(A)$ for a certain oracle A.

To facilitate the proof of Proposition 2(1), we introduce a special circuit, called an $S_1(n)$-circuit, which corresponds to L_1^A. Moreover, it is convenient to deal with a series of Boolean variables as a unique "variable." A *string variable* of length n is a series of n distinct Boolean variables. A *negation* of a string variable x is of the form $\bar{x}_1\bar{x}_2\cdots\bar{x}_n$ if x is of the form $x_1x_2\cdots x_n$, where each x_i is a Boolean variable and $\bar{x}_i$ is a negation of x_i.

Definition 1. *Let $n > 0$. An $S_1(n)$-circuit F consists of 2^n blocks, each of which is indexed $s \in \Sigma^n$ and has fanin n (i.e., has n wires). For each block s, a string variable x_s of length n of positive form is connected to s if $Parity(s) = 1$; otherwise, a string variable of negative form, $\bar{x}$, is connected to s. A restriction ρ is now considered as a map from $\{x_s\}_{s \in \Sigma^n}$ to $\{0, 1, *\}^n$. For a given assignment ρ, F outputs 1 if $\eta_{1,1}^\rho$ is two-to-one with an odd period, and 0 if it is one-to-one, and "?" otherwise.*

Note that an $S_1(n)$-circuit outputs 1 for $2^{n-1} \cdot {}_{2^n}P_{2^{n-1}}$ possible assignments whereas there are $2^n!$ choices that force an $S_1(n)$-circuit to output 0, where ${}_iP_j = i!/(i-j)!$.

The first claim of Proposition 2 follows directly from the following lemma. The main idea of the proof is to kill each candidate for odd periods by assigning a value to an appropriate string variable. We leave its proof to the reader.

Lemma 7. *Let $m, n > 0$. Let F be the $S_1(n)$-circuit and C a $\Pi_1(m)$-circuit. If $2m < 1 + \sqrt{2^{n+3}+1}$, then $F\lceil_{\rho_A} \neq C\lceil_{\rho_A}$ for a certain set A in $\mathcal{A}$.*

To prove the second claim of Proposition 2, we need the notion of an $S_2(n)$-circuit, which is a two-layered circuit of $S_1(n)$-subcircuits.

Definition 2. *An $S_2(n)$-circuit F is a circuit of depth 2 with a top $S_1(n)$-circuit and $n2^n$ bottom $S_1(n)$-subcircuits with all distinct variables. Each bottom $S_1(n)$-subcircuit is labeled (i,s), where $1 \leq i \leq n$ and $s \in \{0,1\}^n$, and is denoted by $F_s^{(i)}$, which is connected to F through the ith wire of block s. We say that a bottom $S_1(n)$-subcircuit is of positive form (negative form, resp.) if it has label (i,s) with $Parity(s) = 0$ ($Parity(s) = 1$, resp.). For a given assignment ρ, F outputs 1 if $\eta_{2,1}^\rho$ is two-to-one with an odd period, and 0 if it is one-to-one, and "?" otherwise.*

Using the test language L_2^A, we show that there is an oracle A relative to which $L_2^A \notin \Pi_2^P(A)$. The following lemma supplies the main piece for showing this claim. The proof uses an argument from Ko's [24] paper on the **BP**-hierarchy.

Lemma 8. *Let $m, n > 0$. Let F be the $S_2(n)$-circuit and C a $\Pi_2(m)$-circuit. If $m < 2^{n/3}$, then $F\lceil_{\rho_A} \neq C\lceil_{\rho_A}$ for a certain set A in $\mathcal{A}$.*

Proof sketch: Let ℓ satisfy $1 \leq \ell \leq 2^m$. Assume that C is an AND of D_i's, $1 \leq i \leq \ell$, where each D_i is an OR of ANDs of fanin $\leq 2^m$ and the bottom fanin is at most m.

Let K be the collection of restrictions ρ_A such that $A \in \mathcal{A}$, $C\lceil_{\rho_A} = 0$, and the condition (*) that exactly the same number of bottom $S_1(n)$-circuits of negative form output 1 and 0. Because of the choice of K, the total number of bottom $S_1(n)$-circuits that output 1 equals $n2^n/2$. Thus, we have $|K| = c_n \cdot (2^{n-1} \cdot_{2^n} P_{2^{n-1}})^{n2^{n-1}} (2^n!)^{n2^{n-1}}$, where c_n is the number of assignments that force the top $S_1(n)$-circuit to output 0 under condition (*). Take a subcircuit D_{i_0} of C such that $|\{\rho \in K \mid D_{i_0}\lceil_\rho = 0\}| \geq |K|/2^m$. Since the number of D_i's is $\leq 2^m$, such D_{i_0} exists.

Initially, set $K_0 = \{\rho \in K \mid D_{i_0}\lceil_\rho = 0\}$, $Q_0 = \emptyset$, and $R_0 = \emptyset$. Inductively, for every $i \in [0, 2^{n-1}] \cap \mathbb{Z}$, we define a triplet (K_i, Q_i, S_i) such that $K_i \subseteq K$, $Q_i \subseteq VAR \times \{0,1\}^n$, and $R_i \subseteq \{0,1\}^n$, where VAR is the set of all string variables (of positive form) that appear in F. We may assume that all the variables in C are in VAR. We also ensure that each triplet satisfies the following four requirements: (i) $|K_i| \geq |K_0| \cdot [\binom{m}{2}^2 2^{2n}]^{-i}$, (ii) $|Q_i| = 4i$, (iii) $|R_i| = 2i$, and (iv) $\rho(w) = d$ for every $\rho \in K_i$ and every pair $(w, d) \in Q_i$.

Notice that the above requirements hold for (K_0, Q_0, R_0). Now, we assume that the requirements hold for $i > 0$. We first claim that there exists a subcircuit G (of depth 1) of D_{i_0} satisfying the following four conditions:

1. Let ρ_i be such that $\eta_{2,1}^{\rho_i}$ is two-to-one with an odd period a and ρ_i is consistent on Q_i (that is, $\rho_i(w) = d$ for any pair $(w,d) \in Q_i$). Since $C\lceil_{\rho_i} = 1$ by our assumption, $D_{i_0}\lceil_{\rho_i} = 1$.
2. G satisfies $G\lceil_{\rho_i} = 1$. Note that $G\lceil_\rho = 0$ for any $\rho \in K_i$ since $K_i \subseteq K$.

3. Since a is a period given by ρ_i, there exist two blocks s_1 and s_2 in the top $S_1(n)$-circuit and an integer k such that $s_1, s_2 \notin R_i$, $s_1 = s_2 \oplus a$, and $F_{s_1}^{(k)} \lceil_{\rho_i} \neq F_{s_2}^{(k)} \lceil_{\rho_i}$.
4. For those $F_{s_1}^{(k)}$ and $F_{s_2}^{(k)}$, G contains two pairs, $\{w_1, w_2\}$ appearing in $F_{s_1}^{(k)}$ and $\{v_1, v_2\}$ appearing in $F_{s_2}^{(k)}$, which are all distinct string variables such that, for any legal ρ, if $\rho(w_1) = \rho(w_2)$ and $\rho(v_1) = \rho(v_2)$ then $F_{s_1}^{(k)} \lceil_\rho = F_{s_2}^{(k)} \lceil_\rho = 1$.

We fix such a subcircuit G. Note that there are at most $\binom{m}{2}^2$ pairs in G satisfying condition (4). Take such two pairs $\{w_1, w_2\}$ and $\{v_1, v_2\}$ satisfying the additional condition that $|\{\rho \in K_i \mid \rho(w_1) = \rho(w_2) \& \rho(v_1) = \rho(v_2)\}| \geq |K_i|/\binom{m}{2}^2$. Notice that these two pairs kill the period a. We then choose two values d_1 and d_2 in $\{0,1\}^n$ such that $|\{\rho \in K_i \mid \rho(w_1) = \rho(w_2) = d_1 \& \rho(v_1) = \rho(v_2) = d_2\}| \geq |K_i|/[\binom{m}{2}^2 \cdot 2^{2n}]$.

At the end, we define $K_{i+1} = \{\rho \in K_i \mid \rho(w_1) = \rho(w_2) = d_1 \& \rho(v_1) = \rho(v_2) = d_2\}$, $Q_{i+1} = Q_i \cup \{(w_j, d_1) \mid j = 1, 2\} \cup \{(v_j, d_2) \mid j = 1, 2\}$, and $R_{i+1} = R_i \cup \{s_1, s_2\}$.

After $e = 2^{n-1}$, we cannot proceed the above procedure because F is determined to output 0. Thus, $|K_e| \geq |K_0| \cdot [\binom{m}{2}^2 \cdot 2^{2n}]^{-e} > |K|/[2^m \cdot m^{2^{n+1}} \cdot 2^{(n-1)2^n}]$. Since the bottom $S_1(n)$-circuit, say F', attached to each block in R_e already has a certain odd period, there are only $_{2^n-1}P_{2^{n-1}-1}$ possible assignments left for F'. Thus, $|K_e| \leq c_n (2^{n-1} \cdot {}_{2^n}P_{2^n-1})^{(n-2)2^{n-1}} ({}_{2^n-1}P_{2^{n-1}-1})^{2^n} \cdot (2^n!)^{n2^{n-1}}$, which equals to $|K|/[2^{n2^n} \cdot 2^{(n-1)2^n}]$. Therefore, we have $2^m m^{2^{n+1}} > 2^{n2^n}$. Since $m < 2^{n/3}$, this is a contradiction. □

7 Almost-BQP Might Collapse To BQP

We turn our attention to random oracles and the complexity classes that they define.

For a relativizable complexity class $\mathcal{C}$, we write Almost-$\mathcal{C}$ to denote the collection of sets S such that $S \in \mathcal{C}^A$ holds relative to a random oracle A with probability 1.

It is easy to see that Almost-**BQP** contains **BQP**. Our goal is to show that Almost-**BQP** collapses to **BQP**. This provides a quantum analogue to the classical result Almost-**P** = Almost-**BPP** = **BPP** [9, 25]. To do this, we import the machinery developed by Nisan and Wigderson [28], who used the idea of pseudorandom generators that are secure against small-depth Boolean circuits. We will adapt their ideas to handle quantum computation.

In what follows, we will focus only on *robust* quantum networks. We call a quantum network U *robust* if for any oracle A and any input x, the acceptance probability of U on x with oracle A is in $[0, 1/3] \cup [2/3, 1]$.

First, we redefine the notion of *hardness* of a Boolean function. We always use X_n, Y_n, etc., as random variables over $\{0,1\}^n$. Let f

be a Boolean function from $\{0,1\}^*$ to $\{0,1\}$ and let U be a robust (m,n,t)-quantum network. Let us define $bias(U,f)(n)$ of U in computing f as follows: $bias(U,f)(n) = |\text{Prob}_{X_n}\left[\|\langle f(X_n)|U^{X_n}|0^m\rangle\|^2 \geq \frac{2}{3}\right] - \text{Prob}_{X_n}\left[\|\langle \bar{f}(X_n)|U^{X_n}|0^m\rangle\|^2 \geq \frac{2}{3}\right]|$, where $\bar{f}(x) = 1 - f(x)$.

Definition 3. *A function f from $\{0,1\}^*$ to $\{0,1\}$ is called* robustly (ϵ, m, t)-hard *if, for every robust (m,n,t)-quantum network U, $bias(U,f)(n) \leq \epsilon$ for all sufficiently large n.*

We then define the notion of a *pseudorandom generator* which is secure against any robust quantum networks with bounded queries.

Definition 4. *Let ℓ be a map from $\mathbb{N}$ to $\mathbb{N}$ and let $G = \{G_n\}_{n \in \mathbb{N}}$ consist of functions G_n from $\{0,1\}^{\ell(n)}$ to $\{0,1\}^n$. The function G is called a* pseudorandom generator *secure against any robust (m,n,t)-quantum network with stretch function ℓ if (i) $\ell(n) < n$ for all n, (ii) G is computed deterministically in polynomial time in $\ell(n)$, and (ii) for every robust (m,n,t)-quantum network U, $|\text{Prob}_{X_n}\left[\|\langle 1|U^{X_n}|0^m\rangle\|^2 \geq \frac{2}{3}\right] - \text{Prob}_{X_{\ell(n)}}\left[\|\langle 1|U^{G(X_{\ell(n)})}|0^m\rangle\|^2 \geq \frac{2}{3}\right]| < \frac{1}{n}$ for all sufficiently large n.*

Definition 5. *Let U be a robust (m,n,t)-quantum network, b a map from $\{0,1\}^n$ to $\{0,1\}$, and f a function from $\{0,1\}^n$ to $\{0,1\}^k$. The prediction probability of U for b given f is $pred(U, b, f) = |\text{Prob}_{X_n}\left[\|\langle b(X_n)|U^{f(X_n)}|0^m\rangle\|^2 \geq \frac{2}{3}\right] - \text{Prob}_{X_n}\left[\|\langle \bar{b}(X_n)|U^{f(X_n)}|0^m\rangle\|^2| \geq \frac{2}{3}\right]|$, where $\bar{b}(x) = 1 - b(x)$ for every x.*

Definition 6. [28] *Let k, ℓ, m, n be positive integers. A Boolean $n \times \ell$ matrix A is called a (k, m)-design if its n rows $v_1, v_2, \ldots, v_n$ satisfy the following conditions: (i) $|S_i| = m$ for all i with $1 \leq i \leq n$, and (ii) $|S_i \cap S_j| \leq k$ for all distinct pair (i,j), where $S_i = \{j \mid \text{the } j\text{th bit of } v_i \text{ is } 1\}$. Let f be a Boolean function from $\{0,1\}^m$ to $\{0,1\}$ and let A be a Boolean $n \times l$ matrix which is a (k, m)-design. We denote by f_A the function from $\{0,1\}^\ell$ to $\{0,1\}^n$ defined as follows: $f_A(x_1 x_2 \cdots x_\ell) = y_1 y_2 \cdots y_n$ such that $y_i = f(x_{i_1} x_{i_2} \cdots x_{i_m})$, where $S_i = \{i_1, i_2, \ldots, i_m\}$ with $i_1 < i_2 < \cdots < i_m$.*

Next, we state two helpful lemmas that are based on Nisan and Wigderson's work [28]. The proofs of these lemmas are left to the avid reader.

Lemma 9. *Let k be an increasing, unbounded function on $\mathbb{N}$ such that $n \in O(2^{k(n)})$. Let c be any sufficiently large positive integer. For almost all $n \in \mathbb{N}$, there exists a Boolean $n \times 2c^2 k(n)$ matrix which is a $(k(n), ck(n))$-design.*

Lemma 10. *Let m, k, ℓ, and s be functions on $\mathbb{N}$ and assume that $0 < m(n) \leq \ell(n) < n$ for all $n \in \mathbb{N}$. Also let f be a function from $\{0,1\}^*$ to $\{0,1\}$. Assume that, for every $(k(n) + 5s(n) + i\log(n) + 1, m(n), 2s(n)(s(n) + 1))$-quantum network U, we have $bias(U, f)(m(n)) \leq 1/n^2$ for all sufficiently large n. For any positive integer n and any Boolean $n \times \ell(n)$ matrix A which is a $(s(n), m(n))$-design, f_A is a function from $\{0,1\}^{\ell(n)}$ to $\{0,1\}^n$ that is pseudorandom to any $(k(n), n, s(n))$-quantum networks.*

In proving our main result, we show a hardness result for PARITY against robust quantum networks. A crucial idea of the proof is to use a result of Nisan and Szegedy [27] on certificate complexity and apply Håstad's hardness result against depth-2 circuits [20].

Lemma 11. *PARITY is robustly* $(2^{-n^{1/k}}, (\log n)^k, (\log n)^k)$-*hard for any positive integer* k.

We call a well-formed, oracle QTM M a *robust* **BQP**-*machine* if (i) M runs in polynomial time and (ii) for all oracles A and all inputs x, the acceptance probability of M on input x with oracle A does not fall into the interval $(1/3, 2/3)$. Let $\mathbf{BQP}^A_{\text{robust}}$ denote the collection of all sets that can be recognized by robust **BQP**-machines with oracle A.

Proposition 3. *If* $\mathbf{BQP}^A = \mathbf{BQP}^A_{\text{robust}}$ *for a random oracle* A *with probability* 1, *then* Almost-**BQP** = **BQP**.

Proof sketch: Since **BQP** $\subseteq$ Almost-**BQP**, we need only to show that Almost-**BQP** $\subseteq$ **BQP**. Let L be any set in Almost-**BQP** and set $\mathcal{A} = \{A \mid L \in \mathbf{BQP}^A_{\text{robust}}\}$. By our assumption, $\mathbf{m}(\mathcal{A}) = 1$.

By Lemma 1, for every ϵ ($0 < \epsilon < 1$), there exists a subset $\mathcal{A}'$ of $\mathcal{A}$ and a single robust **BQP**-machine M such that (i) $\mathbf{m}(\mathcal{A}') \geq \epsilon$ and (ii) $L(M, A) = L$ for all A in $\mathcal{A}'$. We set $\epsilon = 5/6$. Let p be a polynomial such that M^A on input x runs in time $p(|x|)$ for every x and A. As shown by Yamakami [33], without loss of generality, we can assume that M makes the exactly the same number of queries of the same length on all computation paths. Hence, by Lemma 10, M^A can be simulated by a robust $(p(n), 2^{p(n)+1}, p(n))$-quantum network U.

We define the desired well-formed QTM N as follows. By the hardness of PARITY given in Lemma 11, using Lemma 4, we can define a pseudorandom generator, stretching $p(n)$ bits to $2^{p(n)}$ bits, that is secure against U. On input x of length n, N produces $p(n)$ 0's on an extra tape and returns its head to the start cell by applying qubit-wise Walsh-Hadamard transformation. Then, N starts to simulate M on input x except that when M enters a pre-query state with query word z, N scans a string written in the extra tape and generates the zth bit of the outcome of the pseudorandom generator. Since the pseudorandom generator fools U, N accepts the input with almost the same probability as M does. Hence, N recognizes L with bounded error and consequently, L belongs to **BQP**. □

References

1. L. Adleman, J. DeMarrais, and M. Huang, Quantum computability, *SIAM J. Comput.*, **26** (1997), 1524–1540.
2. T. Baker, J. Gill, and R. Solovay, Relativizations of the P=?NP question, *SIAM J. Comput.*, **4** (1975), 431–442.

3. R. Beals, H. Buhrman, R. Cleve, M. Mosca, and R. de Wolf, Quantum lower bounds by polynomials, in *Proc. 39th Symposium on Foundations of Computer Science*, pp.352–361, 1998.
4. R. Beigel, Relativized counting classes: relations among threshold, parity, and mods, *J. Comput. System Sci.* **42** (1991), 76–96.
5. R. Beigel, H. Buhrman, and L. Fortnow, NP might not be as easy as detecting unique solutions, in *Proc. 30th IEEE Symposium on Theory of Computing*, pp.203–208, 1998.
6. P. Benioff, The computer as a physical system: a microscopic quantum mechanical Hamiltonian model of computers as represented by Turing machines, *J. Statistical Physics*, **22** (1980), 563–591.
7. A. Berthiaume and G. Brassard, Oracle quantum computing, *Journal of Modern Optics*, **41** (1994), 2521–2535.
8. E. Bernstein and U. Vazirani, Quantum complexity theory, *SIAM J. Comput.*, **26** (1997), 1411–1473.
9. C.H. Bennett and J. Gill, Relative to a random oracle A, $P^A \neq NP^A \neq coNP^A$ with probability 1, *SIAM J. Comput.*, **10** (1981), 96–113.
10. C.H. Bennett, E. Bernstein, G. Brassard, and U. Vazirani, Strengths and weaknesses of quantum computing, *SIAM J. Comput.*, **26** (1997), 1510–1523.
11. G. Brassard and P. Høyer, An exact quantum polynomial-time algorithm for Simon's problem, in *Proc. 5th Israeli Symposium on Theory of Computing and Systems*, pp.12–23, 1997.
12. H. Buhrman and L. Fortnow, One-sided versus two-sided error in probabilistic computation, in *Proc. 16th Symposium on Theoretical Aspects of Computer Science*, Lecture Notes in Computer Science, Vol.1563, pp.100–109, 1999.
13. D. Deutsch and R. Jozsa, Rapid solution of problems by quantum computation, *Proc. Roy. Soc. London*, Ser.A **439** (1992), 553–558.
14. R. Feynman, Simulating physics with computers, in *Int. J. Theoretical Physics*, **21** (1982), 467–488.
15. S. Fenner, F. Green, S. Homer, and R. Pruim, Determining acceptance probability for a quantum computation is hard for PH, in *Proc. 6th Italian Conference on Theoretical Computer Science*, World-Scientific, Singapore, pp.241–252, 1998.
16. L. Fortnow and J. Rogers, Complexity limitations on quantum computation, in *Proc. 13th Conference on Computational Complexity*, pp.202–209, 1998.
17. M. Furst, J. Saxe, and M. Sipser, Parity, circuits and the polynomial-time hierarchy, *Math. Syst. Theory*, **17** (1984), 13–27.
18. L. Fortnow and T. Yamakami, Generic separations, *J. Comput. System Sci.*, **52** (1996), 191–197.
19. F. Green, Lower bounds for depth-three circuits with equals and mod-gates, in *Proc. 12th Annual Symposium on Theoretical Aspect of Computer Science*, Lecture Notes in Computer Science, Vol.900, pp.71–82, 1995.
20. J.T. Håstad, *Computational Limitations for Small-Depth Circuits*, The MIT Press, 1987.
21. E. Hemaspaandra, L.A. Hemaspaandra, and M. Zimand, Almost-everywhere superiority for quantum polynomial time. Technical Report TR-CS-99-720, University of Rochester. See also ph-quant/9910033.
22. K. Ko and H. Friedman, Computational complexity of real functions, *Theor. Comput. Sci.*, **20** (1982), 323–352.
23. K. Ko, Some observations on the probabilistic algorithms and NP-hard problems, *Inform. Process. Lett.*, **14** (1982), 39–43.

24. K. Ko, Separating and collapsing results on the relativized probabilistic polynomial-time hierarchy, *J. Assoc. Comput. Mach.*, **37** (1990), 415–438.
25. S.A. Kurtz, A note on randomized polynomial time, *SIAM J. Comput.*, **16** (1987), 852–853.
26. H. Nishimura and M. Ozawa, Computational complexity of uniform quantum circuit families and quantum Turing machines, manuscript, 1999. See also LANL quant-ph/9906095.
27. N. Nisan and M. Szegedy, On the degree of Boolean functions as real polynomials, *Computational Complexity*, **4** (1994), 301–313.
28. N. Nisan and A. Wigderson, Hardness and randomness, *J. Comput. System Sci.*, **49** (1994), 149–167.
29. C.H. Papadimitriou, *Computational Complexity*, Addison-Wesley, 1994.
30. D. Simon, On the power of quantum computation, *SIAM J. Comput.*, **26** (1997), 1340–1349.
31. P.W. Shor, Polynomial-time algorithms for integer factorization and discrete logarithms on a quantum computer, *SIAM J. Comput.*, **26** (1997), 1484–1509.
32. T. Yamakami, A foundation of programming a multi-tape quantum Turing machine, in *Proc. 24th International Symposium on Mathematical Foundations of Computer Science*, Lecture Notes in Computer Science, Vol.1672, pp.430–441, 1999. See also LANL quant-ph/990684.
33. T. Yamakami, Analysis of quantum functions, in *Proc. 19th International Conference on Foundations of Software Technology and Theoretical Computer Science*, Lecture Notes in Computer Science, Vol.1738, pp.407–419, 1999. See also LANL quant-ph/9909012.
34. T. Yamakami and A.C. Yao, $NQP_C = \text{co-}C_=P$, *Inform. Process. Lett.*, **71** (1999), 63–69.

Added in proofs. Two corrections were pointed out by Lance Fortnow [1]; we are grateful for his comments.

The proof showing $L_2^A \not\in \Pi_2^P(A)$ is erroneous. Note that Simon's language L_1^A belongs to $\mathbf{AM}^A \cap \text{co-}\mathbf{AM}^A$ for any good oracle A. Since $\mathbf{AM}^A \cap \text{co-}\mathbf{AM}^A$ is closed under polynomial-time Turing reductions, L_2^A should be in $\Sigma_2^P(A) \cap \Pi_2^P(A)$. We still conjecture that the claim itself should be true.

Lately we became aware of the paper by Green and Pruim on the existence of an oracle relative to which $\mathbf{EQP} \not\subseteq \Delta_2^P$. Note that their test language $L(A)$ also belongs to $\mathbf{BPP}^A$ for a target oracle A.

We have shown that $\mathbf{Almost\text{-}BQP} = \mathbf{BQP}$ under the condition that $\mathbf{BQP}^X_{\text{robust}} = \mathbf{BQP}^X$ relative to a random oracle X. This claim can be proven by Fortnow [1] without any assumption (however, the assumption itself is still an open problem). The proof is actually a variant of the classical Bennett-Gill's proof. For any set L in $\mathbf{Almost\text{-}BQP}$, consider a quantum function $f^A \in \#\mathbf{QP}^A$ such that, for almost all x, $|f^A(x) - L(x)|$ is small relative to most oracles A. As in [32 above], we can construct a well-formed QTM M whose final superposition in which the amplitude of configuration $|x\rangle|1^{p(n)}\rangle|1\rangle$ is $f^A(x)$ within polynomial time $p(|x|)$. Consider the following QTM N: on input x of length n, generate $2^{-p(n)/2} \sum_{y:|y|=p(n)} |y\rangle$ and simulate M on input x by reading the ith bit of y as an oracle answer to the ith query of M. Since the acceptance probabilities $f^A(x)$ add up for oracles A, N outputs $L(x)$ with high probability.

Additional References

1. L. Fortnow, Personal communication, October 2000.
2. F. Green and R. Pruim, Relativized separation of EQP from P^{NP}, manuscript, 2000.

Solving NP-Complete Problems Using P Systems with Active Membranes*

Claudio Zandron, Claudio Ferretti, and Giancarlo Mauri

DISCO - Universitá di Milano-Bicocca, Italy

Abstract. A recently introduced variant of P-systems considers membranes which can multiply by division. These systems use two types of division: division for elementary membranes (i.e. membranes not containing other membranes inside) and division for non-elementary membranes. In two recent papers it is shown how to solve the Satisfiability problem and the Hamiltonian Path problem (two well known NP complete problems) in linear time with respect to the input length, using both types of division. We show in this paper that P-systems with only division for elementary membranes suffice to solve these two problems in linear time. Is it possible to solve NP complete problems in polynomial time using P-systems without membrane division? We show, moreover, that (if $P \neq NP$) deterministic P-systems without membrane division are not able to solve NP complete problems in polynomial time.

1 Introduction

The P-systems were introduced in [5] as a class of distributed parallel computing devices of a biochemical type. In the basic model we have a membrane structure composed by several cell-membranes, hierarchically embedded in a main membrane called the skin membrane. The membranes delimit regions and can contain objects. The objects evolve according to given evolution rules associated with the regions. A rule can modify the objects and send them outside the membrane or to an inner membrane. Moreover, the membranes can be dissolved. When a membrane is dissolved, all the objects in this membrane remain free in the membrane placed immediately outside, while the evolution rules of the dissolved membrane are lost. The skin membrane is never dissolved. The evolution rules are applied in a maximally parallel manner: at each step, all the objects which can evolve should evolve.

A computation device is obtained: we start from an initial configuration and we let the system evolve. A computation halts when no further rule can be applied. The output consist in the string obtained by concatenating the objects which leave the system through the skin membrane, taken in the order they are sent out.

---

* This work has been supported by the Italian Ministry of University (MURST), under project "Unconventional Computational Models: Syntactic and Combinatorial Methods".

Many variants are considered in [1], [4], [5], [6], [7], [8] and [10]. In these papers, the membranes are used as separators and as channels of communication, but they participate to the process in a passive way. The whole process is regulated by the evolution rules.

In [3] one consider membrane systems where the membranes play an active role in the computation: the evolution rules are associated both with objects and the membrane; the communication of the objects through the membranes is performed with the direct participation of the membranes; and finally, the membranes can not only be dissolved, but they can multiply by division. There are two different types of division rules: division rules for elementary membranes (i.e. membranes not containing other membranes) and division rules for non-elementary membranes. In [3] and [2] it is shown how to solve the Satisfiability problem and the Hamiltonian Path problem respectively (two well known NP complete problems) in linear time with respect to the input length, using P-systems with active membranes. We show here how to solve these two problems in linear time, using P-systems with active membranes which only make use of division for elementary membranes.

Another interesting question relates to the needed of active membranes: what about the possibility of solving NP complete problems in polynomial time using P-systems without membrane division? In [3] and [9] one points out that P-systems without active membranes already possess a great amount of parallelism (all the rules are applied at the same time on all objects which can evolve) but it seems not to be sufficient to solve complex problems in polynomial time. In fact, we show here that this is the case: if $P \neq NP$, a deterministic P-system without membrane division is not able to solve a NP complete problem in polynomial time; every deterministic P-system without active membranes working in time t, can be simulated by a deterministic Turing Machine working in time polynomial in t.

2 P-systems with Active Membranes

In the following, we refer to [11] for elements of Formal Language Theory. A membrane structure is a construct consisting of several membranes placed in a unique membrane; this unique membrane is called a skin membrane. We identify a membrane structure with a string of correctly matching parentheses, placed in a unique pair of matching parentheses; each pair of matching parentheses corresponds to a membrane. The membranes can be marked with $+$, $-$ or 0, and this is interpreted as an electrical charge (0 is the neutral charge). A membrane identifies a region, delimited by it and the membrane immediately inside it. If we place multisets of objects in the region from a specified finite set V, we get a super-cell. A super-cell system (or P-system) is a super-cell provided with evolution rules for its objects.

A P-system with active membranes is a construct

$$\Pi = (V, T, H, \mu, w_1, \ldots, w_m, R),$$

where:

- $m \geq 1$;
- V is an alphabet (the total alphabet of the system);
- $T \subseteq V$ is the terminal alphabet;
- H is a finite set of labels for membranes;
- μ is a membrane structure, consisting of m membranes, labeled (not necessarily in a one-to-one manner) with elements of H; all membranes in μ are supposed to be neutral;
- $w_1, \ldots, w_m$ are strings over V, describing the multisets of objects placed in the m regions of μ;
- R is a finite set of developmental rules, of the following forms:
  (a) $[_h a \to v]_h^\alpha$, for $h \in H$, $a \in V$, $v \in V^*$, $\alpha \in \{+, -, 0\}$ (object evolution rules),
  (b) $a[_h]_h^{\alpha_1} \to [_h b]_h^{\alpha_2}$, where $a, b \in V$, $h \in H$, $\alpha_1, \alpha_2 \in \{+, -, 0\}$ (an object from the region immediately outside the membrane h is introduced in membrane h),
  (c) $[_h a]_h^{\alpha_1} \to [_h]_h^{\alpha_2} b$, for $h \in H$, $\alpha_1, \alpha_2 \in \{+, -, 0\}$, $a, b \in V$ (an object is sent out from membrane h to the region immediately outside),
  (d) $[_h a]_h^\alpha \to b$, for $h \in H$, $\alpha \in \{+, -, 0\}$, $a, b \in V$ (membrane h is dissolved),
  (e) $[_h a]_h^{\alpha_1} \to [_h b]_h^{\alpha_2} [_h c]_h^{\alpha_3}$, for $h \in H, \alpha_1, \alpha_2, \alpha_3 \in \{+, -, 0\}, a, b, c \in V$ (division rules for elementary membranes)
  (f) $[_{h_0} [_{h_1}]_{h_1}^{\alpha_1} \ldots [_{h_k}]_{h_k}^{\alpha_1} [_{h_{k+1}}]_{h_{k+1}}^{\alpha_2} \ldots [_{h_n}]_{h_n}^{\alpha_2}]_{h_0}^{\alpha_0} \to$
  $\to [_{h_0} [_{h_1}]_{h_1}^{\alpha_3} \ldots [_{h_k}]_{h_k}^{\alpha_3}]_{h_0}^{\alpha_5} [_{h_0} [_{h_{k+1}}]_{h_{k+1}}^{\alpha_4} \ldots [_{h_n}]_{h_n}^{\alpha_4}]_{h_0}^{\alpha_6}$ for $k \geq 1, n \geq 1, h_i \in H, 0 \leq i \leq n$, and $\alpha_0, \ldots, \alpha_6 \in \{+, -, 0\}$ with $\{\alpha_1, \alpha_2\} = \{+, -\}$ (division rules for non-elementary membranes)

These rules are applied accordingly to the principles in [3].

1. All the rules are applied in parallel: in a step, the rules of type (a) (i.e. the rules that do not modify the membranes) are applied to all objects to which they can be applied; all other rules are applied to all membranes to which they can be applied; an object can be used by only one rule, non-deterministically chosen (there is no priority relation among rules), but any object which can evolve by a rule of any form, must evolve.
2. If a membrane is dissolved, then all the objects in its region are left free in the region immediately above it. Because all rules are associated with membranes, the rules of a dissolved membrane are no longer available at the next steps. The skin membrane is never dissolved.
3. All objects and membranes not specified in a rule and which do not evolve are passed unchanged to the next step.
4. If at the same time a membrane h is divided by a rule of type (e) and there are objects in this membrane which evolve by means of rules of type (a), then in the new copies of the membrane we introduce the result of the evolution; that is, we may suppose that first the evolution rules of type (a) are used, changing the objects, and then the division is produced, so

that in the two new membranes with label h we introduce copies of the changed objects. Of course, this process takes only one step. The same assertions apply to the division by means of a rule of type (f): always we assume that the rules are applied "from bottom-up", in one step, but first the rules of the innermost region and then level by level until the region of the skin membrane.
5. The rules associated with a membrane h are used for all copies of this membrane, irrespective whether or not the membrane is an initial one or it is obtained by division. At one step, a membrane h can be the subject of only one rule of types (b) – (f).
6. The skin membrane can never divide. As with any other membrane, the skin membrane can be "electrically charged".

The membrane structure at a given time, together with all multisets of objects associated with the regions defined by the membrane structure, is the *configuration* of the system at that time. The *initial configuration* is $(\mu, w_1, \ldots, w_m)$. We can pass from a configuration to another one by using the rules in R, according to the principles previously described (we call this a *transition*). A *computation* is a sequence of transitions between configurations. A computation *halts* when there is no rule which can be applied to objects and membranes in the current configuration.

During the computation, objects can leave the skin membrane (with a rule of type *(c)*). The terminal symbols which leave the skin membrane are collected in the order of their expelling from the system, so a string is associated to a complete computation. The symbols not in T which leave the skin membrane and all symbols in T which remain in the system at the end of a halting computation are not considered in the generated strings. If a computation never stops, then it provides no output.

Moreover, we will need the extension introduced in [2], about the rules of types *(e)* and *(f)*: a membrane can be divided into finitely many membranes. This type of division is called *bounded membrane division*.

In the following, we do not make use of division for non-elementary membranes, so we give the following definition:

A *P-system with elementary active membranes* is a P-system with active membranes where the rules are of type (a) to (e) only (i.e. a P-system with active membranes not using rules of type (f)).

Finally, we say that a P-system (with or without active membranes) is *deterministic* if, in each moment, there is at most one possible transition. This means that there are no two different rules applicable at the same time in the same region on the same symbol.

More details and examples can be found in [5], [3] and [2].

3 Solving NP Complete Problems Using P-system with Elementary Active Membranes

In this section we show how to solve the NP complete problems proposed in [3] and [2] (Satisfiability and Hamiltonian Path) using P-systems with elementary active membranes: although the systems in [3] and [2] do make use of division for non-elementary membranes, the time we need to execute the computation is of the same orders as theirs. In fact, the systems presented here solve Satisfiability and Hamiltonian Path in a time which is linear with respect to the dimension of the input; this time is comparable with the time needed by the systems in [3] and [2]. Moreover, the structure (i.e. the number of symbols, the number of rules and the number of membranes) of the systems we propose is similar to the structure of the systems proposed in [3] and [2].

Theorem 1. *The SAT problem can be solved in linear time (with respect to the number of variables and the number of clauses) by a P-system with elementary active membranes.*

Proof. Consider a Boolean expression in conjunctive normal form

$$\alpha = C_1 \wedge C_2 \wedge \ldots \wedge C_m$$

with

$$C_i = y_{i,1} \vee y_{i,2} \vee \ldots \vee y_{i,p_i}$$

for some $m \geq 1$, $p_i \geq 1$, and $y_{i,j} \in \{x_k, \neg x_k | 1 \leq k \leq n\}$, for each $1 \leq i \leq m$, $1 \leq j \leq p_i$.

We build the P-system

$$\Pi = (V, T, H, \mu, \omega_0, \omega_1, R)$$

where

- $V = \{a_i, t_i, f_i | 1 \leq i \leq n\} \cup \{r_i | 0 \leq i \leq m\} \cup \{W_i | 1 \leq i \leq m+1\} \cup \{t\} \cup$
  $\cup \{Z_i | 0 \leq i \leq n\}$
- $T = \{t\}$
- $H = \{0, 1\}$
- $\mu = [_0[_1]_1^0]_0^0$
- $\omega_0 = \lambda$
- $\omega_1 = a_1 a_2 \ldots a_n Z_0$

while the set R contains the following rules:
1. $[a_i]_1^0 \to [t_i]_1^0 [f_i]_1^0$, $1 \leq i \leq n$
We substitute one variable a_i in membrane 1 with two variables t_i and f_i. The membrane is divided in two membranes (the charge remains neutral for both membranes). In n steps we get all 2^n truth assignments for the n variables; each truth assignment is in a membrane labelled with 1.

2. $[Z_k \to Z_{k+1}]_1^0$, $0 \leq k \leq n-2$
3. $[Z_{n-1} \to Z_n W_1]_1^0$
4. $[Z_n]_1^0 \to []_1^+ \lambda$

We count n steps, the time needed to produce all the truth assignments. The step $n-1$ introduces the symbol W_1 used in the next steps, while the step n changes the charge of the membranes labelled with 1.

5. $[t_i \to r_{h_{i,1}} ... r_{h_{i,j_i}}]_1^+$, $1 \leq i \leq n$ and the clauses $h_{i,1}, ..., h_{i,j_i}$ contain x_i.
6. $[f_i \to r_{h_{i,1}} ... r_{h_{i,l_i}}]_1^+$, $1 \leq i \leq n$ and the clauses $h_{i,1}, ..., h_{i,l_i}$ contain $\neg x_i$.

In one step, each symbol t_i is replaced with some symbols r_k, indicating the clauses satisfied if we set $x_i = true$; each symbol f_i is replaced with some symbols, indicating the clauses satisfied if we set $x_i = false$ (i.e. $\neg x_i = true$).

After this step, we start the "VERIFY STEPS": we have to verify if there is a membrane, labelled with 1, in which we get all symbols $r_1, r_2, ..., r_m$ (at least one symbol r_k for every k between 1 and m). In fact, this means that there is a truth assignment satisfying all the clauses. To do this, we apply (in a single step) the rules of type 7 in all membranes containing the symbol r_1; then, we apply (in a single step) the rules of types 8, 9 and 10, on all symbols which can evolve. We repeat this 2-steps process for each symbol $r_1, r_2, \ldots, r_m$, thus the whole process takes $2m$ steps.

7. $[r_1]_1^+ \to []_1^- r_1$

We first verify the presence of the symbol r_1 in membranes labelled with 1. Every membrane containing r_1 (i.e. every membrane with a truth assignment satisfying the first clause) sends this symbol outside (in membrane 0) and changes its charge from $+$ to $-$. The membranes not containing this symbol cannot further proceed their computation.

8. $r_1 []_1^- \to [r_0]_1^+$
9. $[r_k \to r_{k-1}]_1^-$, $2 \leq k \leq m$
10. $[W_i \to W_{i+1}]_1^-$, $1 \leq i \leq m$

The computation can only proceed in membranes with negative charge. In one step, each symbol W_i is replaced with a symbol W_{i+1} (the counter of the satisfied clauses is increased), each symbol r_k is replaced with r_{k-1} and the symbols r_1 in membrane 0 are sent back (as r_0) to the negative charged membranes labelled with 1, to change their charge from $-$ to $+$.

After applying rules 8, 9, and 10 (in a single step on all symbols and membranes which can evolve) we are ready to re-apply the rules of type 7. At this time, the symbols r_1 in membranes 1 are the symbols that was r_2 when the verify steps started (the index of each symbol r_k has been decreased). Thus the computation only proceed in membranes which contained both symbols r_1 and r_2 when the verify steps started. After the application of the rules of type 7 we can apply again the rules of type 8, 9 and 10 to prepare the membranes to verify the next clause.

In $2m$ steps we verify the existence of a membrane containing all symbols r_i. It is easy to see that if a membrane does not contain a symbol r_i, the computation in that membrane halts in less than $2m$ steps. As a consequence, that membrane will not contain the symbol W_{m+1}. The membranes containing

this symbol after $2m$ steps are the membranes which contained all symbols r_i when the verify steps started, i.e. the membranes satisfying all the clauses.
11. $[W_{m+1}]_1^+ \to []_1^+ t$
If a membrane labelled with 1 executes all $2m$ verify steps, it contains the symbol W_{m+1}. Thus we send out to membrane 0 the symbol t, indicating that there is a truth assignment satisfying all clauses.
12. $[t]_0^0 \to []_0^+ t$
A symbol t is sent outside membrane 0 and the charge of this membrane is changed from 0 to +. No further computation is possible. Thus, we have to look at the output of membrane 0 after $n + 2m + 3$ steps. If we get the symbol t, it means that there is a truth assignment satisfying α; otherwise the formula is not satisfiable. □

As we have said at the beginning of this section, the time of computation of the system proposed in the previous proof is comparable with the time of computation of the system proposed by Paun in [3], and the same is true for the structure of the systems. In the next table we explicitly show this, by comparing the system presented here with the system proposed in [3]. We compare the systems under a computational complexity point of view as well as under a structural complexity point of view. With n and m we denote respectively the number of variables and the number of clauses of a generic instance of the SAT problem.

	Paun	This
Number of Steps of Computation	$2n + 2m + 1$	$n + 2m + 3$
Number of Symbols	$5n + m + 2$	$4n + 2m + 4$
Max Number of Membranes	$(m + 1) \times 2^n + 1$	$2^n + 1$
Depth	$m + 1$	1
Number of Rules	$3n + 2m + 2nm$	$4n + 2m + 4$

As one can see, for $n \geq 3$ the system we propose in this paper is slightly faster than the system in [3]. Another interesting aspect of the system we propose relates to the depth of the membrane structure: the depth of the system we propose is always one. This indicate a very simple membrane structure, composed by elementary membranes surrounded by the skin membrane.

Theorem 2. *The Undirected Hamiltonian Path Problem can be solved in linear time (with respect to the number of nodes in the given graph) by a P-system with elementary active membranes and bounded membrane division.*

Proof. Consider an undirected graph $G = (U, E)$ with n nodes, $n \geq 2$.
We build the P-system

$$\Pi = (V, T, H, \mu, \omega_0, \omega_1, R)$$

where

- $V = \{a_i, P_i | 1 \leq i \leq n\} \cup \{Z_i | 0 \leq i \leq 2n-1\} \cup \{W_i | 0 \leq i \leq m+1\} \cup$
  $\cup \{r_i, x_i | 1 \leq i \leq n+1\} \cup \{t, R\}$
- $T = \{t\}$
- $H = \{0, 1\}$
- $\mu = [_0[_1]_1^0]_0^0$
- $\omega_0 = \lambda$
- $\omega_1 = RZ_0$

while the set R contains the following rules:
1. $[R]_1^0 \to [a_1]_1^0 ... [a_n]_1^0$
We create n membranes labelled with 1, each containing a symbol a_i corresponding to a node in G.
2. $[Z_k \to Z_{k+1}]_1^0$, $0 \leq k \leq 2n-3$
3. $[Z_{2n-2} \to Z_{2n-1}C_1]_1^0$
4. $[Z_{2n-1}]_1^0 \to []_1^+ \lambda$
We count $2n$ steps, the time needed to produce all paths of length n.
5. $[a_i \to r_i P_i]_1^0$, $1 \leq i \leq n$
6. $[P_i]_1^0 \to [a_{j_1}]_1^0 ... [a_{j_k}]_1^0$, $1 \leq i \leq n$, $1 \leq j_h \leq n$ and (i, j_1), ..., (i, j_k) are edges of G.

In $2n$ steps, we create all paths of length n we can obtain starting from every node in G. At each step, we divide the membranes by replacing the symbol P_i with the symbols corresponding to the nodes directly connected to the node i. Note that the rule of type 4 changes the charge of the membranes labelled with 1 from 0 to +, thus after $2n$ steps, no rule of the previous types can be applied.

7. $[r_1]_1^+ \to []_1^- r_1$
8. $r_1 []_1^- \to [r_0]_1^+$
9. $[r_i \to r_{i-1}]_1^-$, $1 \leq i \leq m$
10. $[C_i \to C_{i+1}]_1^-$, $1 \leq i \leq m$
11. $[C_{m+1}]_1^+ \to []_1^+ t$
12. $[t]_0^0 \to []_0^+ t$

Once the charge of membranes labelled with 1 has become +, what we need to do is to verify the existence of a membrane containing the symbols $r_1, r_2, ..., r_n$. Such a membrane exists if and only if the graph G has a Hamiltonian path. To verify this, we can proceed as in the previous proof, using the rules of types 7 to 12. After $4n + 2$ steps, we look at the output of membrane 0. If we get the symbol t then there is a Hamiltonian Path in G, otherwise no such path exists. □

As for the SAT problem, in the next table we compare the system presented here with the system presented in [2] to solve the Undirected Hamiltonian Path Problem. Given a graph G as the input for the Undirected Hamiltonian Path Problem, we denote with n the number of nodes of the graph.

	Krishna, Rama	This
Number of Steps of Computation	$3n+1$	$4n+2$
Number of Symbols	$4n+1$	$6n+4$
Max Number of Membranes	$(n+2) \times n \times (n-1)^{n-1}$	$2 \times n \times (n-1)^n$
Depth	$n+1$	1
Number of Rules	$14n+8$	$6n+5$

The system we propose to solve the Hamiltonian Path Problem is slightly slower with respect to the system proposed in [2], but the time remains linear with respect to the input length. The depth of the system we propose is still one.

4 P-systems without Active Membranes

One interesting question, pointed out in [3] and [9], concerns the power of the operation of membrane division. Are we able to solve NP complete problems in polynomial time using P-systems without active membranes? In fact, the model without active membranes has a great amount of parallelism, but we do not know if it suffices to solve complex problems in an efficient way.

The systems presented in the previous section are deterministic (like the systems in [3] and [2]): at each step of computation there is at most one possible transition from a configuration to another one. We show now that, under the assumption of using deterministic P-systems, if $P \neq NP$ the amount of parallelism without membrane division is not enough to solve NP complete problems in polynomial time. Membrane division is necessary to get significant speed up of computation.

Theorem 3. *Consider a deterministic P-system, without membrane division and working in time t. We denote with A, B and C respectively the number of membranes, the number of symbols in V and the length of the rules (the number of symbol involved in the rule, on both left and right side) of the P-system. Moreover, we set $D = max\{C, B+2\}$. Such a system can be simulated by a Deterministic Turing Machine (DTM) in time $t' = O(A \times B \times D \times t \times log(A \times B \times C^t))$.*

Proof. Consider a P-system

$$P = (V, T, H, \mu, \omega_0, ..., \omega_k, R)$$

without membrane division working in time t.

We build a DTM $M = (V', K, S_0, \delta)$ with multiple tapes, which simulates P within a number of steps $O(A \times B \times D \times t \times log(A \times B \times C^t))$.

To simulate P with M, we have just to keep track of the quantity of each symbol $z \in V$ in each membrane. In fact, consider the application of a rule $[a \to bc]_i^\alpha$: all the symbols a in membrane i are substituted with two symbols, b and c. We can simulate this rule by adding to the quantities of symbols b

and c, in membrane i, the quantity of symbols a and then by setting the last quantity to zero. In other words, the application of a rule in P corresponds to a modification of the quantities of the symbols involved in the rule of the specific membrane (and eventually, as we will see, in modifying the quantities of all symbols in a membrane, if a membrane placed immediately inside this one is dissolved by the rule). The modification of the quantities is done by adding the quantity of the symbol on the left side of the rule to each quantity of the symbols on the right side of the rule.

For the previous reasons, the DTM M has $2 \times (A \times B) + 3$ tapes:

- $A \times B$ "main" tapes, to keep track of the quantities of each symbol in each membrane (A membranes and B symbols) after every step.
- $A \times B$ "support" tapes, used to make partial sums.
- 1 "polarity" tape, used to keep track of the polarity of each membrane.
- 1 "structure" tape, used to keep track of the structure of membranes.
- 1 tape as output.

Every computation step of P can be simulated with 2 macro steps of M:

1. We simulate the application of a rule on each symbol in each membrane by modifying the quantity written on the tapes containing the partial sums. For example, consider a rule $[a \rightarrow bc]_1^0$, and let the strings written on the tapes corresponding to the symbols a, b and c in membrane 1 contain the quantity 100, 200 and 250. To simulate this rule, we have to put the value 0 in the first string, 300 in the second one and 350 in the third one. We do not write this quantity on the main tapes, because to simulate the application of another rule $[b \rightarrow de]_1^0$ we have to know that the quantity of symbols b when the computation step started was 200. For this reason we need support tapes. The rule to be applied (only one rule per symbol can be applied, because we consider deterministic P-system) is chosen accordingly to the polarity of the membrane; the polarity can be found in the "polarity" tape.

2. When the application of the rules has been executed on all objects, we copy the quantity of each symbol from the support tapes to the main tapes.

We have to repeat these two steps t times. To simulate a single step of computation of P, we have to simulate the application of at most $A \times B$ different rules in R (one rule for each object in each membrane). The number of sums to be executed for each rule in each membrane depends on the type of rule applied. Consider the four types of rules in R:

a) $[a \rightarrow x]_i^\alpha$ where $a \in V$ and $x \in V^*, |x| \leq C - 1$. To simulate such a rule, we have to execute at most C sums.

b) $a[]_i^\alpha \rightarrow [b]_i^\beta$, where $a, b \in V$. To simulate such a rule, we have to sum 1 to the quantity of the symbol b in membrane i and to sum -1 to the quantity of the symbol a in the membrane placed immediately outside membrane i (remember that, accordingly to the rules in [3], at each computation step only one symbol can enter a rule of type $(b), (c)$ or (d) in each membrane). If needed, we change the charge of the membrane in the "polarity" tape.

c) $[a]_i^\alpha \rightarrow []_i^\beta b$, where $a, b \in V$. To simulate such a rule on a symbol, we have to sum 1 to the quantity of the symbol b in the membrane placed immediately outside membrane i and to sum -1 to the quantity of the symbol a in membrane i. If needed, we change the charge of the membrane in the "polarity" tape. If i is the skin membrane, we have to write the symbol b on the output tape.

d) $[[a]_i^\alpha]_k^\beta \to [b]_k^\gamma$, where $a, b \in V$. To simulate such a rule, we have to sum 1 to the quantity of the symbol b in the membrane placed immediately outside membrane i, and to sum -1 to the quantity of the symbol a in membrane i. Then we have to sum the quantity of each symbol in membrane i to the quantity of the same symbol in membrane k. Thus, we have to execute $B + 2$ sums. After these sums, we modify the structure of the membrane in the "structure" tape and we change the charge of the membrane k.

It is easy to see that the most expensive rules (in terms of time needed to simulate them) are those of type (a) and (d). For this reason we set $D = max\{C, B+2\}$. To simulate a single step of computation of P we need at most $A \times B \times D$ sums.

The time required by each sum depends, of course, on the number of digits of the numbers to sum. We can consider membrane system with a maximum number of initial symbols equal to $A \times B$ (i.e. every membrane contains at most one occurrence of each symbol in V). In fact, given a P-systems with an arbitrary finite number of occurrence of each symbol in each membrane, it is easy to build a P-systems with exactly one occurrence of each symbol in each membrane that requires a finite number of steps to reach that same configuration. Thus, after one step, the total quantity of symbols present in the system will be less than $A \times B \times C$. After the second step, this quantity will be less than $A \times B \times C^2$. After t steps, we get no more than $A \times B \times C^t$ symbols. This means that each sum will be made with number involving less than $log(A \times B \times C^t)$ digits.

The time required by a DTM to make a sum is linear in the number of digits of the number involved; every sum requires a time which is $O(log(A \times B \times C^t))$. To execute the sums for each symbols in each membrane, the time needed is $O(A \times B \times D \times log(A \times B \times C^t))$. After the execution of all the sums, we have to copy the obtained results from the support tapes to the main tapes; the time needed is $O(A \times B \times log(A \times B \times C^t))$. The total time requested by M to simulate a single step of P is $O(A \times B \times D \times log(A \times B \times C^t))$.

Of course, we have to simulate t steps of P, thus the total time needed is $O(t \times A \times B \times D \times log(A \times B \times C^t))$. □

If, given a problem, we could build for each instance of size n a P-systems without membrane division such that it gives the solution in time polynomial in n, then we could build a DTM that gives the same output in polynomial time too. From this follows:

Corollary 1. *If $P \neq NP$ then no NP complete problem can be solved in polynomial time, with respect to the input length, by a deterministic P-system without membrane division.*

As one can see, the time needed to simulate a deterministic P-system depends on the number of membranes of the P-system itself (as well as on the number of symbols in V and on the length of the rules). If we use P-systems without membrane division, the number of membrane remains the same (or is decreased, if we dissolve some membranes) during the whole computation process. This is not true, of course, for P-systems with membrane division. With this variant of P-systems, the number of membranes can grow exponentially: if we repeatedly divide A membranes, we can obtain, after t steps, a number of membranes equal to $A \times 2^t$. The simulation of such a system with a DTM (in the same way we have discussed above) would require an exponential number of strings, or an exponential amount of time equal to $O(t \times A \times 2^t \times B \times D \times log(A \times 2^t \times B \times C^t))$.

5 Conclusions

In [3] one points out that a question of a practical importance is to try to implement a P-system either in biochemical media or in electronic media. Moreover, it is underlined that it could be necessary to consider variants of P-systems which are more realistic. This paper goes in this direction, by showing how to solve two NP complete problems (Satisfiability and Hamiltonian Path) without using division for non-elementary membranes (which seems, of course, more complicated with respect to the division for elementary membranes). Moreover, in the same paper and in [9] the question arises if P-systems without membrane division can solve NP complete problems in polynomial time. We have shown that this is not the case: deterministic P-systems without active membranes cannot solve NP complete problems in polynomial time (unless $P = NP$). Every deterministic P-system working in polynomial time, with respect to the input length, can be simulated by a Deterministic Turing Machine working in polynomial time. This proves that P-systems which make use of the operation of membrane division effectively obtain a significant speed-up of computation.

We point out some problems worth further investigations. One problem, already pointed out in [9], is that of simulating P-systems with d-bounded division presented in [2] with P-systems with active membranes presented in [3] (where the division of a membrane always leads to two new membranes). Of course, we know that every NP problem can be reduced to SAT in polynomial time, so we know we can solve every NP problem with a 2-division P-system that works in polynomial time. Nevertheless, it would be interesting to find out if it is possible to simulate every k-division systems with a 2-division system without using the reduction between problems. In [9] one says that the main difficulty appears with the division for non-elementary membranes, but in the view of the results of this paper we need to simulate only division for elementary membranes. Moreover, it can be interesting to follow the direction of finding other ways to increasing the number of membranes; as we have seen, the basic model of P-system does not have sufficient parallelism to solve complex problem in an efficient way, thus we have to find other ways to get such power. Finally, we think another important topic is that of determinism and nondeterminism. An investigation of the power of Deterministic P-systems and Nondeterministic P-systems could be useful from both a mathematical point of view and for a practical implementation of a P-system. Many other open problems and research topics can be found in [9].

References

1. J. Dassow, Gh. Paun, On the power of membrane computing, J. Univ. Computer Sci., 5, 2 (1999), 33-49.
2. S. N. Krishna, R. Rama, A variant of P-systems with active membranes: Solving NP-complete problems, Romanian J. of Information Science and Technology, 2, 4 (1999).
3. Gh. Paun, P-systems with active membranes: attacking NP complete problems, submitted 1999 (see also CDMTCS Research report No. 102, 1999, Auckland Univ., New Zealand, www.cs.auckland.ac.nz/CDMTCS)
4. Gh Paun, Computing with membranes. An introduction, Bulletin of the EATCS, 67 (Febr. 1999), 139-152.
5. Gh. Paun, Computing with membranes, J. of Computer and System Sciences, in press (see also TUCS Research Report No 208, November 1998 http://www.tucs.fi).
6. Gh. Paun, Computing with membranes. A variant, J. of Computer and System Sciences, 11, 167-, 2000 (see also CDMTCS Report No. 0.98, 1999, of CS Department, Auckland Univ., New Zealand, www.cs.auckland.ac.nz/CDMTCS).
7. Gh. Paun, G.. Rozenberg, A. Salomaa, Membrane computing with external output, Fundamenta Informaticae, 41, 3, 2000 (see also TUCS Research Report No. 218, December 1998, http://www.tucs.fi).
8. Gh. Paun, S. Yu, On synchronization in P-systems, Fundamenta Informaticae, 38, 4 (1999), 397-410 (see also CS Department TR No 539, Univ. of Western Ontario, London, Ontario, 1999, www.csd.uwo.ca/faculty/syu/TR539.html).
9. Gh. Paun, Computing with membrane (P-systems): Twenty-six research topics, Research Report, 2000.
10. I. Petre, A normal form for P-systems, Bulletin of EATCS, 67 (Febr. 1999), 165-172.
11. G. Rozenberg, A. Salomaa, eds., Handbook of Formal Languages, Springer-Verlag, Heidelberg, 1997.

Other titles in the DMTCS series:

Combinatorics, Complexity, Logic:
Proceedings of DMTCS '96
D. S. Bridges, C. S. Calude, J. Gibbons,
S. Reeves, I. Witten (Eds)
981-3083-14-X

Formal Methods Pacific '97: Proceedings
of FMP '97
L. Groves and S. Reeves (Eds)
981-3083-31-X

The Limits of Mathematics: A Course on
Information Theory and the Limits of
Formal Reasoning
Gregory J. Chaitin
981-3083-59-X

Unconventional Models of Computation
C. S. Calude, J. Casti and M. J. Dinneen (Eds)
981-3083-69-7

Quantum Logic
K. Svozil
981-4021-07-5

International Refinement Workshop and
Formal Methods Pacific '98
J. Grundy, M. Schwenke and T. Vickers (Eds)
981-4021-16-4

Computing with Biomolecules: Theory
and Experiments
Gheorghe Paun (Ed)
981-4021-05-9

People and Ideas in Theoretical Computer
Science
C. S. Calude (Ed)
981-4021-13-X

Combinatorics, Computation and Logic:
Proceedings of DMTCS'99 and CATS'99
C. S. Calude and M. J. Dinneen (Eds)
981-4021-56-3

Polynomials: An Algorithmic Approach
M. Mignotte and D. Stefanescu
981-4021-51-2

The Unknowable
Gregory J. Chaitin
981-4021-72-5

Sequences and Their Applications:
Proceedings of SETA '98
C. Ding, T. Helleseth and H. Niederreiter (Eds)
1-85233-196-8

Finite versus Infinite: Contributions to an
Eternal Dilemma
Cristian S. Calude and Gheorghe Paun (Eds)
1-85233-251-4

Network Algebra
Gheorge Stefanescu
1-85233-195-X

Exploring Randomness
Gregory J. Chaitin
1-85233-417-7